W9-ACP-623

# VERTEBRATE
# BIOLOGY

*Third Edition*

ROBERT T. ORR, Ph.D.
California Academy of Sciences
San Francisco, California

B. SAUNDERS COMPANY • *Philadelphia* • *London* • *Toronto*

W. B. Saunders Company:     West Washington Square
                           Philadelphia, Pa.   19105

                           12 Dyott Street
                           London, WC1A   1DB

                           1835 Yonge Street
                           Toronto 7, Ontario

Listed here is the latest translated edition of this book together with the language of the translation and the publisher.

Spanish (2nd Edition ) Nueva Editorial Interamericana, S.A., deC.V., Mexico

Vertebrate Biology                                    ISBN    0-7216-7017-2

© 1971 by W. B. Saunders Company.   Copyright 1961 and 1966 by W. B. Saunders Company.   Copyright under the International Copyright Union.   All rights reserved. This book is protected by copyright.   No part of it may be reproduced, stored in a retrieval system, or transmitted in any form or by any means, electronic, mechanical, photocopying, recording, or otherwise, without written permission from the publisher. Made in the United States of America.   Press of W. B. Saunders Company.   Library of Congress catalog number 76-139434.

Print No.:     9     8     7     6     5     4     3     2

# PREFACE

This is the third edition of a book that was essentially a pioneer in its field in 1961. Since that first experimental product, every effort has been made to meet the changing needs of students of vertebrate biology. In each succeeding edition various advances in physiological, behavioral, and systematic knowledge have been incorporated into the text. Likewise, significant current references have been added. More emphasis has been given to environmental factors in this edition than previously, and the chapter on population movements has been completely rewritten so as to present the student with the latest information as well as theories on this subject. The classification of mammals has been updated to concur with more recent concepts on their taxonomy.

The majority of the illustrations used are original. Many of the line drawings were made by Edward Robertson and some were done by Jacqueline Schonewald. The diagram of avian bills was the work of Gene Christman. For permission to use certain illustrations owned or copyrighted by them I am indebted to the following: Academic Press, New York; American Museum of Natural History, New York; Dr. Steven C. Anderson; Dr. David Caldwell; California Academy of Sciences, San Francisco; California Department of Fish and Game; David Cavagnaro; The Clarendon Press, Oxford; Cleveland Museum of Natural History; Dr. Nathan W. Cohen; *The Condor; Copeia;* Arthur D. Cushman; Doubleday and Co., Inc., Garden City, New York; *Ecology;* Dr. W. I. Follett; Dr. Raymond M. Gilmore; Fritz Goro, *Life* magazine; Dr. Bruce Halstead; Dr. and Mrs. G. Dallas Hanna; Hanover House, Garden City, New York; Harper and Brothers, New York; Ed N. Harrison; Holt, Rinehart and Winston, Inc., New York; Houghton Mifflin Co., Boston; Idaho Fish and Game Department; Richard Jennings; *Journal of Mammalogy; Journal of Wildlife Management;* Karl W. Kenyon; Dr. John Koranda; Dr. Alan E. Leviton; Dr. George E. Lindsay; Longmans, Green, and Co., London; Los Angeles County Museum of Natural History; Dr. George H. Lowery, Jr.; Eric Lundberg; The Macmillan Company, New York; Bruce Markham; Marineland of the Pacific, Palos Verdes Peninsula, California; Dr. James A. Mattison, Jr.; McGraw-Hill Book Co., Inc., New York; Dr. Charles F. Mohr; C. F. W. Muesebeck; Dr. Dietland Müller-Schwarze; National Audubon Society, New York; New York Zoological Society; Oxford University Press, New York; Peabody Museum of Natural History; Dr. E. T. Pengelley; the late Dr. Gayle Pickwell; Ronald Press Co., New York; Dr. Edward S. Ross; W. B. Saunders Co., Philadelphia; Dr. Arthur C. Smith; L. L. Snyder; Stanford University Press, Stanford, California; Dr. Tracy I. Storer; the late Cecil Tose; U.S. Bureau of Entomology and Plant Quarantine; U.S. Fish and Wildlife Service; U.S. Forest Service; U.S. National Park Service; U.S. Navy; U.S. Soil Conservation Service; University of California Press, Berkeley; Wildlife Management Institute; John Wiley and Sons, Inc., New York; Wyoming Fish and Game Commission.

Many friends and associates contributed to this work. I wish to mention especially the staff of W. B. Saunders Company, particularly Tyler Buchenau; the late Butler O'Hara; Rollin D. Hemens and Carroll G. Bowen of the University of Chicago Press; my wife, Dorothy B. Orr; Joy Bailey Osborn; the late Toshio Asaeda; Barbara Boneysteele Bortin; Lillian and Robert P. Dempster; Dr. W. I. Follett; Dr. Earl S. Herald; Dr. John Hopkirk; Dr. Alan E. Leviton; Jacqueline Schonewald. To these persons as well as other associates at the California Academy of Sciences, and at the University of San Francisco with which I was formerly affiliated, I wish to express my sincere thanks for helpful advice and constructive criticism. I am most grateful to Charlotte Dorsey for her meticulous editing and proofreading of this text.

ROBERT T. ORR

San Francisco

# CONTENTS

# Chapter One

# INTRODUCTION

Biology as a science has gone through many phases since its beginning several thousand years ago. For centuries its aim was purely descriptive—to describe various forms of life and their component parts. Relationship and function were little understood. Following the rise of Darwinism systematic biology began to develop and replace what had been a nomenclatural science, and descriptive anatomy was gradually supplemented by functional anatomy. These were great advances that led to an understanding of the origin of species and their relationship to each other as well as of the functional performance of the various biological systems within the individual organism.

Today many biologists have gone into the molecular field, trying to understand the minute biochemical differences in animals and plants. Their contributions toward a clarification of the mechanisms by which evolutionary changes occur has been great. Nevertheless, continued study of organisms as a whole, commonly called "organismal biology" as contrasted with "molecular biology," is vital. It means little to know the structure of a protein molecule representing a gene if the organism whose chromosomes contain this gene is not identified. Both organismal and molecular studies in biology are vital and must be carried on together, each complementing the other.

A broader, more modern approach to biology is seen in the study of the relationship of organisms to their environment. This is called ecology, and it is not a new subject. Courses in ecology were given in universities at least as early as the 1920s. However, mankind is now realizing its great importance, not just for the advancement of knowledge but for the very survival of living things, including man himself, on this planet today.

1

In addition to all the basic knowledge that we have acquired on the structure and classification of plants and animals, it is necessary to know their environmental requirements. This involves many things for many kinds of organisms. Environmental temperature, both daily and seasonal, humidity, rainfall, altitudinal limitations, soil characters, water availability, plant and animal associations, food requirements, parasites, enemies, and diseases are just a few of the subjects needing investigation. The ecological requirements of some species are so limited that the changing of one of these environmental factors could spell extinction, as we are rapidly learning.

*Vertebrate Biology* is not meant to be a text on ecology, but its approach to backboned animals is ecological rather than morphological. In the early chapters a brief account is given of the organ systems of each class of vertebrates for the benefit of students who have not taken comparative vertebrate anatomy. Much emphasis in each of these chapters, however, is upon special characters. Most of these are adaptations of the integumentary system and the appendages or are special sensory perceptive devices for particular modes of living. The major part of the book is concerned with the principles of systematic biology, factors governing distribution, methods used by vertebrates to solve environmental problems, reproductive physiology and behavior, and population dynamics.

The demand for persons trained in this field has increased greatly in the last few decades, and the need will be even greater in the future. Vertebrate species have long played a major role in human economy. They have provided man with food, sport, and clothing. Their utilization for these purposes has increased constantly with the enormous growth in human population that continues unabated year after year. How long we can continue the uncontrolled harvest of any living vertebrates is problematical, but surely it is not for very long. More than a century ago it was learned that such animals as fur seals, sea lions, elephant seals, bison, pronghorns, passenger pigeons, and many other species that seemed to be present in limitless numbers could withstand man's ruthless slaughter of their populations only for a relatively short time. The result for many was final or near extermination. Today we are still slaughtering whales to the point where the very existence of the largest creatures to inhabit the earth is threatened. We continue to overfish the sea because most of it is outside of territorial waters and a depletion in its population seems far off to many. It is estimated that since the year 1600, 36 species of mammals have become extinct, and at least 120 are presently in danger of extinction. During the same period of time 94 species of birds have become extinct and 187 have been in danger of complete extermination. Natural causes have accounted for some of this, but extinction of about three quarters of these birds and mammals in a little less than 400 years has been directly or indirectly caused by man.

Biologists are needed to study all aspects of vertebrate life. Many intensive investigations already have been conducted and others are under way, but vastly more information is needed on the environmental requirements, behavior, and reproductive activity of most vertebrates. Man himself, through air, land, and water pollution, is changing many environments to such a degree and so rapidly that native species are being eliminated at an alarming rate. In order to know the effects of pollution we must have a knowledge of the natural composition of the environment. This is becoming increasingly difficult because there are few if any parts of Earth today that have not been affected by human activity to some degree.

More studies have been made on land and freshwater vertebrates than on marine forms because among these are the principal game species as well as those that conflict with agricultural activities. A few investigations are concerned with the transmission of disease. In order to properly manage game fishes, birds, and mammals a very detailed knowledge of their lives is essential. The value of these animals in terms of dollars is enormous. This is obvious when we realize that tens of millions of persons go hunting or fishing in the United States each year. The equipment as well as the cost of travel and lodging to accomplish this runs into billions of dollars annually. Fishing and hunting with their associated activities are big

FIGURE 1.1. A beaver dam and pond in Rocky Mountain National Park.

business. As a consequence national governments as well as individual states and provinces have set up special bureaus to engage in wildlife management and carry on research as well as to recommend legislation in order to protect wildlife adequately. In the United States Department of the Interior the U.S. Fish and Wildlife Service is divided into two bureaus, the Bureau of Sport Fisheries and Wildlife and the Bureau of Commercial Fisheries. Each employs a large number of vertebrate biologists to study the proper methods of management for species under their jurisdiction. Over 30,000 biologists are in the employ of the United States Government alone.

Apart from the commercial and sporting value of vertebrates, they provide an important part of the recreational value of the out-of-doors. These are intangible assets which are difficult to measure in dollars and cents. In our national and state parks, wilderness areas, and wildlife refuges man has a chance to see relatively untouched segments of the land with living things in their natural setting. The sight of trumpeter swans with their young on a lake or bighorn sheep moving across a mountainside is a thrill long to be remembered. Deer, moose, elk, pronghorns, black bears, and even grizzlies are among the more dramatic mammals in such reservations, but smaller species such as beavers with their dams, muskrats, and otters are equally enjoyable to watch in an undefiled environment where they seem to know they have little to fear from this strange tall, two-legged species which elsewhere seems to spell their doom.

# Chapter Two

# FISHES AND FISHLIKE VERTEBRATES

## GENERAL CHARACTERS

For convenience the four recognized classes of fishes and fish-like vertebrates are considered together in this chapter. These include the Agnatha or jawless vertebrates represented by the extinct ostracoderms and the living cyclostomes (the lampreys and hag-fishes); the ancient jawed fishes or Placodermi; the Chondrichthyes or cartilaginous fishes such as sharks, rays, and chimaeras; and the Osteichthyes or bony fishes.

There are actually more fishes in the world today than any other group of vertebrates. The number of kinds of living fishes is estimated to be about 22,000, according to most ichthyologists, and over 20,000 of these are bony fishes. The total number of fishes exceeds that of all other kinds of vertebrates combined. This is not so surprising when we consider that approximately four fifths of the earth's surface is covered with water.

As a group fishes exhibit great range in size. The largest is the whale shark (*Rhineodon typus*), which is said to attain a maximum length of over 50 feet. The smallest known fish is a species of goby found in the Philippine Islands and named *Pandaka pygmea* by Dr. Herre of Stanford University in 1929. It is a little over one third of an inch long. The majority of fishes occur in the sea, but there are about

5

7000 species that are found in fresh water. Among the latter the largest are the paddlefishes, catfishes, and sturgeons. It is reported that the great paddlefish (*Psephurus gladius*) of the Yangtze River in China may attain a length of 23 feet. Some of the Asiatic and European catfishes have been recorded up to 10 feet, while sturgeon in the Volga River may be as long as 14 feet and weigh up to 2250 pounds. Even on the Pacific coast of the United States sturgeon measuring 12½ feet in length have been recorded. No freshwater fishes are known from either Greenland or Antarctica, because of the ice that permanently covers most of these land masses.

While some fishes are strictly confined to fresh water and are classified as *primary freshwater fishes*, there are others that can on rare occasions enter sea water or brackish water for certain periods of time. These are called *secondary freshwater fishes*.

Fishes are tolerant of rather wide extremes in temperature. Some species survive in hot springs where the water may reach 109° F. Others can live at temperatures close to freezing. Within any one species, however, the range of tolerance is generally quite restricted. Like other *poikilothermous* or so-called cold-blooded vertebrates, fishes are *ectothermic*. By this we mean that the body temperature is dependent upon the environment and, consequently, closely approximates it. A trout, therefore, inhabiting a stream where the water is 50° F. has a body temperature essentially the same as that of the stream.

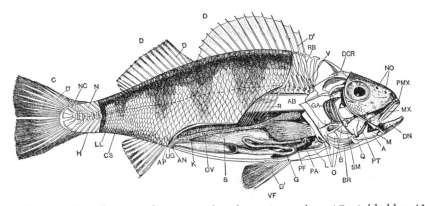

**FIGURE 2.1.**   The general anatomy of a teleost: *A*, angulare; *AB*, air bladder; *AN*, anus; *AP*, abdominal pore; *B*, bulbus arteriosus (conus is rudimentary); *BR*, branchiostegal rays; *C*, caudal fin; *CS*, cycloidal scales; *D*, dorsal fin; *D'*, dermal supports of fins; *DN*, dentary bone; *DCR*, dermal bones roofing cranium; *G*, intestine; *GA*, gill arches; *H*, haemal arch and spine; *K*, kidney (mesonephros); *L*, liver; *LL*, lateral line; *N*, neural arch and spine; *NO*, anterior and posterior nares; *O*, opercula; *OV*, ovary; *PA*, pancreas (pyloric appendices); *PF*, pectoral fin; *PMX*, premaxillae; *PT*, pterygoid; *Q*, quadrate; *R*, ribs, showing accessory supports; *RB*, radial and basal fin supports; *S*, stomach; *UG*, urinogenital opening; *V*, vertebra (centrum); *VF*, ventral fin. (From Dean, Bashford, 1895.)

Although most fishes are restricted to water, there are some, such as the mudskippers, that can walk about on land and even climb the lower branches of trees; others, such as certain lungfishes, may encase themselves in mud cells and live for months without water; and still others, such as flying fish, can glide through the air.

## SKELETAL SYSTEM

In cyclostomes the skull is cartilaginous and very incomplete. In the hagfishes it consists of a plate of cartilage on which the brain rests. The otic capsules which house the inner ears are joined to the posterior part of this plate. The lamprey skull (Fig. 2.2) is more elaborate with sides present, but a roof is lacking so that the brain is protected dorsally only by fibrous connective tissue. Jaws are lacking in cyclostomes. The round mouth is suctorial. The sucking disk, like the tongue, may or may not bear tooth-like structures which enable these animals to rasp off the flesh of prey to which they attach themselves. The gill area is supported by a complicated cartilaginous basket which is difficult to homologize with the gill supports of other vertebrates.

The Chondrichthyes possess a complete cartilaginous skull

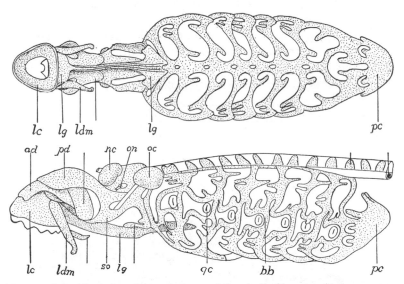

**FIGURE 2.2.** Ventral and lateral views of the skull of lamprey (*Petromyzon marinus*), after Parker. *ad*, anterior dorsal cartilage; *bb*, branchial basket; *gc*, gill cleft; *lc*, labial cartilage; *ldm*, lateral distal mandibular; *lg*, lingual cartilage; *nc*, nasal capsule; *oc*, otic capsule; *on*, optic nerve; *pc*, pericardial cartilage; *pd*, posterior dorsal cartilage. (Kingsley, J. S.: Outlines of Comparative Anatomy of Vertebrates. 3rd Ed., P. Blakiston's Son & Co.)

which is referred to as the *chondrocranium* (Fig. 2.3). Both the olfactory and otic capsules are incorporated into this structure. Although the skull may be partly calcified no true bone is present in living cartilaginous fishes. This is not regarded as a primitive condition. It is generally believed that some of the ancient agnathous vertebrates, represented by the ostracoderms, possessed an exoskeleton of dermal bone. This formed the outer armor which encased these jawless vertebrates. In many respects the outer exoskeleton of ostracoderms was basically similar to the denticles found on the skin of living elasmobranchs. It consisted of a hard outer coat, somewhat like the enamel of the vertebrate tooth, overlying a dentine-like layer. Beneath this there was a layer of spongy bone and under this compact bone.

In the modern cyclostomes all traces of these exoskeletal parts have disappeared. However, in elasmobranchs there are remnants of the outer part of the ostracoderm armor in the form of placoid scales (see p. 31).

The first or mandibular gill arch is modified to form an upper and lower jaw, known respectively as the *palatoquadrate cartilage* and *Meckel's cartilage.* The upper jaw is rather loosely attached to the skull in sharks and rays and is supported posteriorly by the upper or hyomandibular element of the second arch. This type of jaw suspension is referred to as *hyostylic.* In the aberrant chimaeras, the upper jaw is firmly fused to the cartilaginous skull and is, therefore, *autostylic.*

Great variation exists with respect to the degree of ossification in the Osteichthyes. In the lower ganoids such as the sturgeons, the chondrocranium is essentially unossified so that replacement bone is lacking. However, there are a number of dermal or membrane bones, presumably derived from scales that have sunk beneath the surface, that cover the roof of the skull. In the higher bony fishes dermal

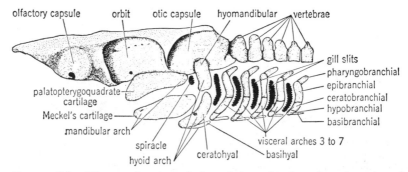

**FIGURE 2.3.**  Diagram showing relation of visceral arches of an elasmobranch to the chondrocranium, vertebral column, and gill slits. (Weichert, C. K.: Anatomy of the Chordates. McGraw-Hill Book Co., Inc., 1958.)

bones are exceedingly numerous, forming an armor around the skull, and parts of the chondrocranium are replaced by bone in the occipital, sphenoidal, otic, and ethmoid regions (Fig. 2.4). Jaw suspension in bony fishes may be either hyostylic as in the Holostei and Teleostei or autostylic as in the Dipnoi.

In bony fishes the mandibular cartilage or its derivatives becomes invested with dermal bones. These are more numerous in ancient fossil forms, especially crossopterygians. The dentary or tooth-bearing bone occupies the anterior and dorsal surface of the jaw with one or more splenials, an angular, a surangular, and a coronoid. Furthermore, the head of Meckel's cartilage may be replaced by a bone called the articular which articulates with the quadrate.

There is no true vertebral column in cyclostomes, the axial skeleton consisting primarily of the notochord. In the hagfishes small cartilaginous elements are present in the caudal region, and in the lampreys two pairs of cartilaginous elements occur on either side of the spinal cord in each body segment. They do not meet above, however, to form a neural arch. ~) arcualia

In some of the lower bony fishes, such as the sturgeon, the notochord persists essentially unchanged and, although neural and haemal arches are present, a centrum is lacking. The chimaeras and lungfishes show but a slight advance over this condition with cartilage beginning to invade the notochordal sheath. In most fishes, however, there is a well developed centrum. Generally fish vertebrae are divisible into two groups, the trunk or preanals and the caudal or postanals. The latter are recognizable by the presence of a haemal arch on the underside of the centrum or the notochord. The ends of the centra in most fishes are concave, a condition referred to as *amphicoelous*.

Paired fins are lacking in cyclostomes, but unpaired median fins, supported by cartilaginous rays, are present. As will be discussed later in this chapter, there is a great deal of variation in both the median and paired fins of fishes. In general, the structure of the front or pectoral appendages is more complicated than that of the hind or pelvic appendages. In sharks the girdle that supports the pectoral fins consists of a curved cartilaginous bar. The ventral part between the bases of the fins is called the *coracoid bar* and the processes that extend dorsal to the bases of the fins on each side are referred to as the *scapular processes*. In bony fishes these structures may remain cartilaginous or may be ossified into two pairs of bones, the coracoids and the scapulars. Furthermore, in this group several dermal bones, presumably derived from dermal scales, may join in the pectoral girdle complex. These are the paired *post-temporals,* the *supracleithrums,* the *cleithrums,* and the *clavicles.* The post-temporals, when

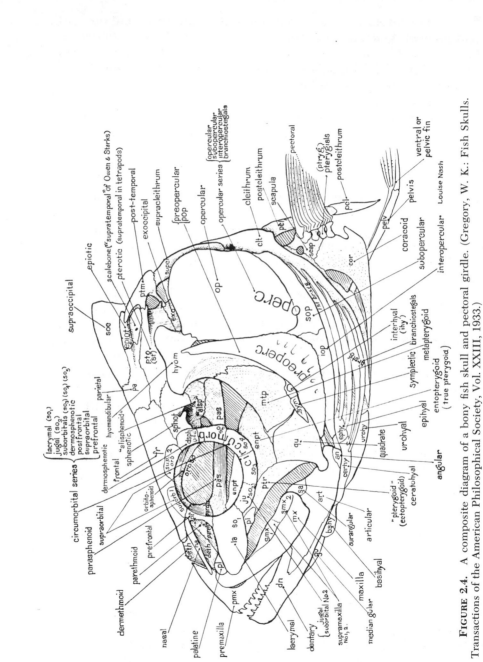

**FIGURE 2.4.** A composite diagram of a bony fish skull and pectoral girdle. (Gregory, W. K.: Fish Skulls. Transactions of the American Philosophical Society, Vol. XXIII, 1933.)

bony fish

present, are dorsalmost and attach on either side to the posterior part of the skull. They extend forward above the gill chamber. This type of pectoral girdle is most highly developed in dipnoans and crossopterygians. In the higher ray-finned fishes some of the dermal bones have been lost (Fig. 2.4).

The pelvic girdle of fishes is less complicated than the pectoral girdle and, unlike that of higher vertebrates, is never attached to the vertebral column. In sharks it consists of a ventral *puboischiac bar* which extends across the body between the bases of the fins, and a small *iliac process* on each side which projects dorsally above the point of articulation with the fins. In higher bony fishes the pelvic girdle is often reduced or missing.

## MUSCULAR SYSTEM

The primary function of the muscular system is to effect movement of various parts of the body. In higher vertebrates, such as birds and mammals, these movements may be very complicated. For example, ordinary conversation with another person requires the use of a large number of muscles in our own body. We must have muscles to move the tongue, the lips, the cheeks, the neck, the larynx, and many other structures, especially if we gesticulate. Voluntary muscular movement in fishes is far less complicated and serves principally to raise and lower the mouth, move the eyes, open and close the gill apertures, move the fins, and produce lateral motion on the part of the body so as to effect locomotion through the water. This requires only a relatively simple muscular system.

In all types of fishes the body muscles tend to maintain their primitive segmental arrangement with septa, known as *myocommata,* present between the successive myotomes. In the cyclostomes there is no separation of the body or *axial musculature* into dorsal and ventral groups. In the cartilaginous and bony fishes the axial musculature is separated by a lateral or horizontal septum into dorsal *epaxial muscles* and ventral *hypaxial muscles* (Fig. 2.5). These two major divisions may also be distinguished by the fact that the epaxial muscles are innervated by dorsal branches of the spinal nerves, while the hypaxial muscles are innervated by ventral branches of the spinal nerves. The epaxial body musculature of fishes consists of an essentially undifferentiated mass of segmented muscle and is referred to as the *dorsalis trunci.*

In the head region, during the embryonic development of fishes as well as higher vertebrates, there are three pairs of somite-like mesodermal blocks called the *premandibular, mandibular,* and *hyoid somites.* The premandibular somite, which is innervated by

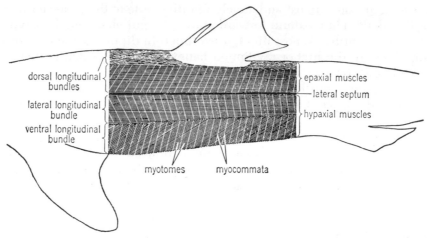

dorsal longitudinal bundles

lateral longitudinal bundle

ventral longitudinal bundle

epaxial muscles

lateral septum

hypaxial muscles

myotomes     myocommata

**FIGURE 2.5.**   Lateral view of portion of body wall of dogfish (*Squalus acanthias*) showing zigzag arrangement of the myotomes. The skin has been removed to expose the muscle fibers beneath. (Weichert, C. K.: Anatomy of the Chordates. McGraw-Hill Book Co., Inc., 1958.)

the third cranial nerve, the oculomotor, gives rise to four of the six *extrinsic muscles* of the eye. These are the superior, inferior, and internal (anterior) rectus and the inferior oblique muscles. The superior oblique muscle, innervated by the fourth cranial nerve, the trochlear, arises from the mandibular somite, and the external (posterior) rectus muscle, which is supplied by the sixth cranial nerve, the abducens, comes from the hyoid somite. These six muscles originate on the orbit and insert on the eyeball. They show relatively little difference in the various classes of vertebrates.

    The closing and opening of the gill apertures and mouth is controlled by what are referred to as the *branchial muscles*. These consist principally of *constrictors* (both dorsal and ventral) and *levators*. These muscles differ from the axial and appendicular muscles in that they are innervated by cranial rather than spinal nerves.

    There is another group of muscles called the *hypobranchial muscles* that extend forward beneath the gills from the coracoid region to the jaw and ventral part of the gill arches. These are true axial muscles, even though they are in the branchiomeric region, and are innervated by spinal nerves. The fin musculature of most fishes consists of dorsal *extensors* and ventral *flexors*.

## CIRCULATORY SYSTEM

    Except in the Dipnoi, the circulatory system of fishes, ranging from cyclostomes to teleosts, is essentially a single system in which only unoxygenated blood goes to the heart (Fig. 2.6). From there it is pumped to the gills, aerated, and then distributed to the body. The heart possesses four chambers, but only two of these (the atrium and the ventricle) correspond to the four chambers (paired atria and paired ventricles) of the higher vertebrates. The first or receiving chamber in the heart of a fish is called the *sinus venosus.* It is thin-walled like the next chamber or *atrium* into which the blood then passes. From the atrium blood passes into the thick-walled *ventricle* and is pumped out through the *conus arteriosus* to the ventral aorta. Blood from the ventral aorta then goes to the gill region to be oxygenated by way of *afferent branchial vessels*, and then from the gills by way of *efferent collector loops* to the dorsal aorta. These vessels are derivatives of the *aortic arches* which connect the ventral and dorsal aortae. In the embryological development of most living vertebrates, six pairs of these arches make their appearance, although they may later be reduced or highly modified.

There are several major venous systems found in fishes. Entering the sinus venosus on each side is a vessel known as the *common cardinal vein* or *duct of Cuvier* which is formed by a fusion of the anterior and posterior cardinals. Blood from the head is collected by the *anterior cardinals,* and blood from the kidneys and gonads is collected by the *posterior cardinals.* Also entering the ducts of Cuvier are the paired *lateral abdominal veins* which receive blood from the body wall and the paired appendages. The *renal portal system* consists of the caudal vein and two renal portal veins which are situated lateral to the kidneys. Blood from the tail region passes from the caudal vein to the renal portals and then into capillaries in the kidneys. The *hepatic portal system* picks up blood from the stomach and intestines and returns it to the liver, where, after passing through

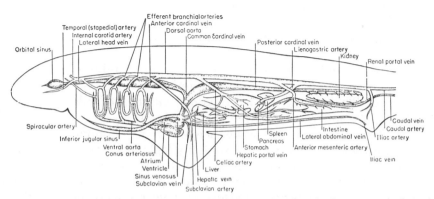

**FIGURE 2.6.** Diagram of the main blood vessels of a shark as seen in lateral view. (Romer, A. S.: The Vertebrate Body, 4th Ed., W. B. Saunders Co., 1970.)

a series of sinusoids, it enters the sinus venosus by way of the paired *hepatic veins.*

Within different groups of fishes there are certain modifications of the circulatory system. In cartilaginous fishes the conus arteriosus has a number of sets of valves which prevent any backward flow of blood into the heart. These are reduced to a single set in bony fishes. In dipnoans we find that a septum dividing the atrium into a right and left chamber is present. This is correlated with the use of the swim bladder as an organ of respiration and represents the first step toward the development of the double-type circulatory system whereby both oxygenated and unoxygenated blood enter the heart and are kept separate. Blood from the right atrium of the lungfish passes into the right side of the ventricle, and from here is pumped to the primitive lung-like gas bladder by way of the *pulmonary arteries* which branch off the sixth pair of aortic arches. The oxygenated blood returns to the left atrium of the heart by way of pulmonary veins very much as in amphibians.

In cyclostomes the number of pairs of aortic arches corresponds to the number of pairs of gill pouches. Although six is considered to have been the basic number of aortic arches for fishes, this number is reduced to five even in sharks and rays with the loss of the first pair. In most teleost fishes the second pair of arches is also gone. Likewise, in teleosts we find that lateral abdominal veins are lacking so that blood from the *subclavians,* draining the pectoral appendages, enters the sinus venosus directly, and blood from the *iliac veins,* draining the pelvic appendages, passes into the postcardinals. Dipnoans, which show many characters intermediate between fishes and amphibians, have a single *ventral abdominal vein,* presumably derived from a fusion of the lateral abdominals. This vein receives blood from the iliacs by way of paired *pelvic veins* and enters the right duct of Cuvier. Furthermore, from the right postcardinal system a new vein that is to be of major importance in higher vertebrates makes its appearance in lungfishes. This is the *postcaval.* It connects with the caudal vein and passes forward through the liver to the sinus venosus.

In fishes and their relatives, as in higher vertebrates, oxygen is carried to the tissues by means of a red blood pigment called *haemoglobin.* This substance is contained in the red blood cells or *erythrocytes* and is transported in the plasma to all tissues of the body. Rather recently it has been found that the antarctic ice fish (*Chaenocephalus aceratus*) is essentially lacking in erythrocytes and, consequently, haemoglobin. These fish live in deep cold water where there is little fluctuation in temperature and food is abundant. They therefore get little exercise and have reduced their need for oxygen. Such oxygen as is required is absorbed slowly from the water.

## DIGESTIVE SYSTEM

Cyclostomes lack teeth in their round suctorial mouths, but they do have horny ridges which serve to tear flesh off their prey. Sharks possess highly developed sets of teeth (see under Scales, p. 29) which enable them to secure and tear other organisms. In rays and chimaeras, teeth have taken on the form of broad plates for crushing mollusks and other hard-shelled organisms which such bottom-feeders live upon (Fig. 2.7). Analogous plates are found in the dipnoans. Ray-finned fishes have well developed teeth which are usually conical. In the higher forms these teeth are often palatal and pharyngeal rather than maxillary and may be joined at the base by a cement-like substance.

Since fishes spend their lives in water, they do not need oral glands to moisten their food. There are, however, some mucous glands present in the mouth. The esophagus in fishes is generally very short and little differentiated from the stomach. Cyclostomes have an intestine that is essentially a straight tube with a spirally arranged, longitudinal flap extending into it. In elasmobranchs, the intestine is divided into a small and large portion. The former is characterized by the presence of a spiral valve which greatly increases the absorptive surface. A spiral valve is present in the small intestines of a few of the more primitive bony fishes, but is lacking in higher forms in which the absorptive surface is increased by the elongation and coiling of the gut.

## RESPIRATORY SYSTEM

An *internal gill* type of respiratory system is characteristic of fishes. Embryologically, gill clefts develop as a result of a series of evaginations from the pharynx which grow outward and meet corre-

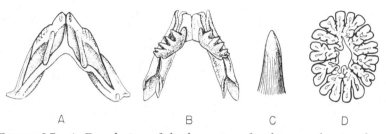

A          B        C      D

**FIGURE 2.7.** *A*, Dorsal view of the lower jaw of a chimaera showing the tooth places; *B*, a dorsal view of the lower jaw of a dipnoan with fan-shaped tooth plates; *C*, an external view of a crossopterygian tooth; *D*, a cross section of a crossopterygian tooth showing the labyrinthine structure. (*A* after Dean; *B* after Watson; *C* and *D* after Bystrow. In Romer, A. S.: The Vertebrate Body. 4th Ed., W. B. Saunders Co., 1970.)

sponding invaginations from the outside. At each point of junction a cleft or aperture is formed so that water coming into the pharynx from the mouth may pass to the outside. The partitions between these passageways contain cartilaginous or bony supports for the gill filaments which are located on each side. Each gill or *holobranch,* therefore, is separated into two parts or *hemibranchs*.

The gill *lamellae* or plates themselves consist of thin, pleated coverings of respiratory epithelium overlying vascular networks associated with the aortic arches so that carbon dioxide in the blood may be exchanged for dissolved oxygen in the water.

There is considerable variation in the gill apparatus in different kinds of fishlike vertebrates (Fig. 2.9). In cyclostomes the gill lamellae are situated in pouches which may number from six to 14 pairs, depending on the species. Each pouch has an internal and an external opening. The former opens either directly into the pharynx or into a tube which connects with the pharynx anteriorly. The external apertures either open directly to the outside or are connected by tubes to a single external opening. When lampreys and hagfishes are attached to prey by means of their suctorial apparatus, they are

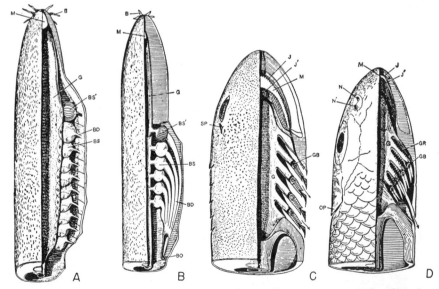

**FIGURE 2.8.** Heads of various fishes to show gill arrangement: *A*, the slime-hag *Bdellostoma; B*, the hagfish *Myxine; C*, a shark; *D*, a teleost. The right half of each is sectioned horizontally through the pharynx. Abbreviations: *B*, barbels around the mouth; *BD*, ducts from gill pouches; *BO*, common outer openings of gill pouches; *BS*, gill sacs; *BS'*, sacs sectioned to show internal folds of gills; *G*, gut (pharynx); *GB*, cut gill arch; *GR*, gill rakers; *J* and *J'*, upper and lower jaws; *M*, mouth; *N* and *N'*, anterior and posterior openings of nasal chamber; *OP*, operculum; *SP*, spiracle. (From Dean, in Romer, A. S.: The Vertebrate Body. 4th Ed., W. B. Saunders Co., 1970.)

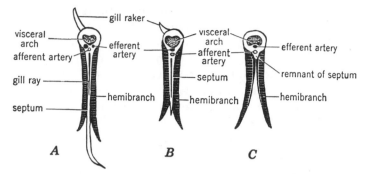

**FIGURE 2.9.** Types of fish gills: *A*, elasmobranch; *B*, chimaera; *C*, teleost. (Weichert, C. K.: Anatomy of the Chordates. McGraw-Hill Book Co., Inc., 1958.)

unable to take in water through the mouth. Consequently, in order to respire at such times, water may be taken into, as well as expelled from, the gill pouches through the external gill openings.

Sharks and rays have from five to seven pairs of functional gill slits plus an anterior pair of nonrespiratory openings referred to as the *spiracles*. The hemibranchs are separated from one another by a well developed interbranchial septum which extends outward from the cartilaginous arch (Fig. 2.9). Distally, each of these interbranchial

**FIGURE 2.10.** The American Pacific leopard shark (*Triakis semifasciata*) possesses five pairs of functional gill slits in addition to the spiracles that are located posterior to each eye. (Photographed at Steinhart Aquarium.)

septa forms a flap or covering for the next posterior gill opening. Thus each gill aperture opens separately to the outside.

In higher bony fishes the interbranchial septum is lacking so that the hemibranchs on the anterior and posterior part of each branchial arch are no longer separated from one another. Furthermore, with the loss of the interbranchial septa we find that in these fishes the gill apertures no longer open separately to the outside. Instead, the gills are enclosed in a single chamber and covered externally by a large bony *operculum* which opens and closes posteriorly to permit water to pass to the outside. In most bony fishes there are four pairs of functional gills, but the number varies from two to six.

Larval lungfishes possess *external gills* very similar to those of amphibian larvae. These are filamentous structures that are attached to the outer edges of the branchial arches and not homologous to internal gills. They disappear, however, in the adults.

It is believed that primitive bony fishes developed lunglike structures from which the tetrapod respiratory system arose. The African bichir (*Polypterus*) (Fig. 2.11), which is a living representation of the ancient chondrosteans, supports this theory. In *Polypterus* paired ventral lungs are present which enable these fishes to survive during periods of drought.

The dipnoans, members of the subclass Sarcopterygii which branched off from the Actinopterygii after bony fishes were established, likewise have a lunglike structure. In all the living lungfishes

**FIGURE 2.11.** The African bichir (*Polypterus ornatipennis*), a relict chondrostean that possesses paired central lungs and fleshy lobed paired fins. (Photographed at Steinhart Aquarium.)

the lung is dorsal to the gut although connected by a tube to the ventral side of the esophagus. It is bilobed in the African (*Protopterus*) and South American (*Lepidosiren*) species but unpaired in the Australian lungfish. The dipnoan lung, unlike that of *Polypterus*, contains internal chambers or pockets to increase the respiratory surface and is heavily vascularized by branches of the pulmonary veins and arteries.

No lunglike structure is present in either the jawless or the cartilaginous fishes. In most bony fishes, apart from *Polypterus*, the dipnoans, and presumably the crossopterygians, the primitive lung has become modified into a gas bladder or *hydrostatic organ* which may or may not be connected with the esophagus by means of a dorsal connection. By means of glands the amount of gas in the bladder may be increased or decreased so as to maintain the body at various levels in the water. Gas bladders are usually lacking or degenerate in species that live on the bottom of the ocean, such as flatfishes. There are some pelagic teleost fishes that lack a gas bladder. Since the bodies of some of these have a greater specific gravity than that of sea water, buoyancy must be effected by swimming. In a number of kinds the pectoral fins are extended to produce a lift similar to that of an airplane. This of course is regularly seen in elasmobranch fishes. In a few kinds of fishes gas bladders may function either for sound reception or for sound production. Some fish, such as carp, receive vibrations from the water by means of the gas bladder and transmit these, through a chain of small bones,

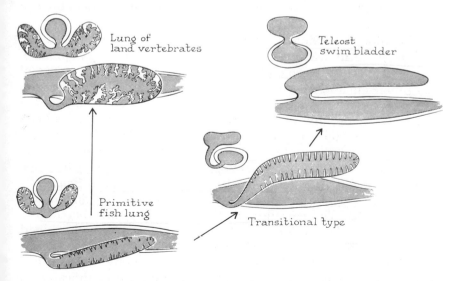

**FIGURE 2.12.** Evolution of lungs and swim bladder. (After Dean, in Villee, Walker, and Smith: General Zoology. W. B. Saunders Co., 1968.)

derived from the anterior vertebrae and known as the *Webberian ossicles,* to the inner ear. Certain fishes make use of the gas bladder directly or indirectly to produce sounds. The croakers and drumfishes, members of the family Sciaenidae, have muscles attached to the bladder which produce a snapping sound that is amplified by the gas-filled chamber. The bladder also serves as a resonance chamber for the grunts (family Pomadasyidae), which produce sounds by grinding their pharyngeal teeth together. These sounds are sometimes audible when the fish are removed from the water, but in water they can readily be detected with the use of a hydrophone.

## UROGENITAL SYSTEM

The excretory system in fishes, as in other vertebrates, serves several purposes. It regulates the water content of the body, maintains the proper salt balance, and eliminates nitrogenous waste resulting from protein metabolism. To accomplish these ends vertebrates have developed three types of kidneys, which are commonly, though perhaps incorrectly, referred to by their embryological names of *pronephros, mesonephros,* and *metanephros.* All three are fundamentally alike, differing principally in their relationship to the blood system, in degree of complexity, and in efficiency.

The pronephros is the most primitive and, while present in the embryonic development of all vertebrates, is functional in the adult of none. Some larval cyclostomes, however, have a kidney, part of which appears to be homologous to the embryonic pronephros of higher vertebrates.

The functional kidney of fishes, sometimes referred to as an *opisthonephros* to distinguish it from the embryonic mesonephros of amniotes, is of the mesonephric type. This consists of a series of renal tubules that in early development exhibit a segmental arrangement which is later lost. Each tubule is coiled or convoluted, both proximally and distally, and leads into a common longitudinal collecting duct called the *archinephric duct.* This in turn leads to the outside, usually by way of the *cloaca* or common chamber receiving products from the digestive and urogenital system. Proximally, each tubule terminates in a hemispherical capsule known as *Bowman's capsule,* into which a capillary knot or *glomerulus* from the circulatory system fits. The capsule and the glomerulus together form a *renal corpuscle.* Water, salts, and waste products from the blood stream pass into the capsule and then may flow through the tubule to the archinephric duct and ultimately out of the body. This system, however, as we shall see, is subject to considerable modification to suit the needs of various groups of fishes. This is particularly evident

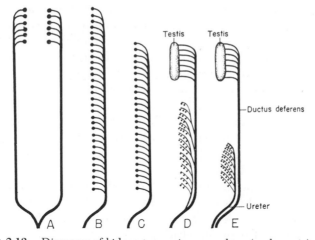

**FIGURE 2.13.**  Diagrams of kidney types: *A*, pronephros (embryonic); *B*, theoretical holonephros (each trunk segment with a single tubule), much as in a young hagfish or apodous amphibian; *C*, primitive opisthonephros: pronephros reduced or specialized, tubules segmentally arranged, as in hagfish; *D*, typical opisthonephros: multiplication of tubules in posterior segments, testis usually taking over anterior part of system, trend for development of additional kidney ducts (most anamniotes); *E*, metanephros of amniotes: an opisthonephros with a single additional duct, the ureter, draining all tubules. In *A*, both sides of the body are included; in *B* to *E*, one side only. (Romer, A. S.: The Vertebrate Body. 4th Ed., W. B. Saunders Co., 1970.)

with respect to the archinephric duct in the male. In sharks it functions as a gonadal duct and the kidneys have developed new, accessory urinary ducts.

The problem of maintaining a proper salt and water balance is an important one in fishes, since they live in a liquid medium which may be salty or fresh. Salt water tends to dehydrate the body, which is only slightly saline, and it also increases the salt concentration in the body. Fresh water has the opposite effect. The kidney of fishes, therefore, must play a major part in maintaining the proper balance within the body. Some marine bony fishes have salt-excreting glands on the gills which help eliminate excessive salt. Most marine fishes, however, other than elasmobranchs, have the renal corpuscles very small so as to reduce water loss. Since the corpuscle is the filter, the smaller it is the less filtrate passes through. In most types of kidney, nevertheless, we find that an excess of water and many valuable products are filtered out through the renal corpuscles. The convoluted tubules here play an important role in conservation. In these parts of the nephric tubules desirable products, including much of the water that would otherwise be lost, are reabsorbed into the blood stream. The renal corpuscles are much larger in freshwater fishes than in marine species. This means more liquid output, which is necessary to prevent overdilution of the body fluids.

Elasmobranchs, unlike most marine fishes, have large renal corpuscles and a relatively large water output as do freshwater fishes.
Two factors are responsible for the maintenance of a proper water
balance in their bodies. First, they have the ability to maintain a very
high concentration of nitrogenous waste in the form of urea in the
blood stream, amounting to about 300 to 1300 milligrams per 100
milliliters of fluid. Thus, while the salt concentration in the body of a
shark is no higher than that of other fishes, the presence of a large
amount of urea raises the osmotic pressure within the body to approximately that of sea water and there is no danger of dehydration. This is
also true of the coelacanth (*Latimeria chalumnae*) and the African
lungfish (*Protopterus aethiopicus*). Second, there is a gland known as
the *rectal gland*, which is attached to the posterior part of the intestine and serves as a salt-excreting organ. It is interesting to note that
recent studies on freshwater bull sharks (*Carcharhinus leucas*) from
Lake Nicaragua have shown that these glands exhibit regressive
changes and are hypofunctional as compared with individuals of the
same species in marine waters. While the urea content in freshwater
sharks in Lake Nicaragua, as well as in Malaysia, is only 25 to 35 per
cent that of marine species, it is far above the level found in other
kinds of vertebrates. An exception among elasmobranchs, however,
has recently been found in rays of the genus *Potamotrygon* occurring
in the Amazon basin of Brazil. In these fishes urea retention is only 2
to 3 milligrams per 100 milliliters of fluid. These freshwater sting-
rays live several thousand miles from the sea, unlike the Nicara-
guan and Malaysian sharks that live only a few miles from the coast,
and no doubt have lost the ability to retain urea because of much
longer isolation from a hypertonic environment.

A bladder for the temporary storage of urine is present in some
fishes, but it generally consists merely of an enlargement of the
lower end of the excretory ducts.

Hermaphroditism is relatively rare among vertebrates, but it
does occur in cyclostomes and, although not common, is rather widespread among bony fishes, occurring in at least 13 families. In
hagfishes the anterior part of the gonad produces ova and the posterior part sperm. In any one individual, however, the gonad at maturity functions only as an ovary or a testis, not as both at once. The
gonads are not paired in cyclostomes and ducts are absent. The ova
or sperm pass directly into the coelom and to the outside through
genital pores. Most bony fishes that are functionally hermaphroditic
are marine, although a few are freshwater species such as the neotropical killifish (*Rivulus marmoratus*) and the Asiatic synbranchid
(*Monopterus albus*). Some are like the hagfishes; they function first as
one sex, and then transform into the opposite sex. Others possess
ovotestes in which the mature ova are fertilized by sperm produced

within the same organ before they are laid (see Hermaphroditism, p. 422).

Cartilaginous and bony fishes have paired gonads and the sexes are generally distinct (Figs. 2.14 and 2.15). The female usually has two oviducts. In elasmobranchs the upper ends of these oviducts are fused so that there is a single, funnel-like opening or *ostium* into which eggs from the ovaries may pass. Also in this group each oviduct possesses a shell gland, since the eggs are fertilized internally and then encased in a horny shell. Some elasmobranchs are *oviparous* and deposit their eggs in the water, whereas others are *ovoviviparous* and hatch them internally in an enlargement of the lower end of the oviduct called a *uterus*.

In most vertebrates the ovary is not directly connected with the oviduct so that theoretically the eggs enter the coelom and then pass into the ostium. Actually, the relationship between these two structures is so close that there is little chance for the ova to enter the body cavity freely. In some bony fishes, however, where prodigious numbers of eggs are produced during the short breeding season, the ovaries are continuous with the oviducts so as to prevent the ova from escaping into the coelom. Likewise, in many teleosts the lower parts of the paired oviducts are fused. Most bony fishes are oviparous, but there are some in which the eggs hatch internally.

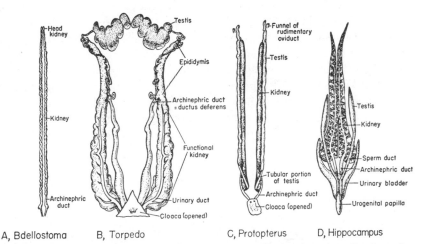

A, Bdellostoma    B, Torpedo       C, Protopterus    D, Hippocampus

FIGURE 2.14.   Urogenital systems in ventral view of males of *A*, the slime hag, *Bdellostoma; B*, the elasmobranch, *Torpedo; C*, the lungfish, *Protopterus; D*, a teleost, the seahorse *Hippocampus*. In *A* the testis, not shown, is pendent from a mesentery lying between the two kidneys and has no connection with them. In *B* the testis has appropriated the anterior part of the kidney as an epididymis, much as in most land vertebrates, and utilizes the entire length of the archinephric duct as a sperm duct. In *C* the testis ducts drain, on the contrary, only into the posterior part of the kidney and thence to the archinephric duct. In *D* the sperm duct is entirely independent of the kidney system. (*A* after Conel; *B* after Borcea; *C* after Kerr, Parker; *D* after Edwards. In Romer, A. S.: The Vertebrate Body. 4th Ed., W. B. Saunders Co., 1970.)

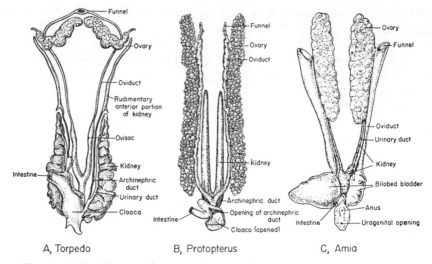

A, Torpedo            B, Protopterus            C, Amia

**FIGURE 2.15.** Urogenital systems in ventral view of females of *A,* the elasmobranch, *Torpedo; B,* the lungfish, *Protopterus;* and *C,* the primitive actinopterygian, *Amia.* (In *Torpedo* the shell gland is not developed.) (*A* after Borcea; *B* after Parker, Kerr; *C* after Hyrtl, Goodrich. In Romer, A. S.: The Vertebrate Body. 4th Ed., W. B. Saunders Co., 1970.)

In all vertebrates above cyclostomes, sperm passes from the testis to the outside by way of a duct. The elasmobranch testis is connected to the anterior part of the archinephric duct by means of some renal tubules so that much of the duct functions for genital purposes. Fertilization is internal in these cartilaginous fishes, the males having developed copulatory organs from the inner parts of the pelvic fins. Each of these fins has a clasper which is grooved medially. When placed together they form a tube which leads from the cloaca. The gonadal duct in male teleosts, unlike that of most other vertebrates, appears to have an independent origin. It may join the archinephric duct posteriorly or may open separately into the cloaca.

## NERVOUS SYSTEM

The vertebrate central nervous system arises embryologically from a dorsal strip of ectoderm that is known as the *medullary plate.* Very early in development this plate, which parallels the long axis of the body, invaginates to form a hollow tube of potential nerve tissue called the *neural tube.* The anterior end of this tube grows much more rapidly than the remainder and soon gives rise to a primitive brain consisting of three primary vesicles. The anterior vesicle is known as the forebrain or *prosencephalon,* behind which is the midbrain or *mesencephalon.* The posterior vesicle is the hindbrain or *rhombencephalon* and is continuous posteriorly with the rest of the neural tube which gives rise to the spinal cord and much of the remainder of the nervous system. Even in the most primitive of living fishlike vertebrates, the forebrain and hindbrain each subdivide into two parts so that there are five brain vesicles. Proceeding from anterior to posterior these are known as the *telencephalon, diencephalon, mesencephalon, metencephalon,* and *myelencephalon.*

The importance of each of these five divisions of the brain varies in different groups of vertebrates and as we proceed from the lower to the higher forms we find a marked increase in nervous tissue, which necessitates folding. These folds, which are required to accommodate the brain within the cranial cavity, are called *flexures.*

In all vertebrates the telencephalon gives rise to the olfactory lobes and the cerebrum. However, the relative importance of these two derivatives of the anteriormost brain vesicle changes markedly as we proceed up the phylogenetic scale. In fishes the telencephalon is primarily olfactory in function and the olfactory lobes are large, especially in elasmobranchs. The cerebral hemispheres are barely indicated in cyclostomes, but are better developed in cartilaginous and bony fishes. The cerebrum in fishes, however, consists of a basal ganglionic mass known as the *corpus striatum* and a thin, dorsal, epithelial layer called the *pallium.* It is the pallium that in higher vertebrates is invaded by gray matter and becomes the center of mental activity. In fishes the pallium is composed essentially of non-nervous tissue and the center of brain activity is back in the mesencephalon.

The diencephalon gives rise to the *thalamus,* which serves primarily as a relay center for olfactory and visual impulses and is associated with several sensory and glandular structures. From the

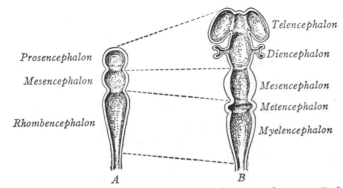

**FIGURE 2.16.** Subdivisions of the brain: *A,* three-vesicle stage; *B,* five-vesicle stage. (Arey, L. B.: Developmental Anatomy. 7th Ed., W. B. Saunders Co., 1965.)

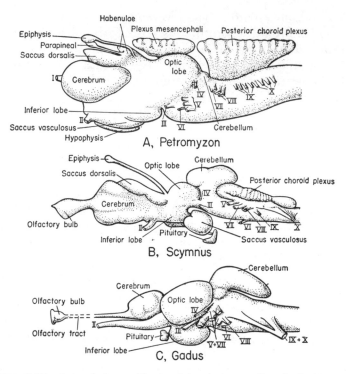

**FIGURE 2.17.** Lateral views of brain of *A*, a lamprey; *B*, a shark; *C*, a codfish. In the lamprey an exceptional condition is the development of a vascular choroid area, the plexus mesencephali, on the roof of the midbrain. (After Bütschli, Ahlborn, in Romer, A. S.: The Vertebrate Body. 4th Ed., W. B. Saunders Co., 1970.)

dorsal wall two median eye-like structures may arise. The anterior one is called the *parietal body* and the posterior one the *pineal body*. Both structures are present in cyclostomes, but in most fishes only the pineal body is present. The ventral part of the diencephalon gives rise to the stalk or infundibulum of the hypophysis as well as the posterior lobe. Since the sensory and pigmental layers of the retina of the eyes arise as outgrowths from the walls of the diencephalon we find that the optic stalks connect ventrally with this part of the brain. Where they meet, certain fibers from each optic nerve cross to the opposite side forming what is termed the *optic chiasma*.

The mesencephalon undergoes relatively little change in the vertebrate series, but its importance is far greater in lower than in higher forms. In fishes the midbrain is the center of nervous coordination. Dorsally, it develops two prominences known as the *optic lobes*, which are concerned with visual reception. In some bony fishes these lobes exceed the rest of the brain in size.

The metencephalon gives rise, dorsally, to the *cerebellum*. Since the cerebellum is a center of muscular coordination, we find that its degree of development is coordinated with bodily activity. Cyclostomes and dipnoans, which are sluggish and slow moving, have a very poorly developed cerebellum. On the other hand, sharks, which are rapid-moving fishes, have a relatively large cerebellum.

The myelencephalon forms the *medulla* of the brain, which, in many respects, resembles the spinal cord with which it is continuous posteriorly. The medulla is extremely important in all vertebrates since it serves as a center for many vital body activities, including respiration, heart action, and metabolism. It also serves as a relay center, and in fishes is a center for the lateral line system and inner ear.

Fishes, like amphibians, possess only 10 cranial nerves. The spinal nerves of cyclostomes differ from those of other vertebrates in that the sensory and motor roots tend not to join together to form a common trunk. In lampreys they remain separate and emerge alternately from the spinal cord, whereas in hagfishes there may be an incomplete union. In all other fishes the sensory root is dorsal and the motor root ventral, and the two join to form a common trunk. This union, however, takes place outside of the vertebral column.

Little is known about the autonomic nervous system in cyclostomes, although sympathetic ganglia are present. The digestive system and heart are supplied by the vagus nerve.

In elasmobranchs an irregular series of sympathetic ganglia lies along the body wall. Fibers from these ganglia connect both to the spinal cord and to the smooth musculature of the digestive and circulatory systems. The autonomic system of bony fishes is more advanced, with the sympathetic ganglia arranged in a chain and extending forward to the trigeminal nerve.

## SENSE ORGANS

Taste buds are stimulated by chemicals in solution; consequently, in terrestrial vertebrates they are necessarily confined to the mucosa of the mouth and pharynx. Fishes, however, since they live in a liquid medium, are not limited as to the location of the organs of taste. Therefore, while taste buds are most often found in the head and mouth region in this group, they may also be widely scattered over other parts of the outer body surface, including even the fins.

In most fishes the olfactory organs consist of a pair of pits lined with folds or ridges of sensory epithelium. The cyclostomes differ from other vertebrates in having a single, median olfactory organ. This is a blind pit in the lampreys, but in the hagfishes it opens into the pharynx. Dipnoans resemble higher vertebrates in possessing paired nasal passages that open by means of choanae into the phar-

**FIGURE 2.18.** A ventral view of the head of a guitarfish (*Rhinobatos productus*) showing the olfactory pits containing folds of sensory epithelium.

ynx. The nasal passages, therefore, have both internal and external openings. Within the canals themselves, the olfactory epithelium appears in the form of folds.

There is no basic difference between the eyes of fishes and those of other vertebrates. Such differences as do occur usually concern methods of accommodation or special adaptations to a particular mode of life, or are the result of degeneration. Accommodation, or the ability of the eye to adjust itself to near and far vision, is accomplished in fishes by moving the lens forward or backward so as to change its distance from the sensitive retina. Sharks, since they are fast-moving predatory animals, usually have distant vision and must move the lens forward or away from the retina to see objects close-up. The reverse is true of most teleosts, whose eyes are normally adjusted to near vision so that the lens must be moved back toward the retina when distant vision is required.

The eyes of some fishes are highly specialized, as an adaptation to a particular mode of life. The most remarkable type of eye, and one of the most famous perhaps, is that of the South American "four-eyed fish" (*Anableps*) (Fig. 2.19). This fish inhabits quiet water where it floats with the upper half of the eyes above the surface. So that it may see both in the air and in water, presumably at the same time, the lens is divided into two parts, each part being a different distance from the retina. Many deep-sea fishes have very large eyes which seem necessary for vision in regions of low-light intensity. Other deep-sea fishes, just like some cave-inhabiting species, have lost the power of sight as a result of degeneration of the eyes.

The ear of a fish, if we may use the term ear in a broad sense, is very different from the ear of a mammal. Our own ear consists of an

**FIGURE 2.19.** The four-eyed fish (*Anableps anableps*). (New York Zoological Society photograph.)

external pinna which funnels sound waves into the external auditory canal. At the inner end of this canal these waves strike the ear drum or tympanic membrane and are then transmitted across the middle ear or tympanic cavity by means of a chain of small bones to the cochlea or ventral part of the inner ear. From this complicated structure sensory impulses pass to the brain by means of the auditory nerve. The dorsal part of our inner ear consists primarily of three semicircular canals which are essentially organs of balance or orientation. The ear of a fish, however, is not suited for the reception of sound with which we usually associate our own ears. There is no external or middle ear, nor is a cochlea present. The inner ear of a fish consists of a dorsal *utriculus*, which is connected with the semicircular canals, and a ventral enlargement known as the *sacculus*. It is from the sacculus that the lagena of amphibians, reptiles, and birds as well as the *cochlea* of mammals arises to permit hearing in the true sense of the word. While certain fishes are definitely able to detect vibrations in the water, some of which may be produced by members of their own species, the ear of a fish is primarily an organ of balance, enabling it to maintain its equilibrium.

Fishes and the larvae of aquatic amphibians possess another type of sensory organ system that is not found in terrestrial vertebrates. This is known as the *lateral line system* and seems to function as a means of determining pressure changes and currents in the water. The receptors consist of sensory papillae or *neuromasts* which are arranged in rows on the body. In most fishes there is a row on each side of the body which branches into three parts on each side of the head. The neuromasts may be on the surface of the skin or they

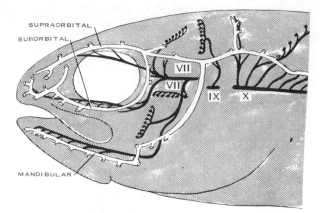

**FIGURE 2.20.** Distribution of lateral line canals and their nerves in the bowfin *Amia.* (Ballard, W. W.: Comparative Anatomy and Embryology. The Ronald Press, 1964. After E. P. Allis, 1888.)

may be in covered canals which open to the outside through small pores.

## ENDOCRINE GLANDS

The endocrine glands are the *ductless glands* whose products enter the blood stream directly. These products, which are called *hormones,* are chemical regulators of the body. They function essentially as catalytic agents by stimulating other glands, regulating growth, controlling metabolic activity, and maintaining chemical equilibrium in the body without undergoing change themselves.

Most of the endocrine glands found in mammals are present in the primitive fishlike vertebrates. The parathyroid glands, however, represent an exception. These glands, which are so important for regulating calcium metabolism in higher vertebrates, are lacking in cyclostomes and in cartilaginous and bony fishes. The pancreas, which is both a duct and ductless gland, is not definitely represented in cyclostomes. It is present in cartilaginous and bony fishes and possesses well marked islands of Langerhans that serve for the production of insulin. The adrenal glands of fishes differ from those of higher vertebrates in that the parts corresponding to the cortex and medulla in mammals are entirely separate from one another.

Fishlike vertebrates have well developed thyroid and pituitary glands. The gonads of fishes, like those of other backboned animals, are partly endocrine in function and are responsible for various secondary sexual characters that usually manifest themselves during the breeding season. Mention has already been made of the various types of hermaphroditism, including sex reversal, that may occur in both cyclostomes and teleost fishes. No doubt such functional changes in gonadal activity are under hormonal control.

## SPECIAL CHARACTERS

### SCALES

External body scales are among the most characteristic features of cartilaginous and bony fishes, although they are lacking in modern cyclostomes. Scales, in general, serve as a protective covering for the body. Their origin, structure, and function, however, vary in different groups of fishes so that they are of importance in classification. (See Chap. 7.)

The skin of a shark or ray feels like sandpaper owing to the presence of numerous tiny scales embedded in the skin. These scales are known as *placoid scales* and are essentially the same as teeth in structure. Each scale is composed of a basal plate of bone. Projecting upward through the skin and then directed backward is a spine that is made of dentine. Just as in a tooth, there is a central cavity, the pulp, in which there are blood vessels. The spine is capped by a layer of harder material that some believe to be homologous with tooth enamel, although there is considerable question as to whether or not this interpretation is correct. There is no essential difference between the teeth and the scales of a shark except that the former are larger (Fig. 2.21). Both are subject to loss and replacement. In rays, however, the teeth are frequently modified into broad, flat plates, set close together so as to enable these fishes to crush the shells of mollusks. Placoid scales vary considerably in shape and arrangement in different kinds of sharks and rays. In the chimaeras they are present in embryonic life, but in the adults they are greatly reduced in number and limited to a few localized areas on the skin.

Some of the more primitive ray-finned fishes possess what are referred to as *ganoid scales*. Scales of this type are best seen on the gars (Fig. 2.22), where they are rhomboidal in shape, pressed tightly against one another, and arranged in diagonal rows on the body. Overlying the basal plate of each scale are layers of a shining enamel-like substance known as *ganoin*. Because of the presence of ganoin on the scales of some of the Chondrostei and Holostei, members of these groups are often referred to as ganoid fishes. Some of the modern, so-called ganoids, however, no longer have ganoin present. The large, plate-like scales of the sturgeons lack ganoin and in the paddlefish the body is essentially lacking in scales.

The scales of most of the higher bony fishes are embedded

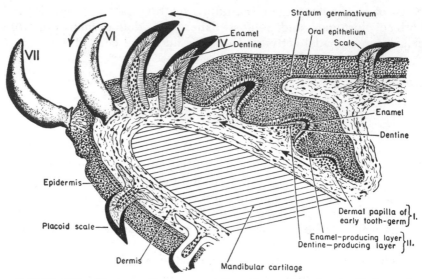

**FIGURE 2.21.** Diagrammatic representation of a section of a shark's lower jaw showing the manner of origin of the teeth and indicating their successive replacement. Roman numerals *I* to *VII* indicate successively older teeth. The continuous movement of the teeth forward and outward is suggested by the arrows. *V* and *VI* are fully developed functional teeth at the edge of the jaw. *VII* suggests a tooth about to be shed. (Rand, H. W.: The Chordates. McGraw-Hill Book Co., Inc., 1950.)

anteriorly in pockets in the skin with the free, posterior edge overlapping the next scale. Unlike placoid scales, these overlapping or imbricate scales are not replaced when lost. They are entirely dermal in structure with no epidermal covering of enamel-like material or ganoin. *Cycloid scales* (Fig. 2.23) are essentially round and they increase in size as the fish grows. As a consequence, in certain species they are marked by growth rings, somewhat comparable to the growth rings of trees. These rings are most noticeable on the embedded part of the scales in species whose growth is retarded during the winter as a result of reduced temperature and food supply.

**FIGURE 2.22.** The scales of the alligator gar (*Lepisosteus spatula*), one of the bony ganoids inhabiting streams and rivers entering the Gulf of Mexico, are covered with a hard outer layer of ganoin and do not overlap. (Photographed at Steinhart Aquarium, San Francisco.)

FIGURE 2.23. Diagram showing (*left*) cycloid scale of a sucker and (*right*) ctenoid scale of a perch.

*Ctenoid scales* (Fig. 2.23) are essentially the same as cycloid scales as regards structure and arrangement in higher bony fishes, but differ in that their free or posterior margins possess a number of comb-like projections. In some species these scales may be reduced to a single spine. Ctenoid scales are most often found on fishes possessing spiny dorsal fin rays.

## COLOR

Some of the most beautiful coloration in the vertebrates is to be found among fishes and birds. In the latter, color is largely confined to the feathers, which, when fully grown, are nonliving integumental structures. Color changes in birds can be brought about only by wear, oxidation, or molt. In fishes, color, to a large extent, results from the presence of pigment cells called *chromatophores* that are located in the dermis. Crystals of guanine, however, located in cells called *iridocytes* that are associated with the scales, may be responsible for iridescence.

FIGURE 2.24. The speckled sand dab (*Citharichthys stigmaeus*) of the North Pacific is often difficult to see against the substratum because of its protective coloring. (Photographed at Steinhart Aquarium.)

Some bony fishes, like certain reptiles, are capable of changing color very rapidly under the control of the nervous system. Much slower color changes occur in some cartilaginous fishes as well as amphibians as a result of hormones released by the pituitary and circulated in the blood stream. In fishes the resulting changes, from either cause, may be effected by movement of the pigment granules within the chromatophores or from a change in the shape of the chromatophores themselves. If pigment granules are evenly distributed throughout these cells, the color is very evident, whereas the cells appear largely colorless if the pigment is concentrated at one point.

The principal pigments found in fishes are the *melanins* which produce brown, gray, or black and the *carotenoids* which are largely responsible for the yellow, orange, and red colors. Chromatophores containing melanins are referred to as *melanophores* and those containing carotenoids are sometimes collectively called *lipophores*. In some kinds of fishes there are marked color changes that occur in the breeding season. These may serve in part as a means of sex recognition. In most fishes, however, color appears to be an environmental adaptation to enable them to blend with their background. Species inhabiting coral reefs are often brightly colored. Those found in kelp beds frequently have alternating bands of light and dark on the body. The upper surface of bottom-dwelling fishes sometimes blends remarkably well with the surrounding substratum.

## APPENDAGES AND LOCOMOTION

There are no indications of paired fins in living cyclostomes. However, unpaired median dorsal and ventral fin folds as well as a caudal fin are present. The latter is rounded and has the notochord extending straight through it to the posterior tip. This type of tail is known as *protocercal* (Fig. 2.25). Both the median and caudal fins in cyclostomes are supported by delicate cartilaginous rays.

Both cartilaginous and bony fishes basically possess paired pectoral and pelvic appendages in addition to unpaired median and caudal fins. In sharks and rays both paired and unpaired fins are supported by cartilaginous rays. All the fins, however, are covered with heavy skin so that the individual supporting elements cannot be seen. In sharks there are usually two dorsal fins but in some species the number is reduced to one. A single anal fin is present. The tail in sharks is generally described as *heterocercal* (Fig. 2.25), consisting of a large dorsal flange, into which the distal end of the axial skeleton extends, and a smaller ventral flange. In the ray-like fishes the pectoral fins are very much enlarged and are attached for their full

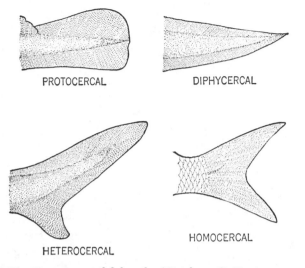

PROTOCERCAL  DIPHYCERCAL

HETEROCERCAL  HOMOCERCAL

**FIGURE 2.25.** Four types of fish tails. (Weichert, C. K.: Anatomy of the Chordates. McGraw-Hill Book Co., Inc., 1958.)

length from the head back to the anterior end of the pelvic fins. In the electric ray they are even confluent around the tip of the snout. Most rays possess two median dorsal fins situated far back on the tail, but these are lacking in the stingrays. No anal fin is present. Although a caudal fin is lacking in most rays it is well developed on the tail of the electric ray. The inner parts of the pelvic fins of male sharks and rays are modified as claspers, which serve as a means of transferring sperm to the female.

In the bony fishes there is considerable diversity in the shape, position, and number of fins (Fig. 2.26). Generally there are paired pectoral and pelvic fins, one or two median unpaired dorsal fins, a median ventral anal fin and a caudal fin. In some species, however, such as the ocean sunfish, the wolf-eel, and some of the blennies, pelvic fins are completely lacking. In other species the pelvic fins are so far anterior that they appear to be directly under the pectoral fins or even anterior to them. In the morays both pectoral and pelvic fins are missing and the dorsal, anal, and caudal fins may be hidden beneath the skin and hardly discernible. The dorsal fin may be small in some fishes or it may extend from the head to the tail and even be continuous with the caudal fin. Frequently there are two dorsal fins. In some species there are two anal fins, although one is the usual number. Like the dorsal fin the anal fin may sometimes be very long as in most of the flatfishes. Some fishes, such as the mackerels, possess a number of median, dorsal, and ventral finlets situated between the dorsal and anal fins and the caudal fin.

In ray-finned fishes the fins are generally supported by either

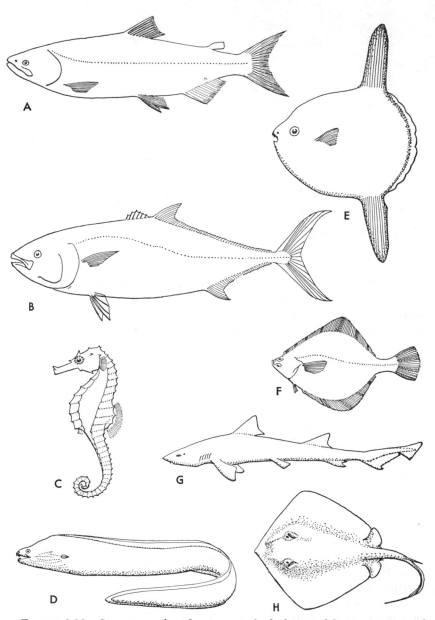

**FIGURE 2.26.** Some examples of variation in body form and fin structure: *A*, salmon; *B*, tuna; *C*, sea horse; *D*, moray; *E*, ocean sunfish; *F*, flounder; *G*, shark; *H*, ray.

soft or spiny rays which are easily seen, since they are not covered by heavy skin as in elasmobranchs. There are some species, however, that possess adipose fins without any internal support. In trout, salmon, smelt, and certain other fishes, the fleshy adipose fin is situated between the dorsal fin and the tail.

Living members of the subclass Sarcopterygii have paired fins that differ markedly from those of the ray-finned fishes. They either are fleshy throughout or have a fleshy base. The latter type of appendage was characteristic of the ancient crossopterygians and is seen on the only known living representative of this group, the coelacanth *Latimeria* (Fig. 2.27). The basal lobe of the crossopterygian fin contains bony elements believed to be homologous to those of the tetrapod limb. Proximally there is a large element which *a o w* seems to represent the humerus or femur and which connects with the girdle. Distal to this are two other bones that correspond to those of the forearm or the shank of higher land vertebrates. Even more distal are series of smaller elements which could be construed as potential wrist and ankle bones. In *Latimeria* the anal fin as well as the more posterior of the two dorsal fins possesses basal lobes.

In the dipnoans the fleshy pectoral and pelvic fins contain a central segmented skeletal axis from either side of which additional supporting elements extend. This is known as an archipterygial fin. In *Neoceratodus* of Australia the archipterygial fins are quite broad, whereas they are more slender and very elongate in the several species of African lungfishes of the genus *Protopterus* (Figs. 2.28 and 2.29). In the South American lungfish (*Lepidosiren*) the pectoral and pelvic appendages are very small and somewhat degenerate (Fig. 2.30). For many years it was thought that the dipnoans were ancestors of the higher land vertebrates. It would be most difficult, however, to derive a tetrapod limb from an archipterygial fin, whereas no

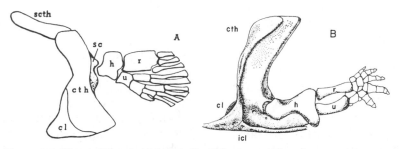

**FIGURE 2.27.** *A*, The shoulder girdle and pectoral fin of a crossopterygian; *B*, the same structures in an ancient fossil amphibian, placed in a comparable pose to show the basic similarity in limb pattern. Abbreviations: *h*, *r*, and *u*, humerus, radius, and ulna of the tetrapod and obvious homologues in the fish fin; *cl*, clavicle; *cth*, cleithrum; *icl*, interclavicle; *sc*, scapula; *scth*, supracleithrum. (*A* after Gregory, in Romer, A. S.: The Vertebrate Body. 4th Ed., W. B. Saunders Co., 1970.)

**FIGURE 2.28.**   The Australian lungfish (*Neoceratodus forsteri*) is the largest of living dipnoans. (Photographed at Steinhart Aquarium.)

such difficulty exists with respect to the crossopterygian paired appendages.

The caudal fin in some of the more primitive bony fishes like the sturgeons is heterocercal as in sharks. In dipnoans the tail is referred to as *diphycercal* with the axial skeleton extending nearly to the tip, much as in the protocercal tail of cyclostomes (Fig. 2.25). It is believed, however, to have possibly originated from the heterocercal type with the loss of the ventral flange. The great majority of bony fishes have a *homocercal* tail in which the supporting rays are distal to the terminus of the axial skeleton and the dorsal and ventral flanges are symmetrical (Fig. 2.25).

It is logical to associate the paired pectoral and pelvic appendages with locomotion, yet in most fishes these structures serve primarily to orient the body. The pectoral fins are used essentially for steering, while the pelvic as well as the dorsal and anal fins effect stability and assist in maintaining the body upright in the water. Rapid forward progress is generally produced by lateral movement of the posterior part of the body and caudal fin. In many fast-swimming fishes such as tuna, albacore, and marlin the tail is crescent shaped, the caudal peduncle narrow, and the body streamlined for speed. There are, however, certain slow-moving fishes that do rely entirely upon their fins for locomotion. Such species are usually protected by a heavy integumental covering or by spines or concealing coloration.

FIGURE 2.29. One of the several known species of African lungfishes belonging to the genus *Protopterus*. Note the large and elongate archipterygial fins. (Photographed at Steinhart Aquarium.)

FIGURE 2.30. The South American lungfish (*Lepidosiren paradoxa*) has very small fleshy fins. (Photographed at Steinhart Aquarium.)

The heavily armored sea horses, for example, swim principally by movement of the dorsal fin with some assistance by the pectoral fins. The same is true of the related pipefishes. Puffers and turkey fishes swim by means of fins, but are protected by spines which in the latter are poisonous. Many of the rays swim by means of undulatory waves on the part of the large pectoral fins. The dorsal coloration in these species is usually similar to the surrounding bottom on which the fish rest, and many have poisonous spines or electric organs which may also serve for protection. Most of the eel-like fishes rely on neither the fins nor the tail for locomotion, but rather on lateral undulations of the entire body as does a water snake.

Some fishes are capable of jumping a considerable distance out of the water. Marlin and sailfish are noted for this. Salmon make remarkable leaps to ascend small waterfalls when they are returning to their home streams to breed. Trout frequently leap out of the water to secure insect food in the air. Other fishes such as mullet leap into the air when frightened. Most remarkable are the flying fishes which are found in most of the warm oceanic waters of the world. These fishes do not truly fly, but are capable of gliding through the air for distances as great as 1200 feet. This is accomplished by attaining great speed under water, as a result of lateral movement of the tail and posterior part of the body, then suddenly rising to the surface. When the fish passes into the air the very large pectoral fins, which were pressed against the body in the water, are spread so as to form a gliding surface. Although flying fishes are principally marine, there is a freshwater fish in Venezuela that is capable of gliding.

**FIGURE 2.31.** A model of the coelacanth (*Latimeria chalumnae*). (Courtesy of David K. Caldwell, Los Angeles County Museum.)

**FIGURE 2.32.** A suckerfish (*Remilegia australis*) showing the dorsal suction disk, which is partly derived from the anterior dorsal fin rays. (Photograph by W. I. Follett.)

There are a number of fishes that use their fins for walking. Some of the lungfishes seem to literally walk on the bottom by means of fins. Mudskippers use their fins to climb up on banks and onto the roots and lower limbs of mangroves. The so-called walking perch may live for a day out of water because of the presence of a supplemental respiratory chamber above the gills which enables it to use air. In Siam they have been known to travel considerable distances overland at night, using their fins as legs.

One of the most interesting fin modifications is to be found in the remoras or suckerfishes (family Echeneidae), which are members of the higher, spiny-rayed teleosts (Fig. 2.32). In these fishes the spiny, anterior dorsal fin rays are separated into two horizontally placed rows which form transverse folds on a suction disk on the top of the head. These fishes use this apparatus to attach themselves to other larger fishes and thereby move around with the host and feed on various leftovers. By raising or lowering the folds the remoras may increase or decrease the vacuum by which they attach themselves to their host.

## VENOMOUS AND POISONOUS FISHES

Serious illness or even death may be produced by certain species of fishes. Such results may be effected by two different means. Injury from *venomous fishes* is the result of an injection of poison into the victim by means of a spine or some other sort of stinging mechanism. *Poisonous fishes*, on the other hand, produce illness or death when their flesh or certain organ systems are ingested.

A number of species of sharks and rays as well as the chimaeras possess dorsal spines that are associated with venom glands (Fig. 2.34). The latter are sometimes located in the epithelium in the

**FIGURE 2.33.**   The California hornshark (*Heterodontis francisci*) derives its name from a protective spine situated along the anterior base of each of the dorsal fins. (Photographed at Steinhart Aquarium.)

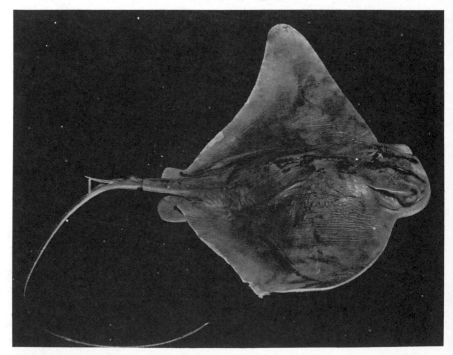

**FIGURE 2.34.**   A bat-ray (*Myliobatis californica*) with the venom spine elevated to show its position on the dorsal surface of the tail.

posterior or ventral grooves of the spine or else in the integumentary sheath surrounding the spine. Some of the most venomous species of fishes are to be found among the sting-rays and bat-rays. Many human deaths have resulted from accidental contact with these elasmobranchs.

Among the bony fishes the most venomous species belong to the family Scorpaenidae. Notable among these are the stonefish (*Synanceja verrucosa*), whose sting may produce death within a few hours, and various species of the genus *Pterois,* including the turkey fish (Fig. 2.35), scorpion fish, and zebra fish as well as others. In these fishes the spines of the dorsal, ventral, and anal fins are stings, and the venom is produced by glandular secretion in the epithelial sheath of the spines. The stonefish, however, is one of the few species in which there is a duct leading from the poison gland to the tip of the glandular groove in the spine. The California scorpion fish or bullhead (*Scorpaena guttata*), so common along parts of the Pacific Coast, has similar poison glands which may inflict painful

**FIGURE 2.35.** The zebra fish or turkey fish (*Pterois volitans*). Members of the genus *Pterois* are found in shallow waters about coral reefs in tropical seas. Their beautiful lace-like fins possess venomous spines which can inflict a very serious and painful injury. (Photograph courtesy of Dr. Bruce Halstead, World Life Research Institute.)

wounds. Certain of the catfishes such as the gafftop (*Galeichthys felis*), a marine species that occurs in fresh water to a limited extent, are able to produce very painful wounds by means of pectoral and dorsal spines.

Poisonous fishes are found in all the warmer waters of the world, but are most abundant in the tropical Pacific Ocean and the Caribbean. Best known perhaps is the puffer fish (*Arothron hispidus*). The mortality rate from ingesting the flesh of this species is estimated to be about 60 per cent and death has been known to occur within 20 minutes. Many hundreds of other species are known to be poisonous at certain times, although the exact cause remains to be determined. Poisoning has been reported from eating such edible species as pompano, red snapper, sea bass, tuna, and mackerel in tropical areas. These fish poisons are most often concentrated in such visceral organs as the liver and gonads and, in some species, their presence seems correlated with the reproductive season. Experiments with the cabezon (*Scorpaenichthys marmoratus*), a large cottid fish that is common along the Pacific Coast of North America, have shown that the roe of ripe females will produce violent illness, whereas the flesh of the same individuals is harmless. Cooking seemingly has no effect on these toxins.

Recent authorities on *ichthyosarcotoxism,* which is the technical term for intoxication resulting from eating poisonous fishes, are of the opinion that fishes may acquire their poisonous qualities from their food. Marine plants may be the source. Under such circumstances almost any fish in a region where poisonous fishes occur could be toxic to human beings.

## BIOLUMINESCENCE

*Bioluminescence* has been described as the emission of light by certain organisms as a result of the oxidation of some substrate in the presence of an enzyme. The substrate is a heat-stable compound known as *luciferin,* and the heat-sensitive enzyme that serves as a catalyst to bring about this oxidation, which results in *cold light,* is called *luciferase.*

Bioluminescence is produced by certain bacteria, fungi, and many invertebrate animals. Among backboned animals, however, only fishes are capable of producing light. Luminescent organs have been found in some sharks, the electric ray (*Benthobatis moresbyi*), and a number of bony fishes, especially those species that live at considerable depths in the ocean.

The presence of light-producing organs would seem to be of value to deep-sea organisms living in total or near total darkness.

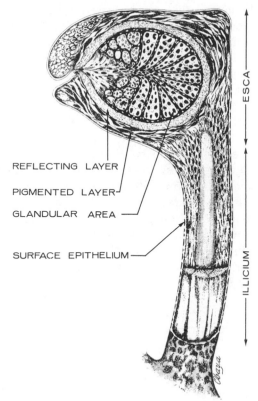

FIGURE 2.36. Theoretical sagittal section through the light organ of *Melanocetus murrayi.* (Courtesy of Hulet and Musil, Copeia, 1968.)

REFLECTING LAYER

PIGMENTED LAYER

GLANDULAR AREA

SURFACE EPITHELIUM

ESCA

ILLICIUM

Some have suggested that they serve to attract food, to confuse enemies, to light the environment, or to attract members of the opposite sex. It seems likely, however, that their principal function is to serve as a means of recognition between individuals of the same species.

Luminescent organs may occur in or on any part of the body (Fig. 2.37). In some species they are on or close to the surface, whereas in others they are deep in the body. In females of all deep sea angler fishes the first dorsal fin ray is modified into a lurelike

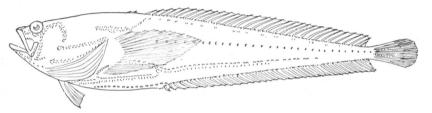

FIGURE 2.37. Drawing of a midshipman (*Porichthys notatus*). Note the rows of photophores on the head and body.

structure that can be extended forward over the head. The bulbous end of the lure contains a light organ which presumably attracts various marine organisms to within reach of the angler's jaws. The color of the light varies in different kinds of fishes. The light may be continuous, it may increase or decrease in intensity, or it may flash off and on.

Bioluminescence is produced in two ways in fishes: either by self-luminous photophores or by special organs in which symbiotic luminous bacteria are cultured. There are several means by which the light intensity of these luminous organs may be increased or decreased. One method depends on the expansion and contraction of overlying chromatophores in the skin. These pigment cells either are directly under the control of the sympathetic nervous system or else react to adrenalin secreted by the adrenal glands, which are also controlled by the sympathetic system. Organs containing luminous symbiotic bacteria may have light emission regulated by muscles which either move the organ itself or move an overlying opaque screen.

## ELECTRIC ORGANS

Among the unique structures developed by vertebrates are the electric organs that are found in several widely separated groups of fishes. In some species the electric current that is produced is quite weak, but in others it is very strong. The maximum recorded discharge for the electric eel (*Electrophorus electricus*) is 550 volts, for the electric catfish (*Malopterurus electricus*), 350 volts, and for the electric ray (*Torpedo nobiliana*), 220 volts.

Electric organs (Fig. 2.38) are composed of *electroplates* or *electroplaxes,* which are aggregations of disklike cells arranged so that they all face in the same direction. As a result, an electric organ appears to be composed of parallel, prismatic columns, each consisting of a large number of electric plates and separated from one another by connective tissue. There is a network of nervous tissue on one side of each of the electric plates and the latter are multinucleate like muscle tissue. The entire electric organ is semitransparent and consequently appears somewhat gelatinous. The direction of the flow of current varies in different kinds of electric fishes. In the electric eel it passes from the tail toward the head while in the electric ray the flow of current is from the ventral to the dorsal surface of the body.

Why electric fishes do not electrocute themselves is not clearly understood. Two partial explanations have been presented. The nervous system in these species is unusually well insulated with layers

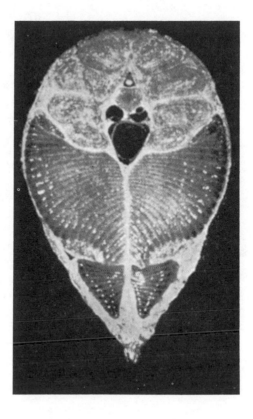

FIGURE 2.38. Cross section of the tail of the electric eel, showing the two major divisions of the paired electric organs and the plates (electroplaxes) of which they are composed. (After Keynes, in Smith, H. M.: Evolution of Chordate Structure. Holt, Rinehart and Winston, Inc., 1960.)

of fat and the lines of current are at right angles to excitable structures.

Since electric organs are limited to a relatively few species of rays and bony fishes, the phylogenetic origin of such structures is somewhat obscure. Embryologically, however, it appears that they may arise from muscle tissue. This is logical in view of the fact that muscular activity is capable of producing electricity. It has even been shown that muscular action potentials can be detected in the water some distance away from non-electric fish. Lissmann (1958) suggests that this represents the evolutionary starting point for electric organs.

Electric organs appear to perform several functions. Those that produce a relatively weak charge have been shown to serve as a means of orientation. Nearby objects whose electrical conductivity differs from that of the water surrounding the fish will distort the flow-pattern of the electric current and thereby be recognized. Even objects of the same shape but with differing electrical conductivity may be distinguished. This is particularly useful to fishes inhabiting turbid water where visibility is poor. It is believed that the lateral line system enables fishes to detect nearby objects as a result of pressure changes, but this would not enable them to distinguish objects of the same shape but composed of different materials as appears to be true of electric organs. Experiments by Lissmann have even shown that *Gymnarchus niloticus* and *Gymnotus carapo* can detect the presence of a magnet on the outside of an aquarium, and individuals of the latter species were conditioned to feed only when a magnet was present. In those species whose electric organs are capable of producing a powerful shock, the function of these structures seems to serve both as a weapon of defense and as a means of stunning prey for food.

Much remains to be learned about the receptor apparatus which enables fishes with electric organs to detect changes in the electrical field around them.

Certain electric fishes have been found to possess special receptor organs on the body called *mormyomasts* which are connected with the nervous system of the lateral line organs. It has been suggested that these function as electroreceptors. Furthermore, fishes with electric organs generally show a marked enlargement of the cerebellum.

## CLASSIFICATION OF FISHES AND FISHLIKE VERTEBRATES

### CLASS AGNATHA (JAWLESS FISHES)

Primitive, fishlike vertebrates lacking true jaws with an endoskeletal support; paired appendages lacking in living species; a single median nasal aperture; Ordovician to Recent; two subclasses.

SUBCLASS OSTRACODERMI (OSTRACODERMS). Ancient agnathous fishes possessing an exoskeleton of denticles or dermal plates often forming a heavy cephalothoracic shield; pectoral-pelvic fin folds present in some forms, with some indication of paired appendages; Ordovician to Devonian; five orders.

SUBCLASS CYCLOSTOMATA (LAMPREYS AND HAGFISHES). Modern agnathous vertebrates with a round suctorial mouth; tongue possessing horny rasping "teeth"; exoskeleton lacking; skin covered with mucous glands; cartilaginous cranium lacking a roof; axial skeleton composed principally of a notochord with some vertebral elements present; Recent; two orders.

ORDER PETROMYZONIFORMES (LAMPREYS). Freshwater, marine, or anadromous cyclostomes with sides on the cartilaginous cranium; nasohypophyseal canal not penetrating roof of mouth; cartilaginous vertebral elements present along sides of entire notochord; coastal parts of Europe and Asia, the Caspian basin, and both coasts of North America; eight genera.

ORDER MYXINIFORMES (HAGFISHES). Marine cyclostomes with a very reduced cranium; nasohypophyseal canal penetrating roof of mouth; cartilaginous vertebral elements present only in caudal region of notochord; temperate and colder waters of the Atlantic and Pacific oceans; three to four recognized genera.

*Note:* Cyclostomata is used by some as the class name for all agnathous fishes, with five subclasses as follows: Pteraspides, Cephalaspides, Paleospondyli, Myxini, and Petromyzones.

## CLASS PLACODERMI (ANCIENT JAWED FISHES)

Fishlike vertebrates with first gill arch modified to form an upper jaw (palatoquadrate cartilage) and a lower jaw (Meckel's cartilage); cranium partly ossified; second or hyoidean arch not modified to support the jaws; Silurian to Permian; five major subclasses.

SUBCLASS ANTIARCHI. Placoderms possessing a pectoral girdle which is essentially the anterior part of the thoracic shield, as in some ostracoderms; dermal jawbones present.

SUBCLASS ARTHRODIRA. Placoderms possessing a pectoral girdle composed of anterior and basal parts of pectoral shield; dermal jawbones present.

SUBCLASS ACANTHODII. Placoderms possessing a dermatocranium and partly ossified endocranium; pectoral girdle present.

SUBCLASS MACROPETALICHTHYIDA. Placoderms possessing a well ossified dermatocranium and endocranium; pectoral and pelvic fins sharklike.

SUBCLASS RHENANIDA. Placoderms possessing broad pectoral fins; body intermediate in shape between sharks and rays.

*Note:* The primitive jawed fishes are classified by some authorities as series Aphetohyoidea (the Acanthodii) and series Placodermi (the Arthrodira, Euarthrodira, Macropetalichthyida, Rhenanida, and Pterichthyes) under the class Pisces.

## CLASS CHONDRICHTHYES (SHARKS, RAYS, AND CHIMAERAS)

Jawed fishes with a cartilaginous skeleton; swim bladder lacking; a spiral valve present in intestine; pelvic fins modified to form claspers in the male; Upper Devonian to Recent; oceans of the world, a few in fresh or brackish water; two subclasses.

SUBCLASS ELASMOBRANCHII (SHARKS AND RAYS). Small to very large cartilaginous fishes with an external skeleton in the form of placoid scales or spines; jaw support amphistylic or hyostylic; five to seven gill clefts, not covered by an operculum; tail usually heterocercal; Devonian to Recent; three orders, only one of which is represented in the world today.

ORDER CLADOSELACHIFORMES (ANCIENT MARINE SHARKS). Pectoral fins not constricted basally as in modern sharks; mouth terminal; vertebral centra absent; notochord persistent; anal fin absent; tail heterocercal; Upper Devonian to Upper Carboniferous.

**ORDER XENACANTHIFORMES (ANCIENT FRESHWATER SHARKS).** Pectoral fins of the archipterygeal type; jaw support amphistylic; tail diphycercal; Carboniferous.

**ORDER SELACHIFORMES (LIVING SHARKS AND RAYS).** Moderately small to very large cartilaginous fishes (0.25 to over 15 m.);[*] paired fins not of the archipterygeal type; jaw support amphistylic or hyostylic; mouth ventral, not terminal except in *Chlamydoselachus;* tail heterocercal; Lower Carboniferous to Recent; oceans of the world, a few estuarine and freshwater species; four suborders.

*Suborder Hexanchoidei (frilled and cow sharks).* Medium to large sharks possessing six to seven pairs of gill clefts; nictitating membrane absent from eye; notochord persistent; vertebral centra simple or degenerate; a single dorsal fin; anal fin present; Jurassic to Recent; bays as well as deep oceanic waters; two living families.

*Suborder Heterodontoidei (hornsharks).* Small sharks possessing five gill clefts; two dorsal fins, each having a heavy spine along the anterior margin; anal fin present; vertebral centra well developed in living species; Carboniferous to Recent; widespread in temperate and tropical waters but absent from the Atlantic and Mediterranean areas; one living family.

*Suborder Selachoidei (modern sharks).* Small to very large sharks possessing five or six gill clefts situated on each side of the head; spiracle present or absent; upper eyelid present; two dorsal fins; notochord absent in adult; vertebral centra well developed; anal fin present or absent; Jurassic to Recent; range, same as that of the order; three superfamilies.

*Note:* There are a number of methods of classifying cartilaginous fishes. Nikol'skii's (1954) system is followed here.

Superfamily Carcharinoidea (Sand Sharks, Basking Sharks, Whale Sharks, Mackerel Sharks, Nurse Sharks, Smooth Dogfishes, Requiem Sharks, Hammerheads, and Allied Groups). Small to very large sharks with asterospondylous vertebrae (verterbral centra strengthened by longitudinal, calcified plates radiating outward from a central cylinder surrounding the notochord); anal fin present; spiracle absent or very small; geographic range, that of order; about 20 living families.

Superfamily Squaloidea (Spiny Sharks and Their Allies). Very small to large sharks with cyclospondylic vertebrae (having a single calcified cylinder surrounding the notochord in each vertebral centrum); anal fin absent; spiracle usually large; widespread in oceanic waters; three or four living families.

Superfamily Squatinoidea (Angel Sharks). Small to medium-sized sharks with enlarged, raylike pectoral fins and a dorsoventrally compressed body; gill openings on sides of

---

[*]Measurements given in this classification represent total length.

body; spiracle well developed; anal fin absent; dorsal fins located on tail; Atlantic and Pacific Oceans; one family.

*Suborder Batoidei (rays and skates).* Small to very large cartilaginous fishes with a dorsoventrally flattened body and greatly enlarged pectoral fins which are attached anteriorly to the sides of the head; five pairs of gill clefts ventral in position; a pair of large dorsal spiracles; upper eyelid absent; Upper Jurassic to Recent; range, that of order; seven to 10 recognized families.

SUBCLASS HOLOCEPHALI (CHIMAERAS OR RATFISHES). Moderately small cartilaginous fishes (up to 1.5 m.) with placoid scales essentially lacking; jaw support autostylic; four gill apertures covered by a single dermal operculum; Carboniferous to Recent; two orders.

ORDER CHONDRENCHELYIFORMES (ANCIENT HOLOCEPHALANS). Vertebral centra well developed; pectoral fins archipterygeal; known only from the Carboniferous.

ORDER CHIMAERIFORMES (MODERN HOLOCEPHALANS). No true vertebral centra, only calcified rings; pectoral fins not of the archipterygeal type; deep oceanic waters; three living families.

## CLASS OSTEICHTHYES (BONY FISHES)

Fishes possessing a skeleton that is at least partly ossified; a swim bladder usually present; a spiral valve present in the intestine of only the most primitive forms; pelvic fin not possessing claspers in the male; gills covered by a single bony operculum; Devonian to Recent; all oceans as well as nearly all rivers, streams, and lakes; two subclasses.

*Note:* There are several divergent views on the classification of bony fishes, especially as regards the higher taxa. Berg (1947) regards the Dipnoi as a class separate from the Teleostomi, which he separates into two subclasses: the Crossopterygii, and the Actinopterygii. Nikol'skii (1954) slightly modifies this system by considering the Dipnoi and Teleostomi as subclasses of the series Osteichthyes. Romer's (1962) classification, down to and including superorders, is followed here.

SUBCLASS ACTINOPTERYGII (RAY-FINNED FISHES). Fishes that generally lack choanae and a fleshy base to the paired fins; five superorders.

*Superorder Paleonisci (paleoniscoids).* Ancient ray-finned fishes with sharklike body; tail usually heterocercal; a single dorsal fin present; vertebrae possessing either very rudimentary centra or none at all; Devonian to Cretaceous, freshwater and marine deposits; about 11 orders.

*Superorder Polypteri (fringe-finned ganoids).* Medium-sized (up to 1 m.), primitive, elongate, ray-finned fishes lacking choanae but possessing a fleshy base to the pectoral fins; caudal fin symmetrical; dorsal fin divided into numerous finlets; ganoid scales present; air bladder functioning as a lung and connected to ventral side of gut; a spiral valve present in intestine; Eocene to Recent; one order.

ORDER POLYPTERIFORMES (BICHIRS). Characters of the superorder; rivers and lakes of Africa; one family, the Polypteridae, represented by two living genera, *Polypterus* and *Calamoichthys*.

**Superorder Chondrostei (cartilaginous ganoids).** Large to very large (up to 9 m.), primitive, ray-finned fishes with sharklike body and a heterocercal tail; mouth subterminal; ganoid scales present only in tail region; much of skeleton cartilaginous; little ossification of chondrocranium, which is surrounded by dermal bones; vertebrae lacking centra; air bladder smooth internally and arising from dorsal side of gut; spiral valve present in intestine; Jurassic to Recent; one order.

ORDER ACIPENSERIFORMES (STURGEONS AND PADDLEFISHES). Characters of the superorder; three families, the extinct Chondrosteidae (Jurassic to Cretaceous), Acipenseridae or sturgeons (larger rivers of the Northern Hemisphere), and the Polyodontidae or paddlefishes (two living genera, *Polyodon* of the Great Lakes and Mississippi Valley of North America and *Psephurus* of the Yangtze River of China).

**Superorder Holostei (bony ganoids).** Medium to large (0.7 to 3 m.), ray-finned fishes with shortened, heterocercal tail and a skeleton that, in many respects, resembles that of the Teleostei; scales ganoid or cycloid; air bladder functioning as an accessory respiratory organ, containing numerous folds on its inner surface, and connected to the dorsal side of the gut; a spiral valve present in intestine; marine and fresh water; four extinct orders (Jurassic to Eocene) and two living orders.

ORDER AMIIFORMES (BOWFIN). Small holostean with nonprotruding jaws and a transversely rounded head; dorsal fin long and spineless; head covered with bony plates; scales cycloid; lakes and streams of eastern United States; one species, *Amia calva*.

ORDER LEPISOSTEIFORMES (GARS). Medium to large holosteans with elongate snout; dorsal fin posterior in position; ganoid scales rhomboidal in shape; vertebrae opisthocoelous, unlike those of other fishes; lakes and rivers of North and Central America and Cuba; one living genus, *Lepisosteus*.

**Superorder Teleostei (higher bony fishes).** Very small to large ray-finned fishes with well developed vertebral centra and considerable cranial ossification; scales lacking ganoin in living species, usually cycloid or ctenoid; tail usually homocercal; air bladder arising from dorsal wall of gut, although sometimes connection disappears by maturity; intestine lacking a spiral valve but generally elongate and coiled; fresh water and marine, first appearing in the Triassic; as many as 41 living orders recognized by some (Berg, 1947).

*Note:* Twenty-five major orders of living teleosts are given here, essentially following Nikol'skii (1954).

ORDER CLUPEIFORMES (HERRING-LIKE FISHES, TARPONS, SALMONIDS, AND THEIR RELATIVES). Small to rather large (0.035 to 2.4 m.), primitive teleosts with soft-rayed fins; vertebral centra with a central aperture; pelvic girdle not attached to pectoral girdle; swim bladder usually connected to gut; widespread, marine and fresh water; a number of suborders and many families.

ORDER SCOPELIFORMES (INIOMOUS FISHES). Mainly small (0.025 to 1.0 m.) fishes with soft-rayed fins; usually lacking an air bladder; many possess light-producing organs; a small adipose dorsal fin usually present on caudal peduncle; living principally at considerable depths in the ocean; widespread.

ORDER SACCOPHARYNGIFORMES (DEEP-SEA GULPER EELS). Elongate (up to 2 m.), eel-like teleosts with enormous jaws, which may be many times the length of the skull; pelvic fins, ribs, scales, and swim bladder absent; photophores present on body; found at great depths in the sea; two or three families.

ORDER GALAXIIFORMES (GALAXIIDS). Small (up to 0.6 m.), scaleless teleosts with posteriorly situated dorsal and anal fins; adipose fin lacking; fresh, brackish, and marine waters of the Southern Hemisphere, occurring in and around southern South America, South Africa, New Zealand, and Australia; one family.

ORDER ESOCIFORMES (PIKE). Small to medium-sized (less than 0.05 to 1.5 m.), soft-rayed fishes lacking a mesocoracoid in the shoulder girdle; swim bladder connected to gut; scales cycloid; freshwater inhabitants of the Northern Hemisphere; three living families. (The Esociformes is considered as a suborder of the Clupeiformes by some.)

ORDER MORMYRIFORMES (MORMYRIDS). Small to medium-large (up to 1.5 m.), soft-rayed fishes with a greatly enlarged cerebellum; mesocoracoid present; snout frequently elongated; electric organs often present on tail; fresh waters of tropical Africa; one or two families.

ORDER CYPRINIFORMES (CYPRINIDS). Small to large, soft-rayed fishes possessing a chain of bony ossicles, called the Webberian apparatus, which extends from the swim bladder to the inner ear; swim bladder usually connected to gut; pelvic fins usually behind pectoral fins; found almost entirely in the fresh waters of the world; four major suborders, numerous families.

*Suborder Characinoidei (characids).* Cyprinids possessing scales, lacking barbels, and usually having an adipose fin.

*Suborder Gymnotoidei (gymnotid eels).* Cyprinids possessing an eel-like body and having the anal opening on the throat.

*Suborder Cyprinoidei (minnows).* Cyprinids having a protrusible mouth as well as pharyngeal teeth; anal opening not on the throat.

*Suborder Siluroidei (catfishes).* Cyprinids having a nonprotrusible mouth, lacking scales, and possessing barbels.

ORDER ANGUILLIFORMES (EELS). Small to medium-sized (up to 2 m.) fishes with elongate body; dorsal and anal fins continuous with caudal fin; pelvic fins lacking; scales small or lacking; mesocoracoid lacking; swim bladder connected to gut when present; marine, except for one family; over 20 families recognized.

**ORDER CYPRINODONTIFORMES (CYPRINODONTIDS OR TOP MINNOWS AND THEIR RELATIVES).** Small (usually less than 0.3 m.), soft-rayed fishes lacking a lateral line; one dorsal fin; mesocoracoid absent; an air bladder present but not connected with the gut in the adult; primarily inhabitants of fresh water in North and South America, Europe, Africa, and parts of Asia, although a few forms occur in brackish or salt water; about seven families.

**ORDER BELONIFORMES (HALFBEAKS, NEEDLEFISHES, AND FLYING FISHES).** Small or medium-sized (up to 1.2 m.), elongate, soft-rayed fishes whose lateral line is situated very low on the body; scales small or of medium size and cycloid; mesocoracoid lacking; swim bladder not connected to gut in adult; largely marine in various oceans of the world, although a few species occur in fresh water; four families.

**ORDER GADIFORMES (CODFISHES AND HAKES).** Small to fairly large (0.1 to 2 m.), soft-rayed fishes whose pelvic fins are anterior to the pectoral fins; scales cycloid; mesocoracoid lacking; swim bladder not connected to gut in adult; marine waters of both the Northern and Southern Hemispheres; four families; one fresh-water species, *Lota lota*, which is holarctic.

**ORDER MACRURIFORMES (DEEP-SEA RATTAILS).** Medium-sized fishes (up to 1 m.) related to the Gadiformes but with the caudal fin continuous with the dorsal and anal fins; a spine occasionally associated with the first dorsal fin; pelvic fins anterior to pectoral fins; scales cycloid or ctenoid; Atlantic, Pacific, and Indian Oceans at considerable depths; one family.

**ORDER PERCOPSIFORMES (TROUTPERCH AND PIRATE PERCH).** Small (0.075 to 0.15 m.) fishes with characters intermediate between the primitive, soft-rayed teleosts and the more advanced spiny-rayed species; pelvic fins posterior to pectoral fins; dorsal, pelvic, and anal fins with one or more spines preceding the soft-fin rays; two of the three known species possessing an adipose fin on the caudal peduncle; mesocoracoid absent; swim bladder not connected to gut in adult; fresh waters of North America; one family (three species).

**ORDER BERYCIFORMES (SQUIRRELFISHES AND THEIR RELATIVES).** Small (0.075 to 0.6 m.) fishes related to the Perciformes but retaining some more primitive characters; pelvic fins under or slightly posterior to pectoral fins; dorsal fin preceded by a series of spiny rays; spines present on pectoral fins; orbitosphenoid usually present; swim bladder may or may not connect with gut in adult; scales ctenoid; temperate and tropical waters of the Atlantic, Pacific, and Indian Oceans; 15 or 16 families recognized.

**ORDER PERCIFORMES (PERCHLIKE FISHES).** Very small to very large (0.01 to 5 m.), spiny-rayed fishes having the pelvic girdle connected to the pectoral girdle by a ligament; pelvic fins either under or anterior to pectoral fins; usually possessing two dorsal fins, the anterior one bearing spiny rays; mesocoracoid and orbitosphenoid absent; swim bladder not connected to gut in adult; scales usually ctenoid; worldwide in occurrence in both marine and fresh waters; about 20 suborders and more than 125 families.

ORDER ECHENEIFORMES (REMORAS). Medium-small, perciform-like fishes (up to 1 m.) whose spiny-rayed anterior dorsal fin is modified to form a series of paired transverse folds on a suction disk situated on the top of the head; swim bladder absent; worldwide in tropical or temperate seas; one family containing about eight species.

ORDER ZEIFORMES (JOHN DORYS AND THEIR RELATIVES). Medium-sized (up to 1 m.), laterally compressed, deep-bodied fishes closely related to the Perciformes; a spinous-rayed dorsal fin present; one to four spines anterior to anal fin; widespread in the seas of the world at mid-depths; three families.

ORDER PLEURONECTIFORMES (FLATFISHES). Small to large (0.15 to over 3 m.), asymmetrical fishes with laterally compressed bodies and both eyes on the same side of the head; one side of body unpigmented and used to rest on the substratum; dorsal and anal fins extending along much of body length; swim bladder usually absent; widespread, benthonic, and largely marine; two suborders and five families.

ORDER GASTEROSTEIFORMES (STICKLEBACKS AND TUBENOSE). Very small (0.04 to 0.15), spiny-rayed fishes with three to 15 free spines anterior to the dorsal fin; one large spine at anterior end of each pelvic fin; cold and temperate marine and fresh waters of the Northern Hemisphere; two or three families, depending upon whether the family Indostomatidae (represented by a single freshwater species in Burma) is included.

ORDER SYNGNATHIFORMES (TUBE-MOUTHED FISHES). Small (0.03 to 0.7 m.) fishes with elongate, tubelike snout; anterior dorsal fin, when present, possessing spiny rays; ribs lacking; swim bladder not connected to gut in adult; principally tropical and subtropical marine waters; two suborders and seven families.

ORDER OPHIOCEPHALIFORMES (SNAKEHEADS). Small to medium-sized (0.15 to 1 m.), perciform-like fishes whose pelvic fins are posterior to the pectoral fins; pectoral and pelvic girdles connected by a ligament; mesocoracoid absent; unpaired fins lacking spines; swim bladder not connected to gut in adult; a suprabranchial chamber present above gills for breathing air; freshwater inhabitants of Asia and Africa; one family.

ORDER MULIGIFORMES (BARRACUDAS, MULLETS, AND SILVERSIDES). Small to moderately large (0.13 to 2 m.), spiny-rayed fishes with pelvic fins situated posteriorly on the abdomen; anterior dorsal spiny-rayed fin separated from posterior dorsal soft-rayed fin; pelvic fin composed of five rays preceded by a spine; pelvic girdle connected to pectoral girdle by a ligament; primarily tropical and subtropical marine waters, but some freshwater species, occurring in such widely separated places as the East and West Indies, New Guinea, and America; two suborders and three families.

ORDER PHALLOSTETHIFORMES (PHALLOSTETHIDS). Very small (usually less than 0.04 m.) fishes with pelvic fins rudimentary or lacking; two dorsal fins present, the anterior one spiny-rayed; swim bladder not connected to gut in adult; anal and genital openings in throat; fresh, brackish, and salt water from the Philippines west to India; one family.

ORDER LOPHIIFORMES (ANGLER FISHES). Small to medium-sized (0.05 to 1.3 m.) fishes whose pectoral fins are so modified that they may be used for walking; pelvic fins located on throat; first ray of dorsal fin modified as a lure to attract other fishes; mesocoroid and ribs absent; swim bladder not connected to gut in adult; Atlantic, Pacific, and Indian Oceans; two suborders and five families.

ORDER TETRAODONTIFORMES (TRIGGER FISHES, PUFFERS, OCEAN SUNFISHES, AND THEIR RELATIVES). Small to very large (up to 3.7 m.) fishes possessing a relatively small mouth and small gill openings; body often covered with plates or spines; tropical to temperate waters of Atlantic, Pacific, and Indian Oceans; four suborders and eight families.

SUBCLASS SARCOPTERYGII (FLESHY-FINNED FISHES). Fishes that usually possess choanae and that have paired fins which at least have a fleshy base over a bony skeleton; two orders (see Note under Osteichthyes, p. 51).

ORDER CROSSOPTERYGII (LOBE-FINNED FISHES). Fishes possessing a fleshy lobe to the base of the fins; usually two dorsal fins; internal choanae generally present; teeth possessing labyrinthine folds of enamel; two groups, the osteolepids and the coelacanths (regarded by some as belonging to separate orders, the Osteolepiformes and the Coelacanthiformes); Devonian to Recent; two closely related living species, *Latimeria chalumnae* and *Malania anjouanae*, known from the Indian Ocean near the Comoro Islands.

ORDER DIPNOI (LUNGFISHES). Fishes possessing fleshy fins of the archipterygeal type; internal choanae present; teeth in the form of two large plates; swim bladder connected to ventral side of gut and functioning, at least in part, as a respiratory organ; living representatives usually divided into two suborders, the Ceratodoidei, containing the genus *Neoceratodus* of Australia, and the Lepidosirenoidei, represented by *Lepidosiren* of South America and *Protopterus* of Africa.

## References Recommended

Barrington, E. J. W, 1961. Metamorphic Processes in Fishes and Lampreys. Am. Zool. 1:97–106.

Berg, L. S. 1947. Classification of Fishes Both Recent and Fossil (English and Russian). Ann Arbor, Mich., J. W. Edwards, Publisher, Inc.

Brown, M. E., Ed. 1957. The Physiology of Fishes. New York, Academic Press, Inc., 2 Vols.

Clemens, W. A., and Wilby, G. V. 1961. Fishes of the Pacific Coast of Canada. Bull. 68, Fisheries Res. Board of Canada, 2nd Ed. pp. 1–443.

Daniel, J. F. 1928. The Elasmobranch Fishes. University of California Press.

Giese, A. C. 1968. Cell Physiology. 3rd Ed. Philadelphia, W. B. Saunders Co.

Greenwood, P. H., Rosen, D. E., Weitzman, S. H., and Myers, G. S. 1966. Phyletic Studies of Teleostean Fishes, With a Provisional Classification of Living Forms. Bull. Amer. Mus. Nat. Hist. *131*:339–456.

Gregory, W. K. 1951. Evolution Emerging. New York, The Macmillan Co., 2 Vols.

Halstead, B. W. 1953. Some General Considerations of the Problems of Poisonous Fishes and Ichthyosarcotoxism. Copeia *1953*:31–33.

Halstead, B. W. 1967. Poisonous and Venomous Marine Animals of the World. Vol. 2, Vertebrates. Washington, D.C., U.S. Govt. Printing Office.

Halstead, B. W., Chitwood, M. J., and Modglin, F. R. 1955. The Venom Apparatus of the California Scorpionfish, *Scorpaena guttata* Girard. Trans. Am. Microscopical Soc. *74*:145–158.

Halstead, B. W., and Lively, W. M. 1954. Poisonous Fishes and Ichthyosarcotoxism. U.S. Armed Forces Med. J. 5:157–175.

Harder, W. 1964. Anatomie der Fische. *In* Wunch, H. H., Ed. Handbuch der Binnenfischerie Mitteleuropas. Berlin, Schweizerbart'sche.

Harvey, E. N. 1952. Bioluminescence. New York, Academic Press, Inc.

Hubbs, C. L., and Lagler, K. F. 1947. Fishes of the Great Lakes Region. Bull. Cranbrook Inst. Sci. *26*:1–186.

Hubbs, C. L., and Wick, A. N. 1951. Toxicity of the Roe of the Cabezon, *Scorpaenichthys marmoratus*. Calif. Fish and Game *37*:195–196.

Hulet, W. H., and Musil, G. 1968. Intracellular Bacteria in the Light Organ of the Deep Sea Angler Fish, *Melanocetus murrayi*. Copeia *1968*:506–512.

Johnson, F. H., Ed. 1955. The Luminescence of Biological Systems. American Association for the Advancement of Science.

Jordan, D. S. 1905. A Guide to the Study of Fishes. New York, Henry Holt and Co., Inc., Vol. 1.

Jordan, D. S., and Evermann, B. W. 1896. The Fishes of North and Middle America: A Descriptive Catalogue of the Species of Fish-like Vertebrates Found in the Waters of North America, North of the Isthmus of Panama. Bull. U.S. Nat. Mus. *47*:1–3313.

Jordan, D. S., Evermann, B. W., and Clark, H. W. 1930. Check List of the Fishes and Fishlike Vertebrates of North and Middle America North of the Northern Boundary of Venezuela and Colombia, Rept. U.S. Comm. Fish. 1928, pp. 1–670.

Kleerekoper, H. 1969. Olfaction in Fishes. Bloomington, Indiana Univ. Press.

Knowles, F. G. W. 1963. Animal Colour Changes and Neurosecretion. Triangle, *6* (No. 1):2–10.

Lissman, H. W. 1958. On the Function and Evolution of Electric Organs in Fish. J. Exper. Biol. *35*:156–191.

Lissman, H. W., and Machin, K. E. 1958. The Mechanism of Object Location in *Gymnarchus niloticus* and Similar Fish. J. Exper. Biol. *35*:451–486.

Magnuson, J. J. 1970. Hydrostatic Equilibrium of *Euthynnus affinis*, a Pelagic Teleost Without a Gas Bladder. Copeia *1970*:56–85.

Myers, G. S. 1963. Fresh-water Fishes. Pacific Discovery, *16* (No. 4):36–39.

Nelson, G. J. 1969. Origin and Diversification of Teleostean Fishes. Ann. New York Acad. Sci. *167*:18–30.

Nickol'skii, G. V. 1961. Special Ichthyology (Chastnaya Ikhtiologia. 1954, Moscow, Gos. Izd. "Sovetskaia Nauka.") Translated by J. I. Lengy and A. Krauthamer. Published by the Israel Program for Scientific Translations for the National Science Foundation. (Available at Office of Technical Services, U.S. Department of Commerce.)

Norman, J. R. 1931. A History of Fishes. London, Ernest Benn Ltd.

Norman, J. R., and Fraser, F. C. 1949. Field Book of Giant Fishes. New York, G. P. Putnam's Sons, Inc.

Oguri, M. 1964. Rectal Glands of Marine and Fresh-water Sharks: Comparative Histology. Science *144*:1151–1152.

Rand, H. W. 1950. The Chordates. Philadelphia, The Blakiston Co.

Romer, A. S. 1945. Vertebrate Paleontology. 2nd Ed. Chicago, University of Chicago Press.

Romer, A. S. 1970. The Vertebrate Body. 4th Ed. Philadelphia, W. B. Saunders Co.

Schrenkeisen, R. 1938. Field Book of Freshwater Fishes of North America, North of Mexico. New York, G. P. Putnam's Sons, Inc.

Schultz, L. P. 1936. Keys to the Fishes of Washington, Oregon and Closely Adjacent Regions. Univ. Wash. Publ. Zool. 2:103-228.

Schultz, L. P., and Stern, E. M. 1948. The Ways of Fishes. Toronto, New York, and London, D. Van Nostrand Co., Inc.

Thorson, T. B. 1967. Osmoregulation in Fresh-Water Elasmobranchs, pp. 265-270. *In* P. W. Gilbert, R. F. Mathewson, and D. P. Rall [eds.] Sharks, Skates, and Rays. Baltimore, Johns Hopkins Press.

Thorson, T. B., Cowan, C. M., and Watson, D. E. 1967. *Potamotrygon* spp.: Elasmobranchs With Low Urea Content. Science *158*:375-377.

Yapp, W. B. 1965. Vertebrates; Their Structure and Life. New York, Oxford University Press.

*Chapter Three*

# AMPHIBIANS

## GENERAL CHARACTERS

There are about 3000 species of amphibians living in the world today. These are commonly grouped into three orders: the Anura (frogs and toads), the Caudata or Urodela (salamanders), and the Gymnophiona or Apoda (caecilians). There are only about 60 species of caecilians and about 200 different kinds of salamanders, so most of the amphibians are various kinds of frogs and toads.

The term "amphibian" is applied to members of this class because most of them spend the early stages of their life cycle in water as gilled, larval forms commonly referred to as tadpoles. Later the larvae metamorphose into lung-breathing terrestrial adults. This is not true of all amphibians, however, as there are some that never leave the water, remaining permanently in the larval state, and others that never enter water at any stage of their life cycle. The young of the latter type emerge from the egg capsule fully formed. Some also lack lungs as adults and respire entirely through the skin. The skin is moist and glandular.

As a group amphibians represent the first vertebrates to live on land. Basically they possess *pentadactyl* (five-toed) limbs, although the number of toes is frequently reduced.

Like fishes and reptiles, they are ectotherms depending on their environment as a source of heat. Since most amphibians lay their eggs in ponds and streams and none can travel far overland, few species live far from water.

## Skeletal System

Amphibians have a proportionately broad, flat skull, in contrast to that of most fishes. The ancient Paleozoic and Mesozoic amphibians, however, retained many of the cranial features of their cross-opterygian ancestors, including a large number of dermal bones. These are greatly reduced in the modern amphibian skull, being principally represented by paired premaxillaries, maxillaries, nasals, frontals, parietals, and squamosals (Fig. 3.2). Much of the upper surface of the head lacks a bony covering. This is especially true of anurans. Parts of the chondrocranium still remain unossified. In the occipital region only the exoccipitals are ossified, and each of these possesses a condyle which articulates with the first vertebra. The basisphenoid, presphenoid, and alisphenoids are cartilaginous.

There is no secondary palate in amphibians; consequently the internal nares open far forward in the roof of the mouth. The underside of the brain case is covered by a dermal bone termed the parasphenoid. This is very broad in caecilians. On either side of this bone there are large vacuities. The epipterygoid, which represents the anterior part of the palatoquadrate cartilage of fishes, is never ossified in amphibians and even the quadrate is incompletely ossified in urodeles. The latter is immovable, being fused to the otic region.

Although in ancient amphibians there were numerous bones in the lower jaw, the number has been markedly reduced in living species. The head of Meckel's cartilage is replaced by the articular bone in caudate amphibians but remains cartilaginous in the anurans. A dentary as well as a posterior splenial and angular may be present.

Teeth may be present on the premaxillary, maxillary, palatine, vomer, parasphenoid, and dentary bones. In some amphibians they are entirely lacking and in others they may be absent from the lower jaw.

The number of vertebrae in amphibians varies from 10 in the Salientia to approximately 200 in the Gymnophiona. Furthermore, there is greater differentiation of the vertebrae than is found in fishes. The skull articulates with a single *cervical* vertebra. Behind this are the *trunk* vertebrae. The pelvic girdle in all except the limbless caecilians is attached to the axial skeleton by a single *sacral* vertebra. The number of *caudal* vertebrae is variable. In the Salientia there is but a single elongate element called the *urostyle* which extends from the sacrum to the posterior end of the pelvis.

Amphibians are the first vertebrates in which a sternum appears. Ribs, however, are poorly developed and in no instance contact the sternum as they do in most reptiles and in birds and mammals.

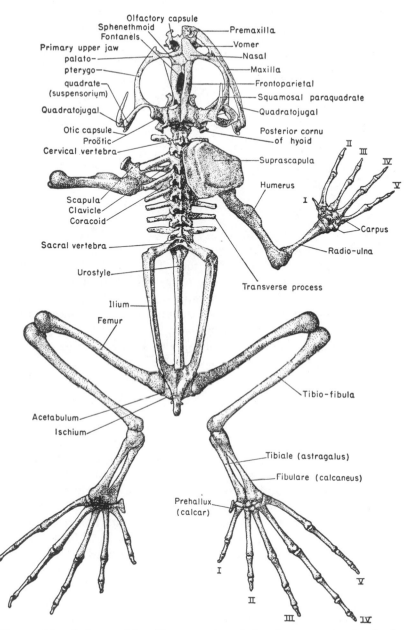

**FIGURE 3.1.** Skeleton of frog (*Rana catesbiana*): dorsal view except that the feet are extended backward so that they are seen in ventral aspect. In the skull most of the dermal bones of the left side have been removed (the quadratojugal remaining), together with a narrow strip of the right frontoparietal, so as to expose the chondrocranium and the primary cartilaginous upper jaw (palato-pterygo-quadrate) continuous with the cartilaginous olfactory capsule in front and the auditory (otic) capsule behind. The proötic bone is the only ossification in the otic capsule and, in the adult frog, is fused with the adjacent exoccipital ossification. On the right side the suspensorium is omitted so as to expose to view the annulus tympanicus, which supports the tympanic membrane of the "middle ear." (Rand, H. W.: The Chordates. McGraw-Hill Book Co., Inc., 1950.)

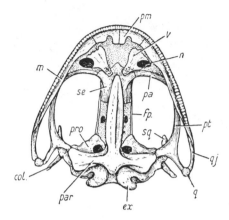

FIGURE 3.2. Ventral view of the skull of the frog: *Col.*, columella auris; *ex.*, exoccipital; *fp.*, frontoparietal; *m.*, maxilla; *n.*, nostril; *pa.*, palatine; *par.*, parasphenoid; *pm.*, premaxilla; *pro.*, proötic; *pt.*, pterygoid; *q.*, quadrate; *qj.*, quadratojugal; *se.*, sphenethmoid; *sq.*, squamosal; *v.*, vomer. Young, J. Z.: The Life of Vertebrates. Oxford University Press, 1950. After Marshall: The Frog, The Macmillan Co., 1912.)

Most amphibians have two pairs of limbs with four toes on the front foot and five on the hind foot. The number of toes may be reduced to as few as two in some forms. The hind limbs are lacking in certain of the eel-like salamanders, and both pairs of limbs are absent in the caecilians. This is a degenerate rather than a primitive condition. Claws are generally lacking. However, there may be horny protuberances or even adhesive disks on the toes. Limb adaptations of amphibians are discussed later in this chapter.

## MUSCULAR SYSTEM

The muscular system of amphibians, like many other organ systems in this group, presents an interesting transition between that of fishes and reptiles. In fishes the muscular system is primarily concerned with lateral body movement, opening and closing the mouth and gill apertures, and relatively simple fin movement. Life on land necessitated some changes in this arrangement. The trunk musculature of amphibians is still primarily metameric as in fishes, but some marked differences are apparent (Fig. 3.3). We find that the horizontal septum that divides the dorsal and ventral body muscles is more dorsal in position. This has resulted in a reduction in the size of the epaxial or dorsal trunk muscles. Parts of this muscle system as well as some of the original gill musculature effect movement of the head, an action which a fish is incapable of doing. The ventral or hypaxial musculature shows even greater modification. Myosepta or myocommata, which are so evident in separating the muscle bundles of each body segment in fishes, are reduced or even absent in the ventral body musculature of many amphibians. Furthermore, the hypaxial muscles are delaminated or split into separate layers, thus forming the external oblique, the internal oblique, and the transversus

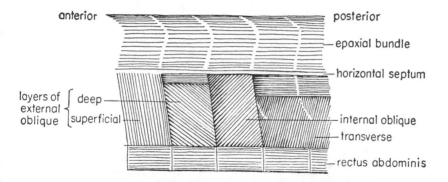

anterior — posterior

— epaxial bundle

— horizontal septum

layers of { deep — internal oblique
external
oblique { superficial — transverse

— rectus abdominis

FIGURE 3.3. Muscles of body wall in a salamander (*Dicamptodon ensatus*), partly dissected, left side. (Eaton, Theodore H., Jr.: Comparative Anatomy of the Vertebrates. Harper & Bros., 1960.)

muscles. The external oblique may even be subdivided into a superficial and deep portion. The midventral rectus abdominus which is present in a few fishes is well developed in amphibians. Dermal muscles are essentially lacking.

To attempt to describe the appendicular muscles is beyond the scope of this work. Even in the most primitive tetrapods locomotion on land involves many factors not encountered by lower vertebrates. In fishes the muscles that move the fins are essentially within the body and are therefore extrinsic to the appendages. They consist simply of dorsal and ventral groups. The former either raises or extends the appendage and the latter either lowers or adducts it. To perform the various motions that different kinds of amphibians must, in order to swim, walk, hop, or climb, involved the development of many other types of muscles. Some of these are located entirely within the limb itself and are, therefore, referred to as *intrinsic* muscles.

## Circulatory System

In fishes that respire entirely by means of gills only unoxygenated blood is received by the heart, which pumps it forward to the gill region. Most amphibians have a somewhat different problem to contend with since the heart receives both unoxygenated blood from the body and oxygenated blood from the lungs. This is also true of the lungfishes. To prevent too much mixing of the two kinds of blood, therefore, we find that both of these groups have started to develop a double circulatory system (Fig. 3.5). This is brought about by the formation of an interatrial septum, deep pockets in the ventricular cavity, and a division of the conus arteriosus into systemic and pulmonary vessels. Blood from the body enters the right atrium from

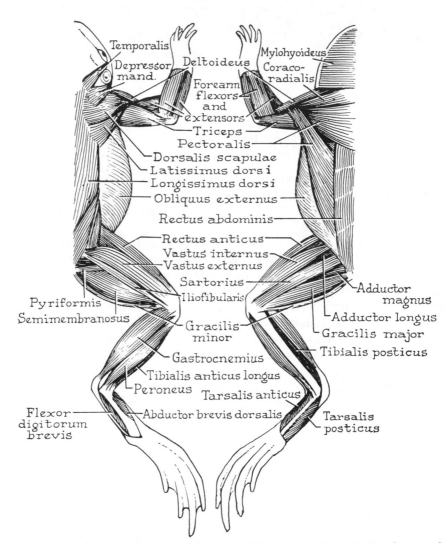

**FIGURE 3.4.** Superficial skeletal muscles of the frog in a dorsal (*left*) and a ventral (*right*) view. (Villee, C. A., Walker, F. W., Jr., and Smith, F. E.: General Zoology. 3rd Ed., W. B. Saunders Co., 1968.)

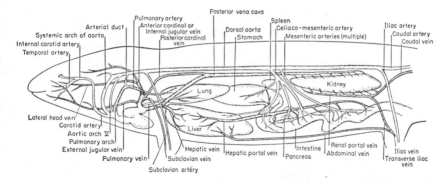

Pulmonary artery
Posterior vena cava
Arterial duct
Anterior cardinal or
Internal jugular vein
Spleen
Systemic arch of aorta,
Posterior cardinal
vein
Dorsal aorta
Celiaco-mesenteric artery
Iliac artery
Internal carotid artery,
Stomach
Mesenteric arteries (multiple)
Caudal artery
Temporal artery,
Caudal vein

Kidney

Lung

Lateral head vein
Carotid artery
Aortic arch Ⅴ
Liver
Pulmonary arch
External jugular vein
Hepatic vein
Intestine
Renal portal vein
Iliac vein
Pulmonary vein
Subclavian vein
Hepatic portal vein
Pancreas
Abdominal vein
Transverse iliac
vein
Subclavian artery

**FIGURE 3.5.** Diagram of the main blood vessels of a urodele amphibian as seen in lateral view. (Romer, A. S.: The Vertebrate Body. 4th Ed., W. B. Saunders Co., 1970.)

the sinus venosus, passes into the right side of the ventricle, and from here is pumped to the lungs. Oxygenated blood from the lungs enters the left atrium by way of the pulmonary veins, goes to the left side of the ventricle, and then is pumped out by way of the systemic part of the conus. Some exceptions occur among lungless salamanders in which the interatrial system is incomplete and pulmonary veins are absent. In most amphibians the first, second, and fifth pairs of aortic arches are missing. The third aortic arches on either side form the bases of the internal carotid arteries, and the fourth aortic arches constitute the systemic arches that are continuous posteriorly with the dorsal aorta. From the proximal part of the sixth pair of aortic arches the pulmocutaneous arteries branch off. They carry blood to the lungs and skin where aeration occurs. In the tailed amphibians a remnant of the distal part of the sixth aortic arches persists on each side. This is called the ductus arteriosus and permits some mixing of nonoxygenated and oxygenated blood.

The venous system of amphibians very much resembles that of lungfishes except for the fact that the abdominal vein enters the hepatic portal system rather than the sinus venosus. The common cardinal veins that enter the sinus venosus on either side are referred to as the precaval veins, and the postcaval vein, which was derived in a complicated manner from several cardinal systems, assumes greater importance.

## DIGESTIVE SYSTEM

Aquatic amphibians have little need for oral glands, since their food is generally secured in the water. They do, however, have a few mucous glands scattered about the mouth. These glands are more numerous in terrestrial amphibians, especially on the tongue, which is used by frogs and toads to capture food. Terrestrial amphibians also possess an intermaxillary gland on the roof of the mouth which produces a sticky secretion. There are some amphibians in which the tongue is either lacking or immovable. Most amphibians, however, have a protrusible tongue and in some frogs and toads it may even be folded back on itself when not in use. A short esophagus, which is usually distinguishable from the stomach, is present. The intestine shows considerable variation. In the caecilians it exhibits little coiling and is not differentiated into a small and large tract. In frogs and toads, on the other hand, there is a relatively long, coiled small intestine and a short, straight large intestine which open into the cloaca.

## RESPIRATORY SYSTEM

During the larval or tadpole stage most amphibians respire by means of gills. These gills are not of the internal type found in most fishes, but are external, filamentous structures covered with a ciliated epithelium and partly enclosed by a large, fleshy operculum which later encases them completely (Fig. 3.6). They are similar in many respects to the gills possessed by larval lungfishes and are either resorbed or degenerate during metamorphosis. In a few of the caudate amphibians the larval gills are retained throughout life.

The lungs of amphibians are simple, saclike structures (Fig. 3.7).

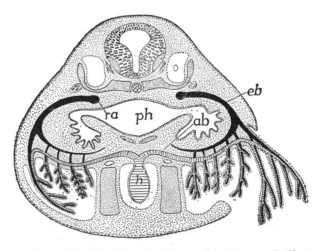

**FIGURE 3.6.** Diagram of the relations of external and internal gills in the anuran tadpole: *ab* & *eb*, afferent and efferent branchial arteries; *h*, heart; *o*, ear cavity; *ph*, pharynx; *ra*, radix aortae. (After Maurer, courtesy of Kingsley: Comparative Anatomy of Vertebrates. McGraw-Hill Book Co., Inc., 1926.)

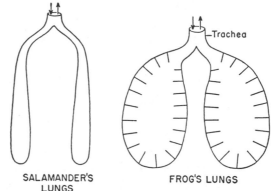

FIGURE 3.7. Lung of a caudate amphibian (*Necturus*), and an anuran (*Rana*). (Villee, C. A.: Biology. 5th Ed., W. B. Saunders Co., 1967.)

SALAMANDER'S LUNGS

FROG'S LUNGS

In some aquatic forms the inner surface may be smooth, but in most frogs and toads the walls contain numerous folds lined with alveoli so as to increase the respiratory surface. Many caudate amphibians possess a short trachea, supported by cartilages,which bifurcates into two bronchi that open into the lungs. The upper end of the trachea is enlarged, especially in the frogs and toads, to form the *larynx* or voice box, in which the vocal cords are located. The opening from the pharynx into the larynx is called the *glottis*. Air is generally pumped into the amphibian lungs by a simple swallowing process.

Most amphibians respire to some extent through the skin and some, e.g., the plethodontid salamanders, which lack both lungs and gills as adults, obtain oxygen entirely through the skin and the oral epithelium. This means that the skin must be kept moist. In terrestrial amphibians this is accomplished by the numerous mucous glands distributed over the surface of the body.

## UROGENITAL SYSTEM

The amphibian kidneys, like those of fishes, are of the opisthonephric type. In tailed amphibians the kidneys are rather elongate structures as in elasmobranchs, but in anurans there is a tendency for these structures to be short and compact. Since many amphibians live partly or entirely in fresh water, we find that they have developed large renal corpuscles to assist in the elimination of water and thus prevent excessive dilution of the body fluids.

In some amphibians the archinephric duct is both genital and excretory in nature in the male, whereas in others the archinephric duct serves only for the transport of sperm, and the kidney is drained by a new duct, somewhat comparable to the ureter of higher vertebrates.

Amphibians have developed a bladder which differs from the

type frequently found in fishes. In the latter the bladder generally results from an enlargement of the lower end of the archinephric duct. The amphibian bladder represents a new structure that develops as an evagination from the floor of the cloaca. Urine, therefore, must first pass from the ducts of the kidneys into the cloacal chamber. From here it is then forced into the bladder for storage. In certain terrestrial amphibians some of the water from urine so stored is resorbed into the system at certain times to compensate for moisture lost through the skin.

The amphibian ovaries are paired and contain a cavity within them that is filled with lymph. The oviducts also are paired although in some forms the lower ends are fused together. Frequently, the lower end of each oviduct is enlarged into a uterus-like structure or ovisac which serves as a temporary storage space for ova before they are shed or, in a few species that are ovoviviparous, for the development of the young. Glands that secrete a jelly-like covering for the eggs are usually situated along the oviduct.

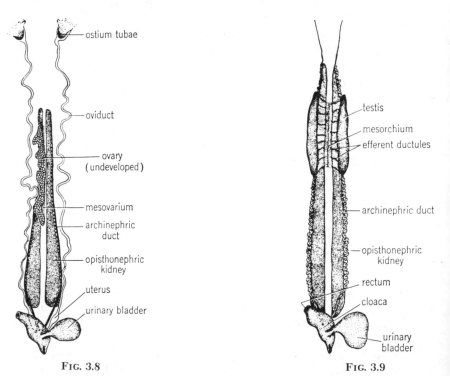

FIG. 3.8                                   FIG. 3.9

**FIGURE 3.8.** Urogenital system of female *Necturus* during the nonbreeding season, ventral view. The left ovary has been removed to show the opisthonephros above. (Weichert, C. K.: Anatomy of the Chordates. McGraw-Hill Book Co., Inc., 1958.)

**FIGURE 3.9.** Urogenital system of male *Necturus*, ventral view. (Weichert, C. K.: Anatomy of the Chordates. McGraw-Hill Book Co., Inc. 1958.)

The testes, like the ovaries, are paired and are connected either directly or by way of mesonephric tubules to the archinephric ducts which in turn open into the cloaca. No special copulatory organs are present. In some toads there is a structure called Bidder's organ located anterior to each testis. Under certain circumstances this may develop into an ovary. Even stranger, perhaps, is the fact that old females at times may produce sperm.

## NERVOUS SYSTEM

The nervous system of amphibians is still basically like that of fishes. The center of brain activity remains in the dorsal part of the midbrain, where the gray cells are concentrated in what is called the tectum. The telencephalon is largely olfactory in nature, but, for the first time in vertebrates, we find nerve cells invading the pallium. Although these cells are internal in location, the result is a definite enlargement of the cerebral hemispheres. Both parietal and pineal bodies are present although, in living amphibians, neither penetrates the cranial roof. Since members of this vertebrate class are notably slow and sluggish in their movements, we find that the cerebellum is very small. Except in the caecilians, flexures are lacking in the amphibian brain. There are only 10 cranial nerves. The dorsal and ventral roots of the spinal nerves unite in their passage through the intervertebral foramen rather than outside as in most fishes or within the neural canal as in amniotes.

## SENSE ORGANS

The taste buds of amphibians, unlike those of many fishes, are restricted to the roof of the mouth, the tongue, and the mucosa that lines the jaws. Since internal choanae are present, the nasal apertures no longer are strictly olfactory in function but also serve as air passages. In amphibians the olfactory epithelium is generally smooth and is restricted to the upper part of the nasal passages. Another

FIGURE 3.10. Lateral view of brain of a frog (*Rana*). (Romer, A. S.: The Vertebrate Body. 4th Ed., W. B. Saunders Co., 1970.)

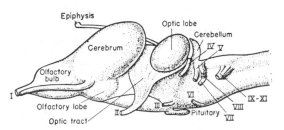

olfactory structure that appears for the first time in amphibians is _Jacobson's organ_ or, as it is also known, the _vomeronasal organ._ This arises as an evagination from the nasal passage and is lined with olfactory epithelium. It is believed to be an aid in tasting food. It has been found, as a result of applying colored liquids to the nasolabial grooves of the salamander *Ensatina,* that fluids rapidly pass up to the external nares and from there over chemoreceptors of Jacobson's organ and down into the pharynx by way of the internal nares. *Ensatina* is one of the lungless plethodontid salamanders; consequently these grooves appear to play an important role in olfaction. They may also be important in reproductive behavior, since the first act of the male in courtship is to nose the female's head and neck.

The amphibian eye is basically like that of other vertebrates. The lens is fixed for relatively distant vision but may be moved forward toward the cornea for seeing close objects, by means of small muscles of accommodation. The pupillary aperture may be vertical, horizontal, three-cornered, or four-cornered. Eyelids are poorly represented in aquatic forms but are well developed in many terrestrial species. The lower lid is usually more movable than the upper lid. Since the cornea of land vertebrates would dry up as a result of evaporation, it is necessary that it be bathed by a liquid. _Lacrimal_ or tear glands are present in many amphibians, although they are poorly developed. The eye, however, is kept moist by the oily secretion of another type of gland known as the *Harderian gland.*

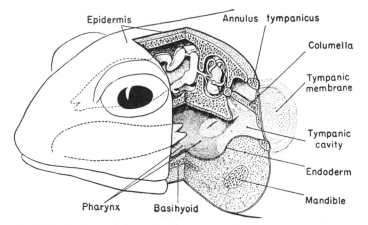

FIGURE 3.11. Auditory apparatus of an anuran. The stereogram exposes a narrow vertical band of the animal's sagittal plane and the left half of a transverse section which has been swung several degrees backward on its sagittal axis to bring it into a favorable angle for the observer. Both sections cut through the pharynx. The transverse section bisects the tympanic membrane and shows the left auditory nerve (VIII) passing from the medulla oblongata to the internal ear (membranous labyrinth). The sagittal view shows the forked anterior tip of the retracted tongue pointing backward into the pharynx. (Rand, H. W.: The Chordates. McGraw-Hill Book Co., Inc., 1950.)

There is considerable variation in the auditory apparatus of amphibians. Salamanders and their relatives lack any middle ear, although it is believed that they can detect vibration. Most frogs and toads, however, have a middle ear and an external ear drum (Fig. 3.11). Sounds are transmitted from the ear drum across the tympanic cavity to the inner ear by means of a bone called the *columella* which is homologous with the hyomandibular element of the gill arches of cartilaginous fishes. The tympanic cavity, since it is derived embryologically from the first pharyngeal or gill pouch, is connected with the pharynx by means of the eustachian tube. This permits equalization of pressure on both sides of the tympanic membrane. From the ventral part of the sacculus of the inner ear there is a ventral outpocketing called the *lagena* which is the forerunner of the cochlea of mammals. This is believed to be concerned with the reception of sound vibrations.

A lateral line system, similar to that of fishes, is present in larval amphibians and even in the adult stage of certain aquatic species.

## ENDOCRINE GLANDS

The amphibian endocrine system is basically like that of higher vertebrates. Parathyroid glands, which are lacking in fishes, are present and serve to regulate the calcium in the system. The two components of the adrenals, the cortex and medulla, are joined together rather than being separate as in fishes. The thyroid gland, which is paired in adult amphibians, not only regulates metabolic activity in the body but is believed to be important in influencing the periodic molting of the outer layer of the skin.

## SPECIAL CHARACTERS

### SKIN AND GLANDS

The amphibian skin is most important for respiration and protection. In some of the aberrant caecilians there are small scales embedded beneath the surface, but in all other amphibians, scales are lacking.

The skin is kept soft and moist by the presence of numerous mucous glands. Even in aquatic species the mucus serves to lubricate the body. There are some amphibians that develop glands on the snout and back before hatching, and the secretion serves to break down the egg capsule at the time of hatching.

Most amphibians possess granular glands in addition to mucous glands. Although the two are quite similar in many respects, the former may produce obnoxious or poisonous secretions which protect these animals from enemies. The so-called warts and the large parotoid glands situated on either side of the neck in toads represent aggregations of poison glands (Fig. 3.12).

The toxicity of various amphibian poisons varies greatly. In the marine toad (*Bufo marinus*) of the American tropics the poison is reported to be sufficiently toxic to cause the death of dogs. The secretion from the glands of some frogs and toads may produce irritation of the skin if one handles these animals. Studies on neotropical frogs of the family Dendrobatidae, many of which are brilliantly colored and poisonous, have shown that the skin toxins are steroidal alkaloids that affect nerve and muscle activities. The coloration in such species is probably a warning. Other types of amphibian poisons are neurotoxins, hallucinogens, vasoconstrictors, hemolytic agents, and merely local irritants. Occasionally when different kinds of amphibians are placed together in a small container some species will succumb in a short while to the poison secreted by others.

Both the mucous and granular or poison glands are classified as *alveolar glands*. By this it is meant that they terminate in blind, spherical sacs whose walls are secretory in nature. In addition to alveolar glands, however, some amphibians possess *tubular glands*.

**FIGURE 3.12.** Western toad (*Bufo boreas*). Numerous warty glands are scattered over the body. The large, oval parotoid gland is posterior to the eye and the tympanic membrane is immediately below its anterior end. (Photograph by Edward S. Ross.)

These are often found on the thumbs of some frogs and toads and occasionally on the chest. They become functional during the reproductive season and produce a sticky secretion which assists the male in clinging to the female during amplexus. Even in some salamanders there are tubular glands on the chins of males whose secretions seem to attract females during the breeding season.

## Coloration

Amphibians exhibit great diversity in color. Brilliant shades of green, yellow, orange, and gold are not uncommon. Red and blue are less frequently seen. There are, of course, many dull colored species as well. Just as is true of the colors of avian feathers, some amphibian colors are structural, while others are produced by pigment or are the result of both structure and pigment. In amphibians, as in fishes, pigment is located in chromatophores in the skin, and these pigment cells are usually named according to the type of pigment that they contain. Melanophores contain various black and brown pigments and lipophores contain red, yellow, and orange pigments (Fig. 3.13). Amphibians also possess cells called *guanophores,* which are somewhat comparable to the iridocytes of fishes and contain crystals of guanine that may produce either iridescence or a white effect. Generally the lipophores are closest to the surface of the skin. Beneath these are the guanophores; the melanophores are deepest.

**Figure 3.13.** Head of an arboreal salamander (*Aneides lugubris*) showing a reticulum of melanophores surrounding the numerous lighter colored glands in the skin.

Chromatophores are rather ameboid in shape with protoplasmic processes extending outward from the main body of the cell. Furthermore, the pigmented cytoplasm within a chromatophore is capable of migrating so as to concentrate or to dilute the color. Some pigment cells, especially the lipophores, are capable of ameboid movement and can migrate toward or away from the surface of the skin. Frequently, as in the tree frogs, a change from green to yellow is the result of contraction of the melanophores and a migration of the lipophores to a position between or below the guanophores.

It has recently been found that the lightening in body color by many amphibians when placed in a dark environment is the result of stimulation of the pineal gland by the lack of sufficient quantities of light or of light of the proper wave lengths. This results in the production of a hormone-like substance called *melatonin.* The latter counteracts the hypophyseal chromatotrophic hormone which would cause melanophore expansion. As a result the melanophores contract and produce a lighter body effect. This occurs even in amphibians whose vision has been impaired. Pinealectomized individuals, however, do not undergo body lightening when placed in a dark envi-

**FIGURE 3.14.** Both color and shape enable certain amphibians like the Surinam toad (*Pipa pipa*) to blend with the background and thereby avoid detection by enemies. (Photograph by Edward S. Ross.)

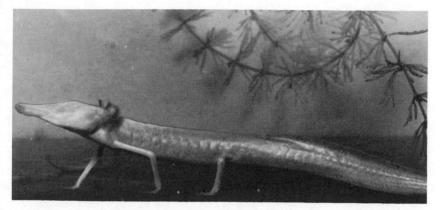

**FIGURE 3.15.** Certain cave inhabiting salamanders, such as the Texas blind salamander (*Typhlomolge rathbuni*), lack functional eyes and pigment in the skin. (Photograph by Edward S. Ross.)

ronment. Many amphibians exhibit marked protective coloration, whereas some cave-dwelling species have lost all trace of pigment (Figs. 3.14 and 3.15).

## MOLT

The entire outer surface of the skin is shed periodically by amphibians. This process of molt appears to be under hormonal control, since it is inhibited in hypophysectomized individuals. The outer layer of the skin does not come off in one piece as in certain reptiles but rather in fragments, although that of the limbs usually remains intact and peels off so that it is inside out when shed. The frequency of the molt varies with different species. The green tree frog (*Hyla cinerea*) of central and southern United States is said to shed its skin every day. In other amphibians a month or more may elapse between molts.

## APPENDAGES

Although it is believed that the primitive, ancestral amphibians possessed two pairs of pentadactyl limbs, there is considerable diversity to be found in modern forms as a result of their adaptation to terrestrial, aquatic, arboreal, and subterranean life. The caecilians, which are confined to tropical parts of the world today, are entirely limbless. Their bodies are superficially wormlike and modified for burrowing in humus and rotten wood.

Most modern caudate amphibians possess four relatively weak limbs which are not suited to rapid travel on land. Generally the front feet possess four digits and the hind feet five digits. Some species, however, show marked limb reduction. Members of the genus *Siren* in southeastern United States and the Mississippi Valley possess only pectoral limbs. The closely related genus *Pseudobranchus* not only lacks hind limbs but has the front toes reduced from four to three. In the so-called "conger eels" of southern United States, the limbs are extremely small in proportion to body size and the toes may be reduced to two or three in number. Although most salamanders are terrestrial or aquatic, some species are arboreal to some extent. At least two members of the genus *Aneides* in western North America can really climb trees. In these species the toes are broadly expanded and truncate to enable them to cling to the trunks, and the tail is somewhat prehensile.

The appendages of frogs and toads in general are much more specialized than those of salamanders. The usual number of toes on the front limbs is four. The hind limbs are elongate and adapted to jumping. Most frogs and toads possess five hind toes plus an additional digit known as the prehallux on the inner side of the foot. In the spadefoots this prehallux forms the bony base of the sharp edged

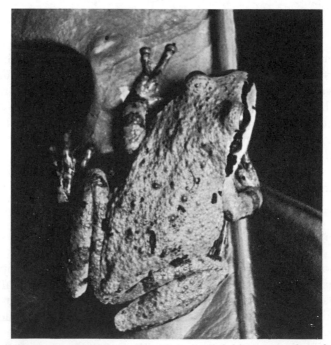

**FIGURE 3.16.** The Pacific tree frog (*Hyla regilla*), like many arboreal or semiarboreal anurans, has adhesive disks on the tips of the toes.

tubercle used for digging. It is interesting to note that these amphibians burrow into the ground backwards.

Many of the arboreal types of frogs have the tips of the toes enlarged and adhesive disks present on the undersurface (Fig. 3.16). Although there are a number of amphibians that have horny, epidermal growths on the tips of the toes, claws comparable to those of lizards are very rare. They do occur on the African clawed frog and in a few salamanders inhabiting mountain streams.

There is considerable variation as regards webbing on the hind feet of anurans. In some species the webbing is complete and extends to the tips of the toes, whereas in others it may be present in varying degrees or even entirely absent. Anurans are unable to regenerate lost limbs or toes although salamanders can do this.

## CLASSIFICATION OF AMPHIBIANS

### CLASS AMPHIBIA (AMPHIBIANS)

Cold-blooded vertebrates respiring either by means of lungs, gills, the skin, or the lining of the mouth and passing through a larval stage, either in water or within the egg in a moist situation; skull incompletely ossified (except in most of the ancient forms) and possessing paired occipital condyles (except in the most ancient forms); skin usually moist and containing mucous glands; scales generally lacking in recent forms or, if present, small and embedded in skin; Devonian to Recent.

SUBCLASS APSIDOSPONDYLI (APSIDOSPONDYLIDS). Ossified vertebral centra, when present, derived from cartilaginous arches or blocks and composed of intercentra anteriorly and pleurocentra (if present) posteriorly; Devonian to Recent; two superorders generally recognized.

Superorder Labyrinthodontia (labyrinthodonts). Ancient apsidospondylous amphibians; braincase well ossified; an otic notch present; appendages usually well developed.

ORDER TEMNOSPONDYLI (TEMNOSPONDYLS). Labyrinthodonts possessing vertebral centra in which the pleurocentra, although often present, are never dominant and never form complete disks; intercentra constituting the dominant vertebral elements; tabulars and parietals not contacting each other; skull often possessing paired occipital condyles; Devonian to Triassic.

ORDER ANTHRACOSAURIA (ANTHRACOSAURS). Labyrinthodonts possessing vertebral centra in which the pleurocentra are either equal in size or larger than the intercentra and form complete disks enclosing the notochord; tabulars and parietals in contact; skull possessing a single occipital condyle; Carboniferous to Permian.

*Superorder Salientia (salientians).* Mostly modern specialized amphibians essentially lacking true vertebral centra; body of vertebra generally formed from ventral parts of neural arches which
encase the notochord; intercentra and pleurocentra vestigial, if
present; braincase not well ossified; ribs generally lacking; Triassic to Recent.

ORDER PROANURA (PROANURANS). Ancient salientians possessing a
tail and ribs; skull resembling modern anurans; limbs not specialized for jumping; Triassic.

ORDER ANURA (FROGS AND TOADS). Amphibians lacking a tail and
possessing hind limbs modified for jumping; ribs usually absent; Jurassic to Recent; 18 families.*

*Family Leiopelmidae (leiopelmids).* Small (about 50 mm.)† frogs
possessing two pairs of free, bony ribs; nine presacral vertebrae; vertebral centra amphicoelous; northwestern North
America and New Zealand; two genera, *Leiopelma* and
*Ascaphus,* and four species.

*Family Pipidae (tongueless frogs).* Small to moderately large (50
to 200 mm.) frogs possessing ribs (free in larva, fused to
transverse processes in adult); vertebral centra opisthocoelous; tongue lacking; pectoral girdle firmisternal or partly
arciferal; South America and Africa; five genera.

*Family Discoglossidae (fire-belly and midwife toads).* Small (50
to 75 mm.) toads possessing ribs on vertebrae 2 to 4; vertebral
centra opisthocoelous; tongue disklike and nonextensible;
pectoral girdle arciferal; Europe, Asia (including the Philippines), and northern Africa; four genera.

*Family Rhinophrynidae (burrowing toad).* A moderately small
(up to about 60 mm.) toadlike anuran lacking free, ossified
ribs; vertebral centra procoelous; tongue free anteriorly; pupil vertical; external ear and teeth lacking; pectoral girdle
arciferal; Mexico and Guatemala; one genus and one species,
*Rhinophrynus dorsalis.*

*Family Pelobatidae (spadefoot toads).* Small (35 to 75 mm.),
toadlike anurans lacking free, ossified ribs; eight presacral
vertebrae; centra usually procoelous; pectoral girdle arciferal;
Europe, northern Africa, southern Asia, and North America;
four genera.

*Family Leptodactylidae (leptodactylids).* Very small to very
large (20 to 215 mm.) anurans lacking free, ossified ribs;
vertebral centra procoelous; intercalary cartilage absent; teeth
present; pectoral girdle arciferal; tropical America, the West
Indies, Australia, New Guinea, and Tasmania; about 43 genera.

*Family Bufonidae (toads).* Very small to very large (30 to 200
mm.), stout-bodied anurans lacking free, ossified ribs; verte-

---

*Only living families, genera and species are included here.
†Measurements given for the Anura represent body length.

bral centra procoelous; parotoid glands present; teeth absent; tongue free posteriorly; pupils horizontal; pectoral girdle arciferal; temperate and tropical parts of the world except Australia (*Bufo marinus* introduced) and Madagascar; five genera.

*Family Rhinodermatidae (mouth-breeding frog).* A small (30 mm.) anuran lacking free, ossified ribs; vertebral centra procoelous; snout elongate; vocal pouch of male greatly enlarged to hold larvae; pectoral girdle firmisternal; southern South America; one genus and one species, *Rhinoderma darwinii.* (*Note:* Referred by some to the family Dendrobatidae.)

*Family Dendrobatidae (dendrobatids).* Very small to small (12 to 50 mm.) anurans lacking free, ossified ribs; vertebral centra procoelous; teeth present or absent; a pair of platelike dermal scutes on ends of digits; skin containing numerous poison glands; pectoral girdle partly or completely firmisternal; Central and South America; nine genera.

*Family Atelopidae (atelopids).* Small (up to 50 mm.) anurans lacking free, ossified ribs; vertebral centra procoelous; toes without disks or dermal scutes; skin containing numerous poison glands; pectoral girdle firmisternal; Central and South America; one genus, *Atelopus*, numerous species.

*Family Hylidae (tree frogs).* Small to medium-sized (25 to 115 mm.), slender, long-legged anurans lacking free, ossified ribs; vertebral centra procoelous; intercalary cartilage present; teeth usually present; toes possessing enlarged pads at tips; pupils vertical, horizontal, or triangular; pectoral girdle arciferal; nearly worldwide in temperate and tropical regions, except Indo-Malayan region, Africa south of the Sahara, and Madagascar; about 27 genera.

*Family Centrolenidae (centrolenids).* Small (20 to 60 mm.), hylid-like anurans lacking free, ossified ribs; vertebral centra procoelous; tarsal bones fused into a single structure; pectoral girdle arciferal; southern Mexico to Ecuador and Brazil; three genera.

*Family Heleophrynidae (heleophrynids).* Small (up to 65 mm.), toadlike anurans lacking free, ossified ribs; vertebral centra procoelous; possessing vomerine teeth; pectoral girdle arciferal; South Africa; one genus, *Heleophryne.*

*Family Pseudidae (pseudids).* Small (50 to 75 mm.) frogs lacking free, ossified ribs; vertebral centra procoelous; digits possessing an additional phalanx; thumbs opposable; larvae disproportionately large in contrast to adult; pectoral girdle arciferal; South America and the Island of Trinidad; one genus, *Pseudis.*

*Family Ranidae (true frogs).* Very small to very large (12 to 250 mm.), long-legged anurans lacking free, ossified ribs; centra of vertebrae 1 to 7 procoelous; eighth vertebra amphicoelous; intercalary cartilage lacking; tongue free posteriorly; maxillary teeth present; ethmoid usually single; pupils horizontal or vertically oval; pectoral girdle firmisternal; hind feet generally webbed; essentially worldwide except New Zealand and the West Indies; about 32 genera.

*Family Rhacophoridae (rhacophorids).* Small (30 to 75 mm.) anurans lacking free, ossified ribs; centra of vertebrae 1 to 7 procoelous; eighth vertebra amphicoelous; intercalary cartilage present; ethmoid usually single; toes possessing enlarged pads at tips; pectoral girdle arciferal; Africa, Madagascar, and southeastern Asia; about 17 genera.

*Family Microhylidae (microhylids).* Small (8 to 100 mm.) anurans lacking free, ossified ribs; centra of vertebrae 1 to 7 procoelous; eighth vertebra amphicoelous; intercalary cartilage absent; ethmoid paired; teeth present or absent; pupils usually horizontal; pectoral girdle firmisternal; North and South America, Africa, Madagascar, and Asiatic and Indo-Australian regions; about 53 genera.

*Family Phrynomeridae (phrynomerids).* Small (maximum about 38 mm.) anurans lacking free, ossified ribs; centra of vertebrae 1 to 7 procoelous; eighth vertebra amphicoelous; ethmoid paired; possess an intercalary cartilage; tips of toes expanded; pectoral girdle firmisternal; Africa; one genus, *Phrynomerus.*

SUBCLASS LEPOSPONDYLI (LEPOSPONDYLS). Vertebral centra resulting from direct deposition of bone around notochord and not preceded by cartilaginous elements; Carboniferous to Recent; four orders.

ORDER AISTOPODA (AISTOPODS). Extinct lepospondyls with elongate bodies that lack limbs; vertebrae very numerous, sometimes numbering more than 100; Carboniferous; two families.

ORDER NECTRIDIA (NECTRIDIANS). Extinct lepospondyls possessing elongate bodies and limbs that are reduced or absent; caudal vertebrae possessing fan-shaped neural and haemal processes; Carboniferous to Permian; three families.

ORDER CAUDATA OR URODELA (SALAMANDERS). Small to large (40 to 1500 mm.)[*] amphibians with elongate bodies terminating in a tail and limbs that are not specialized for jumping; ribs present; Cretaceous to Recent; eight living families.

*Family Hynobiidae (hynobiids).* Moderately small (up to 210 mm.) caudate amphibians lacking functional gills as adults; lungs present or absent; eyelids present; vomerine and maxillary teeth present; vertebrae amphicoelous; females lacking a sperm receptacle; fertilization external; eggs pigmented; Asia; five genera.

*Family Cryptobranchidae (giant salamanders).* Large (500 to 1500 mm.) caudate amphibians lacking gills as adults; lungs present; eyes lacking lids; vomerine and maxillary teeth present; vertebrae amphicoelous; female lacking a sperm receptacle; fertilization external; eggs pigmented; Japan, northeast-

---

[*]Total length.

ern Asia, and eastern United States; two genera, *Cryptobranchus* and *Megalobatrachus*.

*Family Ambystomidae (ambystomids).* Small to moderately large (40 to 325 mm.) caudate amphibians with lungs present in transformed adults, though greatly reduced in *Rhyacotriton;* gills present in neotenic adults; eyelids present; vomerine teeth in transverse rows; vertebrae amphicoelous; costal grooves well defined; female possessing a sperm receptacle; fertilization internal; North America; six genera.

*Family Salamandridae (newts).* Small (100 to 165 mm.), stout-bodied caudate amphibians possessing lungs and lacking gills as adults; eyelids present; vomerine teeth in longitudinal rows that diverge posteriorly; vertebrae opisthocoelous; female possessing a sperm receptacle; fertilization internal; North America, Europe, North Africa, and Asia; six genera and about 43 species.

*Family Amphiumidae (amphiumas).* Large (900 to 1000 mm.), elongate caudate amphibians possessing lungs, with reduced limbs and lacking functional gills as adults; eyelids lacking; vomerine teeth in parallel, longitudinal rows; vertebrae amphicoelous; female possessing a sperm receptacle; fertilization internal; North America; one genus and two species, *Amphiuma means* and *A. tridactyla.*

*Family Plethodontidae (lungless salamanders).* Small (40 to 215 mm.) caudate amphibians lacking lungs and gills as adults; eyelids present; vomerine teeth present; vertebrae amphicoelous or opisthocoelous; costal grooves well defined; female possessing a sperm receptacle; fertilization internal; North America, northern South America, and southern Europe; 26 genera.

*Family Proteidae (mudpuppies and olm).* Moderately small (200 to 300 mm.) caudate amphibians possessing both lungs and gills as adults; eyelids lacking; vomerine teeth present; vertebrae amphicoelous; female possessing a sperm receptacle; fertilization internal; eastern North America and southern Europe; two genera, *Necturus* and *Proteus.*

*Family Sirenidae (sirens).* Medium-large (up to 500 mm.) caudate amphibians possessing gills and lacking lungs at maturity; eyelids lacking; maxillaries absent; teeth absent; hind limbs missing; females lacking a sperm receptacle; fertilization probably external; eastern North America; two genera, *Siren* and *Pseudobranchus.*

**ORDER GYMNOPHIONA OR APODA (CAECILIANS).** Amphibians with elongate, wormlike bodies that lack limbs; some genera have scales embedded in skin; eyes small or vestigial; Recent; one family.

*Family Caeciliidae (caecilians).* Characters of the order. Mexico to South America, Asia, and Africa except Madagascar; 16 genera.

## References Recommended

Bagnara, J. T. 1960. Pineal Regulation of the Body Lightening Reaction in Amphibian Larvae. Science *132*:1481–1483.

Bishop, S. C. 1943. Handbook of Salamanders. The Salamanders of the United States, Canada and Baja California. Ithaca, N.Y., Comstock Publishing Co.

Brown, B. C. 1950. An Annotated Check List of the Reptiles and Amphibians of Texas. Waco, Tex., Baylor University Press.

Brown, C. W. 1968. Additional Observations on the Function of the Nasolabial Grooves of Plethodontid Salamanders. Copeia *1968*:728–731.

Carl, G. C. 1950. The Amphibians of British Columbia. British Columbia Prov. Mus. Dept. Educ. Handbook *2*:1–62.

Conant, R. 1952. Reptiles and Amphibians of the Northeastern States. 2nd Ed. Philadelphia, Zoological Society of Philadelphia.

Cope, E. D. 1889. The Batrachia of North America. Bull. U.S. Nat. Mus., No. 34.

Daly, J. W., and Myers, C. W. 1967. Toxicity of Panamanian Poison Frogs (*Dendrobates*): Some Biological and Chemical Aspects. Science *156*:970–973.

Dunn, E. R. 1926. The Salamanders of the Family Plethodontidae. Northampton, Mass., Smith College Anniversary Series.

Frieden, E. 1961. Biochemical Adaptation and Anuran Metamorphosis. Am. Zool. *1*:115–149.

Gordon, K. 1939. The Amphibia and Reptilia of Oregon. Oregon State Monographs, No. 1.

Kollros, J. J. 1961. Mechanisms of Amphibian Metamorphosis: Hormones. Am. Zool. *1*:107–114.

Lynn, W. G. 1961. Types of Amphibian Metamorphosis. Am. Zool. *1*:151–161.

Moore, J. A. 1964. Physiology of the Amphibian. New York and London, Academic Press, Inc.

Noble, G. K. 1931. The Biology of the Amphibia. New York, McGraw-Hill Book Co., Inc.

Oliver, J. A. 1955. The Natural History of North American Amphibians and Reptiles. Princeton, N. J., D. Van Nostrand Co., Inc.

Pickwell, G. 1947. Amphibians and Reptiles of the Pacific States. Stanford, Calif., Stanford University Press.

Romer, A. S. 1947. Review of the Labyrinthodontia. Bull. Mus. Comp. Zool. 99:1–368.

Ruthven, A. G., Thompson, C., and Gaige, H. T. 1928. The Herpetology of Michigan. Mich. Handbook Ser. Univ. Mich., No. 3.

Schmidt, K. P. 1953. A Check List of North American Amphibians and Reptiles. 6th Ed. Chicago, University of Chicago Press.

Slevin, J. R. 1928. The Amphibians of Western North America. Occasional Papers Calif. Acad. Sci. *16*:1–152.

Stebbins, R. C. 1951. Amphibians of Western North America. Berkeley, Calif., University of California Press.

Stebbins, R. C. 1954. Amphibians and Reptiles of Western North America. New York, McGraw-Hill Book Co., Inc.

Stebbins, R. C., and Hendrickson, J. R. 1959. Field Studies of Amphibians in Colombia, South America. Univ. Calif. Publ. Zool. 56:497–540.

Storer, T. I. 1925. A Synopsis of the Amphibia of California. Univ. Calif. Publ. Zool. 27:1–342.

Taylor, E. H. 1943. Skin Shedding in the Salamander *Amphiuma means*. Univ. Kansas Sci. Bull. *29*:339–341.

Weichert, C. K. 1945. Seasonal Variation in the Mental Gland and Reproductive Organs of the Male *Eurycea bislineata*. Copeia *1945*:78–84.

Wright, A. A., and Wright, A. H. 1949. Handbook of Frogs and Toads of the United States and Canada. Ithaca, N.Y., Comstock Publishing Co.

*Chapter Four*

# REPTILES

## GENERAL CHARACTERS

Living members of the class Reptilia, estimated to number about 6000 species, are descendants of a great group of vertebrates that were dominant during the Mesozoic Era. Their success at that time has generally been attributed to the development of a new method of embryonic protection. Their amphibian ancestors, like modern amphibians, were dependent upon water or at least a moist environment to prevent their eggs from becoming desiccated after they had been laid. Reptiles circumvented this problem partly by developing a substantial shell around an egg that was heavily laden with yolk. The shell was sufficiently porous to permit the passage of respiratory gases, yet solid enough to afford protection against the environment, and the large yolk provided food for the growth of the embryo. Most important, however, was the development of an embryonic membrane called the *amnion* which enclosed a liquid-filled chamber in which the developing individual was protected from injury and desiccation. Once this was accomplished reptiles expanded into many different habitats that previously could not be used by land vertebrates because of the scarcity or lack of water. Why they ultimately went into a decline at the end of the Mesozoic has been the subject of much interesting speculation. Some have attributed it to climatic changes which were directly or indirectly unfavorable to reptiles, while others have suggested that reptiles were unable to compete with mammals. In any event the decline of reptiles was correlated with the rise of numerous mammalian forms.

Members of this class in the past, as in the present, have been limited to the warmer parts of the world because of their lack of internal thermoregulating mechanisms. Being ectotherms they are largely dependent on their external environment for body heat and, therefore, do not thrive in regions where the temperatures are low. During periods of activity, however, many reptiles are capable of maintaining a relatively high body temperature by making use of solar radiation and radiation from the substratum. By controlling the period of exposure to such sources of heat, body temperature may be kept fairly constant. If the ambient temperature is high there is less need for dependence on radiation. Bartholomew (1966) found that the Galápagos marine iguana (*Amblyrhynchus cristatus*), a unique species that feeds in the sea, prefers a body temperature of between 35° and 37° C., which is 10° C. higher than that of the water in which it secures its food. For this reason it shows great reluctance to go into the water unless it is necessary to secure food. Much of its day is spent on dark lava rocks where there is exposure to intense solar radiation. Overheating is prevented by postural adjustments which expose a minimum amount of its body surface to the sun and also permit it to take advantage of relatively cool trade winds.

Desert iguanas (*Dipsosaurus dorsalis*) in western North America were found by De Witt (1967) to often have a body temperature as high as 42° C. although the preferred level was 38.5° C. This level is maintained by appropriate movements in and out of the sun if the ambient temperature is lower than the preferred level. If the ambient temperature is higher the iguanas may prolong their periods of activity above ground by permitting the body temperature to rise above the preferred level. Some species of lizards use panting as a means of lowering the body temperature.

Although reptiles are rather closely related to birds, they exhibit far greater structural diversity than do members of the class Aves. As a group they have lost the aquatic specialization of the lower vertebrates, including respiratory gills, lateral line organs, and external mucous glands.

## Skeletal System

The reptile skull shows greater ossification than that of the amphibian. Considerable variation has occurred in the temporal region in reptiles during their evolutionary history (Fig. 4.1). The ancient stem reptiles, represented by the cotylosaurs of the Carboniferous and Permian periods, lacked any special temporal openings in the skull. This primitive solid-roof condition of the skull is referred to as the *anapsid* type. Among living reptiles it is characteristic only of turtles. The plesiosaurs and their relatives developed a *parapsid* type of skull, in which a single supratemporal fenestra developed on

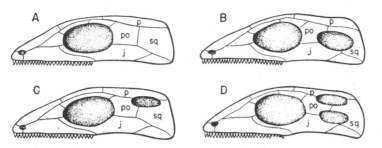

FIGURE 4.1. Diagram of the temporal region in the four major types of reptiles: A, anapsid; B, synapsid; C, parapsid; D, diapsid. (Romer, A. S.: The Vertebrate Body. 4th Ed., W. B. Saunders Co., 1970.)

either side of the skull. The mammal-like reptiles of the Permian to Jurassic possessed a pair of infratemporal openings which represented a *synapsid* type of skull. The ruling reptiles of the Mesozoic had a *diapsid* skull, in which there were both supra- and infratemporal openings. This is also characteristic of most living reptiles apart from the chelonians.

The roof of the braincase itself is arched rather than broadly flattened as in amphibians. A parietal foramen for the pineal, or third eye, is present in the tuatara (*Sphenodon*) and in a few lizards, but is lacking in other living reptiles (Fig. 4.2). Most reptiles, other than snakes, possess a bony interorbital septum. The beginning of a *secondary palate* or shelf, which moves the internal nares to the back

FIGURE 4.2. The tuatara (*Sphenodon punctatum*) of New Zealand, the only living representative of the order Rhynchocephalia. (Photograph by Edward S. Ross.)

of the mouth by increasing the length of the nasal passages, is evident in the turtles and their allies. It is well developed in crocodilians, but lacking in other reptiles. A single occipital condyle is present. The quadrate is solidly fused to the skull in turtles, crocodilians, and the tuatara (Fig. 4.3). In snakes and lizards, however, it is movable and serves somewhat as a hinge between the upper and lower jaws. Teeth are lacking in turtles, which possess a horny bill. They are present in other reptiles and, while usually restricted to the premaxillaries and maxillaries of the upper jaw and the dentary of the lower jaw, they may also occur on the palatines, vomers, and pterygoids. The lower jaw is composed of a number of bones, both dermal and replacement. The connection between the two halves of the lower jaw ranges from a firmly fused condition seen in turtles to a loose connection by means of ligaments in snakes.

Except in the snakes and limbless lizards, the reptilian vertebral column shows considerable differentiation into cervical, thoracic, lumbar, sacral, and caudal regions. The single occipital condyle articulates with the first cervical or *atlas*. The second cervical or *axis* bears an anterior projection known as the *odontoid process* which is believed to represent the centrum of the atlas. The *thoracic* vertebrae bear ribs which, except in limbless reptiles and chelonians, meet the sternum ventrally. Between the thoracics and the two sacral vertebrae are the relatively flexible *lumbar* vertebrae. The number

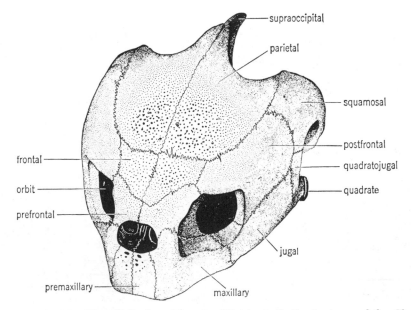

**FIGURE 4.3.** The skull of a sea turtle. (Weichert, C. K.: Anatomy of the Chordates. 2nd Ed., McGraw-Hill Book Co., Inc., 1958.)

of caudal vertebrae is extremely variable. In turtles all vertebrae except the cervicals and caudals are fused to the dermal plates of the carapace. In cobras, members of the family Elapidae which possess fixed front fangs (proteroglyphs), a hood is produced when the snakes are alarmed. This results from the lateral movement of hinged ribs which extend the loose flesh of the neck region into the form of a shelf (Fig. 4.4).

Most reptiles have vertebral centra that are *procoelous*. Thus there is a ball-and-socket type of articulation in which the anterior end of the centrum is hollow and the posterior face rounded. In some reptiles the situation is reversed (*opisthocoelous*), with the anterior end rounded and the posterior end concave.

There is great variability in the appendicular skeleton of reptiles. This, of course, is correlated with the widely divergent types of locomotion used by these animals. In snakes and limbless lizards appendages are lacking. The boas and a few others have vestiges of the hind limbs appearing as spurs on either side of the vent. In the marine turtles the limbs are modified as flippers for swimming, while large land tortoises have feet that superficially resemble those of an elephant so as to bear the body weight. Most lizards have five clawed toes on each foot and some species are capable of running very rapidly for short distances. There are certain lizards, however, in which there is a reduction in the number of toes or even the elimination of the limbs so as to present a snakelike appearance.

**FIGURE 4.4.**   The king cobra (*Ophiophagus hannah*) of Asia with its head raised and hood spread ready to strike. (Photographed at Steinhart Aquarium.)

Most limbless lizards burrow in sand or loose soil. In some of the crocodile-like reptiles the toes may be partly joined together by webbing as an adaptation to aquatic life.

## MUSCULAR SYSTEM

Owing partly to the greater development of ribs and partly to the more advanced types of locomotion in this group, the axial or trunk muscles of reptiles begin to show some of the complexities that lead to the condition found in mammals. In fishes these muscles serve primarily for lateral body movement. Amphibians have necessarily advanced somewhat beyond this as a result of coming onto land. Most reptiles, however, are capable of far greater movement of the vertebral column, and this is brought about by specialization of some of the trunk musculature. This is especially true of snakes in which the limb musculature has disappeared and movement is effected by the trunk muscles. In turtles and their allies the presence of an immovable shell has reduced the need for well developed axial musculature except in the neck region.

The ventral trunk musculature of reptiles, in addition to consisting of the rectus abdominus, the external oblique, internal oblique, and transversus layers found in amphibians, is further subdivided so as to have an intercostal layer that connects the ribs together. Dermal or skin muscles which are lacking in fishes and barely represented in amphibians come into prominence in reptiles, especially snakes. In the latter group, they assist in locomotion by raising and lowering the scales on the ventral side of the body so as to alternately increase and decrease friction with the ground.

The limb musculature of reptiles shows great variation, depending upon the type of locomotion employed. However, it is perhaps most highly advanced in swift running and climbing species of lizards.

## CIRCULATORY SYSTEM

Since reptiles are more perfectly adapted to life on land than amphibians, we find that their circulatory system is correspondingly more efficient. Most of the advances over the amphibian blood system are associated with the elimination of functional gills and with the development of the most advanced type of vertebrate kidney known as a metanephros.

The atrium of the heart is always completely separated into a right and left chamber, and in many forms the sinus venosus is

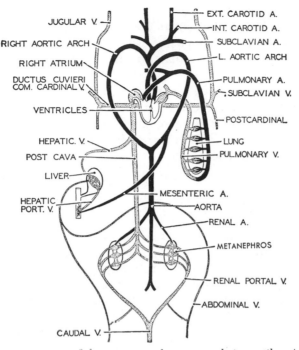

**FIGURE 4.5.** Diagram of the main circulatory vessels in reptiles. Arteries black, veins stippled. It will be noted that some blood vessels persist throughout the entire series. In general features the series may be interpreted as representing stages in the phylogenesis of the mammalian circulatory system. (After Stempell, in Neal and Rand: Chordate Anatomy. McGraw-Hill Book Co., Inc., 1939.)

incorporated into the wall of the right atrium. The ventricle is also partly divided by a septum in most reptiles, and in the alligators and crocodiles is completely two-chambered. This means that oxygenated blood coming from the lungs to the left side of the heart is essentially separated from the nonoxygenated blood returning from the body to the right side. Although the separation of the two types of blood within the heart is complete in the crocodilians and nearly complete in other reptiles, some mixing does occur in other parts of the circulatory system. Strangely enough in reptiles, the embryonic conus arteriosus splits into three instead of two vessels. One of these becomes the pulmonary stem which takes blood to the lungs from the right side of the ventricle. A second becomes the base of the main systemic aorta which carries blood from the left side of the ventricle to the body by way of the right fourth aortic arch. A third stem, however, comes from the right ventricle to the left fourth aortic arch. At its point of contact with the systemic aorta from the left ventricle even in crocodilians, there is a small opening between the two vessels known as the *foramen of Panizzae* where there may be some mixing of the types of blood. Most of the blood from the left aortic arch goes to the left subclavian and coeliac arteries, but there is still a small connection posteriorly where the two aortae join to form a single dorsal aorta so that a further mixture of oxygenated and unoxygenated blood occurs.

The principal advances in the venous system of reptiles over that of amphibians involve the greater development of the pulmonary veins and postcaval vein and the reduction in the importance of the renal portal system which brings blood to the kidneys from the posterior part of the body.

## DIGESTIVE SYSTEM

Some reptiles are herbivorous, while others are carnivorous. Small carnivorous reptiles, e.g., many lizards, feed principally on insects and other invertebrates, while the larger carnivores prey principally on other vertebrates, ranging from fishes to mammals. The reptilian digestive system, therefore, is adapted to suit the food habits of the species involved.

In general we find that nonaquatic reptiles have the glands of the mouth more highly developed than do amphibians. This is associated with the necessity of moistening dry food to reduce friction in swallowing, a problem with which fishes and most amphibians do not have to contend. These oral glands include a *palatine* as well as *labial, lingual,* and *sublingual* glands. The poison glands of reptiles are derived from some of these oral glands. In the Gila monster and Mexican beaded lizard, the poison glands are modified sublingual glands.

The tongue in lizards and snakes is often highly developed. In the chameleons of Africa and India, it is very extensible and is used to capture insects. The tip is thickened and sticky so that the prey adheres to it. The forked tip of the tongue in snakes serves as a means of transferring chemical stimuli from the external environment, such as the ground, to the paired vomeronasal organs on the roof of the mouth. In turtles and crocodilians, however, the tongue cannot be extended. The esophagus of reptiles is more elongate than that of fishes and amphibians and is clearly distinct from the stomach. The crocodilian stomach resembles that of a bird's in that a part of it is gizzard-like and encased in a thick layer of muscle. The small intestine of reptiles is generally more coiled than in amphibians. This serves to increase its absorptive surface. For the first time in vertebrates we find that a caecum or blind diverticulum arises at the point of junction of the small and large intestines. This is not present, however, in all reptiles. The large intestine is straight and empties into a cloaca.

## Respiratory System

The lungs of reptiles are somewhat intermediate between those of adult pulmonate amphibians and higher vertebrates. Generally they are more complex than those of amphibians with an increase in the number of internal chambers and alveoli so as to prove more efficient as respiratory organs. In some lizards one lung is considerably larger than the other, and in snakes the left lobe is reduced or even absent in some species. This reduction or elimination of the lung on one side seems to be correlated with the elongate shape of the body in these reptiles. Crocodilians possess lungs that are quite similar to those of mammals, while a few lizards possess diverticula, extending posteriorly from the lungs, that bear a certain resemblance to the air sacs associated with the lungs of birds. The trachea and bronchi may be short and simple as in amphibians or they may be considerably more complicated. In long-necked reptiles such as turtles, the trachea is not only long, but it is sometimes convoluted. Some lizards have the bronchi subdivided into primary, secondary, and tertiary bronchi. Tracheal cartilages are fairly well developed, sometimes forming complete rings. Bronchial cartilages may also be present.

Reptiles, in addition to swallowing air as amphibians do, also make use of rib and abdominal muscles to suck air into the lungs.

## Urogenital System

The reptilian kidney, like that of birds and mammals, is of the most advanced vertebrate type, known as a metanephros. During embryonic life, however, both pronephros and mesonephros make their appearance, perhaps reflecting reptilian ancestry. The metanephros fundamentally resembles the mesonephros of lower vertebrates, but arises more posteriorly in the body, is more compact, and contains a far greater number of renal units. Furthermore, the renal tubules, instead of draining into the archinephric duct, open into larger collecting tubules which ultimately lead to a new excretory duct called the *ureter*.

The development of a more efficient type of excretory system in reptiles, birds, and mammals is generally believed to have been necessary in order to meet the demands imposed by higher metabolic activity. It is also associated with a more efficient circulatory system in this part of the body. In reptiles the renal portal system has begun to lose its importance and some of the blood from the caudal region goes directly through the kidneys instead of filtering slowly through the capillary network. This is compensated for by the development of renal arteries which bring blood directly from the dorsal aorta to the kidneys. Blood from the latter returns to the postcaval vein by means of renal veins.

Many reptiles possess a urinary bladder which, like that of amphibians, is an outgrowth from the ventral wall of the cloaca. Such a structure, however, is lacking in crocodilians, snakes, and some lizards.

Recent studies have shown that a number of reptiles possess salt-secreting glands located on the head. The purpose of these glands is to eliminate salt rather rapidly from the system. The excretion passes by means of a duct or ducts to the nasal chamber. These glands are very highly developed in the marine iguanas (*Amblyrhynchus cristatus*) of the Galápagos Islands that live on marine algae, which they secure by swimming and diving in the ocean (Fig. 4.6). After feeding they come ashore to rest on rocks, where a concentrated salt solution excreted by the salt glands is frequently expelled during exhalation as a cloud of vapor from the nasal chamber. In certain terrestrial reptiles, such as the common iguana (*Iguana iguana*) and the desert crested lizard (*Dipsosaurus dorsalis*), the elimination of salt by this extrarenal method is thought to be associated with water conservation. By reducing the salt concentration in the urine more of the urine may be reabsorbed in the cloaca. This would be highly advantageous to a reptile living in a warm environment.

The ovaries and testes are paired in reptiles. The former may contain cavities filled with lymph, as in amphibians, or may be solid, as in birds and mammals. Reptilian eggs, though few in number as

**FIGURE 4.6.**   Marine iguanas (*Amblyrhynchus cristatus*) on the shore of Abingdon Island, Galápagos Archipelago. (Photograph courtesy of David Cavagnaro.)

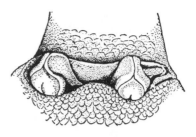

**FIGURE 4.7.** Hemipenes of lizard, *Platy-dactylus*. (After Unterhössel.)

contrasted with lower vertebrates, are relatively large owing to the presence of a large amount of yolk necessary for the growth of the young prior to hatching. They are often surrounded by albumen and encased in a leathery or calcareous shell. The albumen and shell are produced by glands located along the paired oviducts. The latter open separately into the cloaca.

The archinephric or Wolffian duct degenerates in female reptiles but in males becomes the functional genital duct. The upper end is greatly coiled and forms a compact structure called the *epididymis*. Since the eggs of reptiles either hatch internally or are surrounded by a tough shell when laid, fertilization is necessarily internal. Consequently, the males of many species have developed special copulatory organs for the transference of sperm to the females. In lizards and snakes there is a pair of extrusible structures in the cloaca called the hemipenes which serve this purpose, while crocodiles and turtles possess a structure that may be homologous with the mammalian penis (Fig. 4.7).

## NERVOUS SYSTEM

Since mammals, with their highly developed brain, arose from reptilian stock, it is not surprising that we find many features characteristic of the mammalian central nervous system making their appearance in reptiles. In all the anamniotes the midbrain is the center of brain activity, but in reptiles for the first time there is a shift in the nerve center to the cerebrum. This is correlated with a marked increase in the size of the cerebral hemispheres as a result of the invasion of the pallium by many nerve cells so as to form what is generally termed the *neopallium*.

The reptilian cerebellum is relatively larger than that of amphibians. However, it does not attain the size of that of some of the fishes, such as sharks, nor is it anywhere near the size of that of birds or mammals. Again this appears to be correlated with relatively limited locomotive powers of most reptiles. For the first time in vertebrates we find that 12 cranial nerves are present.

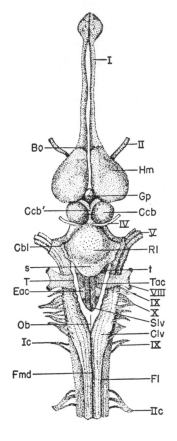

**FIGURE 4.8.** Dorsal view of alligator brain, Roman numerals corresponding to cranial nerves; *Hm*, telencephalon; *Ccb*, midbrain; *Rl*, cerebellum; *Tac*, fourth ventricle. (Redrawn from Bronn.)

## SENSE ORGANS

In most reptiles the taste buds are restricted largely to the pharyngeal region. Within each of the nasal passages, which are elongate in higher reptiles because of the development of a secondary palate, there is a shelf or concha which serves to increase the surface for the olfactory epithelium. Jacobson's organ reaches its peak of development in snakes and lizards and is connected with the roof of the mouth rather than the nasal canal.

In the various lower vertebrate groups accommodation to near and distant vision is accomplished by moving the lens forward or backward so as to change the distance between the lens and the sensitive retina. Accommodation in reptiles and most other amniotes, however, is accomplished not by movement of the lens but by changing its shape. It may be flattened for distant vision or rounded for near vision through the action of the muscles of the ciliary body.

Eyelids are present in many reptiles and generally are more movable than those of amphibians. In snakes and some of the burrowing lizards, however, movable lids are absent, and the eye, which itself is capable of moving, is covered with immovable transparent skin. These so-called "eye plates" are believed to have been derived from the lids. They are continuous with the scales covering the body and, like the scales, are shed periodically. Many reptiles possess a third eyelid called the *nictitating membrane.* This is situated beneath the upper and lower lids and closes the eye by moving backward or outward from the anterior or median margin of the eye aperture. It is more or less transparent. The blood that a horned toad squirts from the eye comes from blood vessels in the nictitating membrane which rupture easily as a result of muscular contraction.

Mention has already been made of the pineal eye which is present in the tuatara and a few lizards. Experimental studies by Stebbins and Eakin (1958) on this structure in some American lizards of the genera *Sceloporus, Uta,* and *Uma,* have shown that it is an aid in regulating the amount of exposure to sunlight (Fig. 4.9). This, of course, is important in ectotherms. Lizards with the parietal eye removed tend to stay longer in sunlight or artificial light than they otherwise would.

There is considerable variation in the structure of the ear in

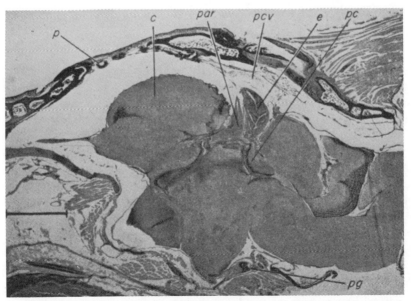

FIGURE 4.9.  Median sagittal section through the brain of *Sceloporus occidentalis,* showing the relative positions of the parietal eye and epiphysis. The length of the epiphysis may be 2 cm. in a lizard, 70 mm. in snout to vent length. Abbreviations: *c,* cerebrum; *e,* epiphysis; *p,* parietal eye; *par,* paraphysis; *pc,* posterior commissure; *pcv,* sinus of posterior cerebral vein; *pg,* pituitary gland. Horizontal line represents 1 mm. (Stebbins, R. C., and Eakin, R. M.: American Museum Novitates, No. 1870, 1958.)

reptiles. The lagena is more elongate than in amphibians, and in the crocodilians actually forms a cochlear duct rather similar to that of birds. In snakes the tympanic membrane, middle ear, and eustachian tube are lacking. Vibrations received by snakes are transmitted by way of the quadrate to the columella and then to the inner ear. Lizards, on the other hand, have a well developed middle ear and in some forms the ear drum has sunk into a depression. Sound waves, therefore, must pass through a short canal known as the *external auditory meatus* in order to impinge upon the tympanic membrane. This is the first indication of an external ear in vertebrates.

Lateral line organs are lacking in reptiles and other amniotes.

## ENDOCRINE GLANDS

The parathyroid glands are usually situated posterior to the thyroid gland, which is unpaired, at least in adult reptiles. The two parts of the adrenal glands — the interrenal bodies, which are homologous with the cortex in mammals, and the chromaffin bodies, which are homologous with the mammalian adrenal medulla — are mixed together in the reptiles. The other endocrine glands of reptiles do not differ markedly from those of most other higher vertebrates.

## SPECIAL CHARACTERS

The reptilian body is typically covered with dry scales which afford protection much as do the scales of fishes. These scales fall into two categories, epidermal and dermal. Those of the former type are superficial and in most reptiles are shed periodically. Dermal scales are permanent plates of bone embedded in the skin and are retained for life. Also present in the dermis of many reptiles are chromatophores, similar to those of fishes and amphibians, which are responsible for the color pattern. Concentration and dispersion of the pigment granules in these chromatophores is responsible for the ability of certain lizards, such as the chameleon, to change their colors in response to environmental stimuli.

## EPIDERMAL SCALES

These are the scales that are most noticeable on lizards and snakes. They are continually being produced by the permanently growing layer of the epidermis known as the *stratum germinativum* and are generally folded so as to overlap one another. When they are fully grown they become separated from the stratum germinativum and appear as nonliving, cornified structures. Snakes and lizards shed their scales, a process known as *ecdysis*. Before ecdysis takes place, however, the new scales that will replace the old ones are formed. Most snakes shed their skins in essentially one piece. The old epidermal covering becomes loosened first in the head region. This skin, including even the transparent plates over the eyes, is turned back and the snake finally crawls out of the old covering, leaving the latter inside out. The number of times a year that a snake sheds its skin is dependent upon the rate of growth. Rapidly growing snakes may shed every two months.

The rattles of rattlesnakes represent horny remnants of the skin which adhere to the base of the tail and are not lost during ecdysis. Each molt adds a new rattle, and the number of rattles, therefore, does not represent the age of the snake in years. Furthermore, especially in older rattlesnakes, the terminal rattles are often lost, so that the total number of rattles does not even indicate the total number of molts.

The epidermal covering of most lizards, unlike that of snakes, is usually shed in pieces.

The terms *carapace* and *plastron* are applied to the dorsal and ventral shells that protect the bodies of most turtles and tortoises (Fig. 4.11). Although these structures are composed largely of bony,

**FIGURE 4.10.** The endemic species of rattlesnake (*Crotalus catalinensis*) on Santa Catalina Island in the Gulf of California, Mexico, has essentially lost its rattles, although it still vibrates its tail as a warning when alarmed.

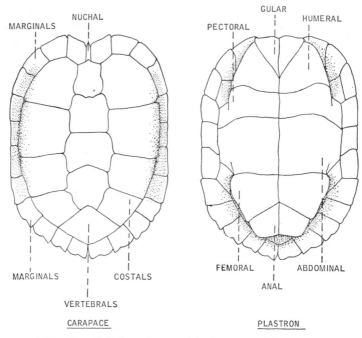

**Figure 4.11.** Dorsal (*left*) and ventral (*right*) view of the shell of a Pacific pond turtle (*Clemmys marmorata*) showing the epidermal plates or shields on the outer surface of the carapace and plastron.

dermal plates, they are covered externally with cornified epidermal scales. Unlike the epidermal scales of snakes and lizards, these scales are not shed regularly, although the older, outer ones do get worn off. Growth is from below and, as a result of expansion of the stratum germinativum, each new scale is larger than its overlying predecessor. As a consequence, the large epidermal plates are marked by growth rings because of the accumulation of layers of cornified scales. A few turtles lack scales and have a leathery skin instead. The bodies of alligators and their relatives are also covered with epidermal scales which are not regularly shed, but gradually wear off and are replaced.

The epidermal scales of reptiles are arranged in very definite patterns and exhibit considerable diversity in shape and structure in different groups. In snakes and lizards, the scales on the body may be arranged in longitudinal, diagonal, or transverse rows. The scales on the head generally differ markedly in appearance from the body scales and are named according to their location. For example, those along the margin of the upper lip are called the upper labials, those around the eyes are the oculars, those between the eyes, the interoculars, just to mention a few. Differences in the size, shape, and

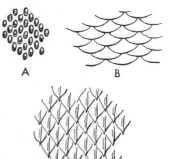

FIGURE 4.12.   Diagram showing some selected examples of the scales of lizards: *A*, granular; *B*, cycloid; *C*, keeled mucronate; *D* and *E*, quadrangular.

number of these scales provide important characters used in classification.

The body scales of snakes are generally cycloid or quadrangular in shape and they may or may not possess a ridge or keel. Those of lizards may usually be classified as granular, cycloid, quadrangular, or mucronate (terminating posteriorly in a sharp point) (Fig. 4.12). They may be smooth or keeled. In some species, such as the horned toads, certain scales may be highly modified into elongate spines (Fig. 4.13). Elongate, spinelike scales are also present in some of the iguanas.

In most snakes, the ventral scales are greatly enlarged into broad, transverse bands called *scutes* that extend the width of the body. The presence of scutes on the undersurface of the body is generally thought of as a character distinguishing snakes from lizards. There are some species of snakes, however, such as the worm snakes, that lack ventral scutes. The large dermal plates on the carapace and plastron of turtles and tortoises show a very definite pattern of arrangement.

## DERMAL SCALES

These are most highly developed in the turtles and tortoises, in which they form the bony carapace and plastron which are joined together by a bridge on each side of the body. Internally, parts of the

FIGURE 4.13.   Many of the scales on the bodies of horned toads (*Phrynosoma*) are enlarged into spinelike structures. (Photograph by Edward S. Ross.)

FIGURE 4.14.   Photograph showing the heavy scales on the head of an Australian stump-tailed skink (*Tiliqua rugosa*).

FIGURE 4.15. The American alligator (*Alligator mississippiensis*) possesses dermal plates beneath some of the epidermal scales.

FIGURE 4.16. Note the scale pattern on the head as well as the keeled body scales on the western garter snake (*Thamnophis elegans*). (Photograph by Edward S. Ross.)

skeleton, including the thoracic, lumbar, and sacral vertebrae as well as the ribs, are fused to the carapace in these reptiles.

Crocodilians possess some bony plates or *osteoderms* beneath certain of the epidermal body scales, and small dermal scales underlie the outer scales of some lizards, notably the Gila monster, the Mexican beaded lizard, the skinks, and alligator lizards. They are absent in snakes.

## TEETH

Teeth are entirely lacking in turtles and tortoises. In these reptiles we find that the upper and lower jaws are encased in horny sheaths resulting in a beak somewhat comparable to that of a bird.

Well developed teeth, however, are present in all other groups of reptiles and are replaced when they are lost. In crocodilians the teeth are of a rather uniform conical shape, and are situated in sockets as in mammals. This type of tooth attachment is referred to as *thecodont* and reduces the chance of tooth loss when fighting or securing prey (Fig. 4.17).

Most lizards have uniform or *homodont* dentition like crocodilians, but there are a few in which the teeth are specialized into incisors, canines, and molars, much as in mammals. This type of dentition is referred to as *heterodont.* There are a few lizards that possess some teeth on the roof of the mouth, but in most species they are attached to the jaws. The usual method of attachment is to the upper surface of the jaw rather than being encased in a socket. This type of attachment is called *acrodont* (Fig. 4.17). There are some lizards, however, in which the tooth rests on a ledge on the inner side of the jaw and is attached on the side as well as the base. This is a *pleurodont* type of attachment.

The only known venomous lizards are the Gila monster and Mexican beaded lizard, closely related species belonging to the genus *Heloderma* (Fig. 4.18). The Gila monster ranges from southwestern United States south into Mexico, and the Mexican beaded lizard occurs from the State of Sonora, Mexico, south to the Isthmus of Tehuantepec. The lower teeth in members of the genus *Heloderma*

FIGURE 4.17. Diagram showing three methods of tooth attachment. (Drawn by G. Schwenk, in Weichert, C. K.: Anatomy of the Chordates. McGraw-Hill Book Co., Inc., 1958.)

ACRODONT   PLEURODONT   THECODONT

FIGURE 4.18. The only known venomous lizards are the Gila monster (*Helo-derma suspectum*) shown above and the Mexican beaded lizard (*Heloderma horri-dum*). Both are inhabitants of parts of western North America. (Photographed at Steinhart Aquarium.)

are pleurodont and deeply embedded in the gums. Each tooth has a groove on its anteromedial surface and occasionally on the posterior surface. There are also grooves on the anterior surfaces of the upper teeth. The poison which is secreted by labial glands in the lower jaw does not pass through any hollow fang but flows into wounds through the grooves in the teeth as these lizards hold onto their victims.

Most snakes have recurved pleurodont teeth arranged in rows on the upper and lower jaws and in many species an additional pair of longitudinal rows is present on the roof of the mouth. As in most reptiles the teeth are replaced when lost. Some venomous snakes have grooved teeth, somewhat like those of lizards of the genus *Heloderma,* on the back of the upper jaw. Snakes of this type are called *opisthoglyphs.* Most of the dangerous venomous snakes, how-ever, have a pair of hollow fangs that are generally situated on either side of the anterior part of the upper jaw. The fang is somewhat like a hypodermic needle and the base is connected to a small sac into which secretion from the poison gland passes through a duct. Muscular contraction around the poison gland at the time a snake strikes is responsible for the injection of poison through the fang into the victim. Fangs, like other teeth, are replaced if lost. In some snakes, referred to as *proteroglyphs,* the fangs are rigid. This is true

**FIGURE 4.19.** The puff adder (*Bitis arietans*) of Africa is a venomous solenoglyph with hinged, erectile fangs.

of the poisonous coral snake of North America, which is related to the cobras and mambas. On the other hand, the rattlesnakes and other pit vipers are *solenoglyphs* and have hinged erectile fangs (Fig. 4.20).

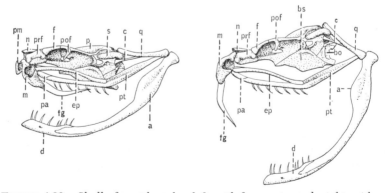

**FIGURE 4.20.** Skull of a rattlesnake: *left*, with fangs retracted; *right*, with mouth partly opened and fangs extended. *a*, Articular; *bo*, basioccipital; *bs*, basisphenoid; *c*, columella; *d*, dentary; *ep*, ectopterygoid; *f*, frontal; *fg*, fang; *m*, maxillary; *n*, nasal; *p*, parietal; *pa*, palatine; *pm*, premaxillary; *pof*, postfrontal; *prf*, prefrontal; *q*, quadrate; *pt*, pterygoid; *s*, squamosal. (After Klauber.)

## Appendages and Locomotion

During their evolutionary history we find that reptiles have made use of essentially every type of locomotion used by vertebrates. In the world today there are reptiles that can swim in the sea or in fresh water, others that can run on land, burrow, climb, and even glide through the air. Back in the Mesozoic the pterosaurs were capable of sustained flight, and the ichthyosaurs were fishlike in form with their limbs modified into finlike structures for life in the sea. There was even a group of reptiles known as the plesiosaurs that had paddle-like appendages.

The basic type of reptilian limb is best exemplified by typical lizards, in which there are paired pectoral and pelvic appendages. Generally each limb possesses five toes and each toe is clawed. Many lizards can run over the ground with great rapidity. Some species use only the hind limbs when running. At such times the front part of the body is elevated. This bipedal type of locomotion was used by a primitive group of dinosaurs that was believed to have given rise to birds.

In addition to rapid locomotion over the ground many lizards can climb vertical surfaces readily. Sharp claws are of considerable assistance in climbing and some lizards, like certain species of geckos, have ridges possessing tiny hooklets on their toes which are an additional aid. Some of the most remarkable lizards are the so-called "flying dragons" of India and the Malay belonging to the genus *Draco*. These reptiles have a large weblike extension of skin that extends out from each side of the body between, but not attached to, the limbs. These winglike structures are supported by extensions of the ribs and permit the lizards to glide through the air for short distances.

Not all lizards have the typical two pairs of pentadactyl limbs. One or both pairs of limbs may be missing. The limbless lizards of the family Anniellidae and the so-called "glass snakes" belonging to the family Anguidae are superficially snakelike in appearance with external limbs lacking. Such species are largely subterranean in habit.

Crocodilians are capable of walking on land as well as swimming in the water. Although there may be webbing between the toes in some forms, rapid movement in the water is accomplished by lateral undulatory body movement. Among living reptiles those best adapted to aquatic life are the marine turtles. In these reptiles the limbs are modified into flippers (Fig. 4.21) and the nails are reduced or absent. Land tortoises (Fig. 4.22), on the other hand, have very strong legs and feet that are able to raise and move the proportionately heavy body. However, they are unable to travel with any degree

**FIGURE 4.21.**   The limbs of marine turtles, such as the loggerhead turtle (*Caretta caretta*) which occurs along both coasts of North America, are modified as flippers for swimming. (Photographed at Steinhart Aquarium.)

**FIGURE 4.22.**   The desert tortoise (*Gopherus agassizi*) has strong limbs adapted to digging and to travel over land. (Photographed at Steinhart Aquarium.)

of rapidity. An impervious shell rather than locomotion is depended on for protection against enemies.

Both marine and freshwater turtles can change the specific gravity of their bodies so as to stay at certain levels in the water. A pond turtle may float on the surface or move to the bottom of the pond. This is probably achieved by changing the volume of air in the lungs and by increasing or decreasing the amount of water stored in the cloaca.

Locomotion in snakes, which of course lack functional limbs, has been a subject of interest to many.

Although we may simply say that snakes get about by crawling, careful study shows that they crawl in four different ways. These four types of locomotion are described as *horizontal undulatory, rectilinear, concertina,* and *sidewinder progression.* The track left by a snake that is moving by horizontal undulatory movement is a wavy line, since the body glides along in a series of waves and each part of the body passes along the same track. In soft dust or sand it is apparent that the dirt or sand is raised or pushed up on the side opposite the axis of travel so as to form a pivot.'

When progressing by rectilinear motion, the axis of the body is essentially straight and movement is effected by alternate movements of the ventral scutes and the body itself. An undulatory movement raises the scutes, carries them forward and permits them to become anchored to the ground. Following this the body literally slides forward until it is reoriented with the skin. By a series of rhythmic undulatory movements on the part of the ventral skin and the body, the snake appears to move along a straight path without any lateral motion.

Concertina progression consists of alternate curving and straightening of the body. After the body is drawn up in a series of curves, the tail serves as an anchor and progress is effected by pushing the rest of the body forward until it is straight. The head and neck are then used as anchors and the remainder of the body is drawn up in a series of curves. This mode of progression is not as rapid as rectilinear motion, but it is used sometimes when snakes are stalking their prey.

Sidewinding is employed by certain desert snakes. In North America it is characteristic of the sidewinder rattlesnake (*Crotalus cerastes*), which derives its name from this peculiar type of locomotion that is difficult to describe. It is essentially a series of lateral looping movements in which only a vertical force is applied and no more than two parts of the body contact the ground at any one time. Figure 4.23 best illustrates the sidewinding movement. The resulting tracks are a series of parallel, diagonal, J-shaped marks. The curve of the J, as can be seen, is made by the head and neck. This type of progression seems to be advantageous in soft sand.

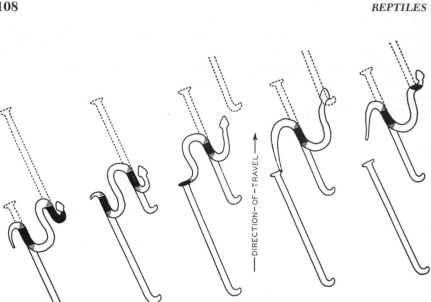

DIRECTION—OF—TRAVEL

**FIGURE 4.23.** How a sidewinder makes its tracks. Consecutive positions of a sidewinder's body in relation to the tracks. The solid track-outlines have already been made; the dotted outlines are yet to be made. Only the solid-black sections of the snake's body are in contact with the ground; the rest of the body is raised sufficiently to clear the ground. (After Mosauer with modifications, in Klauber, L. M.: Rattlesnakes. University of California Press, 1956.)

## RADIATION ORGANS

Rattlesnakes and other members of the pit viper family possess special sensory pits on each side of the head (Figs. 4.24 and 4.25). The presence of these pits was known back in the seventeenth century, but it was not until 1937 that Noble and Schmidt presented a satisfactory theory to account for their function. Experimenting with rattlesnakes that had all their other major sensory organs destroyed or blocked, they discovered that these reptiles could accurately locate and strike at heated objects, that is, objects whose temperature was higher than that of the surrounding environment, or those that were colder. These facial or loreal pits are anteroventral to the eye and have the opening directed forward. The sensory tissue in the pit is innervated by fibers from the ophthalmic and supramaxillary branches of the fifth cranial nerve.

NOSTRIL

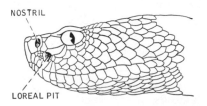

LOREAL PIT

**FIGURE 4.24.** Head of a rattlesnake showing the position of the loreal pit.

FIGURE 4.25.   Western diamondback rattlesnake (*Crotalus atrox*) of North America ready to strike. Note the loreal pit below the nostril.

Under natural conditions it appears that these radiation organs aid pit vipers in locating and accurately striking at small warm-blooded vertebrates at night when the environmental temperature is lower than that of the prey. In captivity, observations made on a rattlesnake that had its vision obscured because it was molting showed that it accurately struck at a mouse.

It has recently been demonstrated that there are infrared receptors in the facial pits of the Australian python, *Morelia spilotes.* Radiation organs are also believed to be present in the supranasal cavities of certain species of vipers occurring in the Old World.

## CLASSIFICATION OF REPTILES

### *CLASS REPTILIA (REPTILES)*

Cold-blooded vertebrates possessing lungs and not passing through a larval stage; skull well ossified and possessing a single occipital condyle; skin dry, covered by epidermal scales, and essentially lacking glands; Carboniferous to Recent; six subclasses.

SUBCLASS ANAPSIDA (ANAPSID REPTILES).   Reptiles lacking temporal openings in the skull; quadrate immovable; two orders.

*Order Cotylosauria (cotylosaurs).* Ancient stem reptiles; skull lacking an elongate snout; a parietal foramen present; teeth situated on margin of jaws as well as on palate; Carboniferous to Triassic.

*Order Chelonia (turtles and tortoises).* Terrestrial, aquatic, or marine reptiles whose bodies are enclosed within two bony shells, a dorsal carapace and a ventral plastron, which are connected with each other laterally; teeth absent; jaws developed into a horny beak; tongue not extensible; eyelids present; neck usually retractible and possessing eight cervical vertebrae; limbs basically pentadactyl; Permian to Recent; temperate and tropical regions; 12 living families and about 240 species.

*Family Chelydridae (snapping turtles).* Large (up to 200 pounds), predaceous, freshwater turtles; carapace ridged and joined to the relatively small plastron by cartilage; five toes on each foot; toes joined by small webs; eastern and central North America south to northern South America; two genera.

*Family Kinosternidae (mud turtles and musk turtles).* Small freshwater turtles; carapace smooth and arched; plastron either much reduced (musk turtles) or with hinged anterior and posterior lobes (mud turtles); five toes on each foot; toes webbed; eastern and central North America south to Brazil; four genera.

*Family Dermatemydidae (Central American river turtle).* Large freshwater turtle; carapace about 300 mm. long, heavy, and connected to plastron by a broad bridge possessing a row of small plates; coastal rivers from Vera Cruz to Guatemala; one genus and one species, *Dermatemys mawi.*

*Family Platysternidae (big-headed turtle).* A small freshwater turtle with a proportionately large head which cannot be retracted into the shell; carapace up to 150 mm. long and very flat; plastron broad and with unhinged lobes; tail relatively long; southeastern Asia; one genus and one species, *Platysternon megacephalum.*

*Family Emydidae (common freshwater turtles).* Small to medium-large, freshwater or terrestrial turtles; carapace 90 to 600 mm. in length, usually oval and arched; plastron possessing 12 plates and united by bone with carapace; five toes on front foot, four or occasionally three on hind foot; webbing usually present between toes; primarily North America, Europe, and Asia, barely extending south to North Africa and northern South America; 25 genera and more than 75 known species.

*Family Testudinidae (land tortoises).* Medium to large terrestrial turtles with a carapace that is usually high and arched; plastron solidly joined to carapace; top of head possessing distinct shields; legs very stout, strong, and covered by hard protective scales; toes short, unwebbed; claws thick; North and South America, southern Europe, southern Asia, Africa, Madagascar, the Galápagos Islands, and islands of the Indian Ocean; seven genera and about 40 species.

Family Dermochelidae (leatherback sea turtle). Very large marine turtle (up to 2.5 m.) lacking an external shell; dorsal surface marked by seven longitudinal ridges on the leathery skin which overlies the bony plates representing the carapace; front and hind limbs modified as flippers; tropical and subtropical seas; one genus and one species, *Dermochelys coriacae.*

Family Cheloniidae (true sea turtles). Medium to large (shell 600 to 1200 mm. long) marine turtles possessing horny plates overlying the bony plates of the carapace and plastron; head incapable of being retracted into shell; limbs modified as flippers; tropical and subtropical seas, coming ashore occasionally on remote islands; four genera.

Family Trionchidae (soft-shelled turtles). Medium-sized (carapace up to 600 or more mm. long) freshwater turtles lacking horny plates on carapace or plastron; shell covered by leathery tissue; head narrow; snout elongate; beak fleshy rather than horny; toes joined by webs; North America, Africa, and eastern and southern Asia; seven genera.

Family Carettochelidae (New Guinea plateless turtle). A medium-sized (carapace 450 mm. long), freshwater, side-necked turtle lacking horny plates on the carapace or plastron, which is covered with a soft skin; neck folded sideways rather than withdrawn into shell by vertical folding; limbs flipper-like but possessing two claws; New Guinea; one genus and one species, *Carettochelys insculpta.*

Family Pelomedusidae (hidden-necked turtles). Medium to large (carapace up to 750 mm. long), freshwater, side-necked turtles; head capable of being completely retracted into shell by lateral rather than vertical bending of neck; carapace and plastron covered by horny shields; plastron generally hinged anteriorly; South America, Africa, and Madagascar; three genera.

Family Chelidae (snake-necked turtles). Small to moderately large (carapace 125 to 400 mm. long), freshwater, side-necked turtles; head and neck occasionally exceeding carapace in length; carapace and plastron possessing external horny shields; toes webbed; South America, Australia, and New Guinea; 10 genera.

SUBCLASS SYNAPTOSAURIA (PLESIOSAURS AND THEIR RELATIVES). Extinct reptiles possessing a parapsid type of skull with the postorbital and squamosal elements meeting below the temporal opening; parietal foramen present; quadrate fused to skull; secondary palate lacking; all major girdle and limb elements present, although adapted to aquatic life in some forms; Permian to Cretaceous; two orders.

Order Protosauria (protosaurs). Small lizard-like reptiles; skull high and relatively narrow; external nares anterior; vertebrae generally amphicoelous with cervical centra sometimes quite elongate; cer-

vical ribs single-headed; trunk ribs two-headed; Permian to Triassic.

*Order Sauropterygia (sauropterygians).* Large amphibious or marine
reptiles; skull low and broad; external nares posteriorly situated;
vertebrae generally amphicoelous; trunk ribs with a single head;
limbs and girdles modified for aquatic life; Triassic to Cretaceous.

SUBCLASS ICHTHYOPTERYGIA (ICHTHYOSAURS). Extinct marine reptiles with dolphin-like bodies and a unique parapsid type of skull
with the postorbital and squamosal elements not bordering the supratemporal fossa; rostrum of skull greatly elongate; nares posterior
in position; eyes large and surrounded by well developed sclerotic
bones; pectoral and pelvic limbs modified into finlike paddles; a
nonskeleton-supported dorsal fin present; caudal vertebrae extending into ventral lobe of vertical tail fin; Triassic to Cretaceous; one
order.

*Order Ichthyosauria (ichthyosaurs).* Characters of the subclass.

SUBCLASS LEPIDOSAURIA (LEPIDOSAURIANS). Diapsid or derivatives of
diapsid reptiles that lack bipedal specializations; Permian to Recent; three orders.

*Order Eosuchia (eosuchians).* Extinct, primitive lepidosaurians with
a diapsid type of skull; teeth thecodont, pleurodont, or acrodont;
limbs slender and lizard-like; Permian to Eocene.

*Order Rhynchocephalia (rhynchocephalians).* Primitive, lizard-like
lepidosaurians with two temporal openings; teeth acrodont; premaxillaries somewhat beaklike; palate primitive; a pineal eye
present; vertebrae amphicoelous; Triassic to Recent.

   *Family Sphenodontidae (tuatara).* Characters of the order; one
   living species, *Sphenodon punctatus*, known from New Zealand.

*Order Squamata (lizards and snakes).* Terrestrial, fossorial, arboreal,
aquatic, and marine reptiles with a modified diapsid type of skull;
lower bony border of infratemporal opening missing; quadrate
movable; teeth present; body covered by epidermal scales which
sometimes overlie osteoderms; copulatory organs paired; Jurassic
to Recent; two suborders.

   SUBORDER LACERTILIA OR SAURIA (LIZARDS). Small to medium-
   large (40 to 3000 mm.)* reptiles with elongate bodies; paired
   limbs usually present; rami of mandible usually joined by a
   suture; eyelids and external auditory meatus generally present;
   worldwide except Antarctica but most abundant in warmer regions; about 20 living families and about 3000 species recognized.

   *Family Gekkonidae (geckos).* Small (75 to 150 mm.) lizards that
   are primarily nocturnal and arboreal; eyes very large; pupils
   usually vertical; eyelids usually lost; toes often possessing
   clinging pads for adhering to smooth surfaces when climbing;
   some capable of loud vocalization; widespread in warmer
   parts of world; about 79 genera.

---

*Total length.

*Family Pygopodidae (flap-footed lizards).* Small to medium-sized (150 to 750 mm.) lizards in which the only remnant of appendages consists of a pair of small flaps on sides of body near vent; eyelids generally immovable; pupils generally vertical; Australia, Tasmania, and New Guinea; eight genera.

*Family Agamidae (agamids).* Small to rather large (125 to 900 mm.) lizards which, for the most part, are unspecialized, although some species are arboreal or aquatic; scales smooth or keeled; teeth acrodont; eyelids movable; limbs usually possessing five toes; range, complementary to that of the Iguanidae, that is, Europe, Asia, and Africa but absent from Madagascar and the Pacific islands; 34 genera and about 300 species.

*Family Chameleonidae (chameleons).* Very small to medium-sized (40 to 600 mm.) lizards adapted to an arboreal life; eyes large, bulging, and mostly covered by thick lids; each eye capable of independent movement; tongue very extensible; tail prehensile; skin capable of rapid color changes; primarily Africa and Madagascar with a very few species found in southern Europe and Asia; about 80 species.

*Family Iguanidae (iguanids).* Small to large (100 to about 2100 mm.) lizards that are largely unspecialized and complement the agamids in range; teeth pleurodont; tongue nonretractible; body either dorsoventrally or laterally compressed; tail usually long, rarely prehensile, and not easily separated from body; limbs possessing five clawed toes; North and South America, the West Indies, Madagascar, and the Fiji Islands; about 44 genera and about 700 species.

*Family Xantusidae (night lizards).* Small (125 to 150 mm.) nocturnal lizards; scales granular; pupils vertical; lower eyelids immovable, transparent, and covering eyes; tail slightly longer than head and body; limbs unspecialized with five toes on each foot; southwestern United States, Mexico, Central America, and Cuba; three or four genera and 11 species.

*Family Scincidae (skinks).* Small to medium-sized (150 to 650 mm.) lizards with cylindrical bodies, and legs that are short or absent; scales generally smooth and cycloid; body scales with bony cores; eyelids often transparent; mostly terrestrial or fossorial; North and South America, Europe, Asia, Africa, Madagascar, Australia, and some of the Pacific islands; about 600 species.

*Family Anelytropsidae (anelytropsids).* Small (200 mm.) blind lizards that inhabit rotting logs; external ears absent; limbs absent; Mexico; one genus and one species, *Anelytropsis papillosus.*

*Family Feylinidae (feylinids).* Moderately small (300 mm.) limbless lizards with rudiments of a pectoral girdle; blind; earless; equatorial Africa; one genus, *Feylinia*, and four species.

*Family Dibamidae (dibamids).* Moderately small (250 to 300

mm.) blind lizards; the male only possesses rudimentary, flaplike hind limbs that lack digits; head conical; external ears absent; Philippines south to New Guinea; one genus, *Dibamus*, and three species.

*Family Cordylidae (girdle-tailed lizards).* Moderately small to medium-sized terrestrial lizards; limbs ranging from normal with five digits on each foot to a condition in which the front limbs are entirely lacking; tail not fracturing easily, possessing large keeled scales which encircle it and sometimes developed as an organ of defense; Africa; four genera and 23 species.

*Family Gerrhosauridae (gerrhosaurids).* Fairly small (200 to 450 mm.) terrestrial lizards; head shields fused to skull; osteoderms present; limbs varying from a normal five-toed condition to complete elimination of the front limbs; Africa and Madagascar; six genera and 25 species.

*Family Teiidae (teiids).* Very small to large (75 to 1200 mm.) lizards that are mostly terrestrial and exhibit great variability in structure; scales smooth, keeled, or granular and very variable in shape and arrangement; head shields not attached to skull; limbs well developed to rudimentary; teeth lacking a hollow base; North and South America and the West Indies; about 40 genera.

*Family Lacertidae (lacertids).* Small to medium-sized (175 to 750 mm.) lizards; scales generally keeled; head shields attached to skull; teeth possessing a hollow base; limbs well developed with five clawed digits; Europe, Asia, and Africa; about 150 species.

*°Family Amphisbaenidae (worm lizards).* Small to medium-sized (300 to 700 mm.) subterranean dwellers with cylindrical bodies and blunt heads and tails; external ears and eyes not apparent; hind limbs absent; front limbs generally absent; North and South America, southern Europe, southwestern Asia, and Africa; about 120 species.

*Family Anguidae (anguids).* Small to medium-sized (350 to 1200 mm.) terrestrial lizards with very elongate tails; osteoderms underlying the epidermal scales; limbs well developed and possessing five toes in some species, but with a reduction in number of toes or even complete absence of limbs in others; tail readily separating from body; North and South America, Europe, Asia, and North Africa; about 60 species.

*Family Aniellidae (limbless lizards).* Small (200 to 250 mm.) burrowing lizards; scales smooth; osteoderms present; external ears lacking; limbs lacking; western North America from central California south to Baja California, Mexico; one genus, *Aniella*, and two species.

---

°This family is placed here tentatively. There is considerable question as to its actual ordinal position.

*Family Xenosauridae (xenosaurids).* Small (250 to 375 mm.), stocky lizards with short tails; osteoderms reduced; limbs normal and sturdy; eastern Asia and Central America; two genera (considered by some to be two families) and four species.

*Family Helodermatidae (Gila monsters).* Medium-sized (400 to 750 mm.), heavy-bodied terrestrial lizards; scales beadlike with osteoderms present and attached to skull; teeth grooved and venom glands present in mouth; tongue thick, but bifid and extensible; limbs normal; tail short and blunt; southwestern United States and adjacent Mexico; one genus, *Heloderma*, and two species.

*Family Varanidae (monitor lizards).* Very small to very large (200 to 3000 mm.) terrestrial lizards; osteoderms reduced or absent; head and neck, as well as tail, relatively long; limbs of the normal terrestrial type; southern Asia, Africa, Australia, and the East Indies; one genus, *Varanus*.

*Family Lanthanotidae (earless monitor).* A moderately small (400 mm.) terrestrial lizard with a fairly long body and thick neck; osteoderms poorly developed; tongue very protrusible; lower eyelids transparent; tail not separable from body; Borneo; one genus and one species, *Lanthanotus borneensis*.

SUBORDER SERPENTES OR OPHIDIA (SNAKES). Small to very large (0.1 up to 9 m.)[*] reptiles with very elongate, limbless bodies; rami of lower jaw usually joined by an elastic ligament rather than a suture; immovable eyelids; external ear opening absent; tongue extensible and forked distally; worldwide except Antarctica, but most abundant in Tropical to Temperate Zones; 13 living families and about 3000 species.

*Family Boidae (boas).* Medium-small to very large (0.5 to 8 m.) terrestrial or arboreal snakes; supraorbital bones absent; no premaxillary teeth; vestiges of ilium, ischium, pubis, and femur present; hind limb vestige indicated externally by a small clawlike spur on either side of vent; paired lungs present; viviparous; temperate and tropical parts of North and South America, Asia, southeastern Europe, Africa, Madagascar, and the Mascarine Islands; about 16 genera.

*Family Pythonidae (pythons).* Medium-sized to very large (1 to 9 m.) arboreal, terrestrial, or burrowing snakes; supraorbital bones present; maxillary teeth present; vestiges of hind limbs apparent; paired lungs present; oviparous; Africa, Asia, East Indies, and Australasia; six genera.

*Family Typhlopidae (blind snakes).* Mostly small (100 to 750 mm.) subterranean snakes; maxilla movably attached to skull; teeth generally lacking in lower jaw; remnant of single pair of pelvic bones present; no external vestige of hind limb; North and South America, Africa, Madagascar, southern Europe, Asia, Australia, and the East Indies; five genera.

---

[*]Total length.

*Family Leptotyphlopidae (slender blind snakes).* Small (100 to 300 mm.) subterranean snakes; maxilla solidly attached to skull; teeth present only in lower jaw; remnant of three pairs of pelvic bones present; external vestige of hind limb sometimes present; Africa and North and South America; one genus, *Leptotyphlops*, and about 40 species.

*Family Anilidae (anilids).* Moderately small (up to 750 mm.) burrowing snakes; premaxillary teeth present; dentary immovable; eyes very small; tail short; vestiges of hind limb present with small external spur on each side of vent; viviparous; southeastern Asia and South America; three genera.

*Family Uropeltidae (shield-tailed snakes).* Small (up to 350 mm.) burrowing snakes; premaxillary teeth absent; dentary immovable; scales at tip of tail flattened and modified in some species into a shield; no vestiges of hind limb; viviparous; Indian peninsula and Ceylon; nine genera.

*Family Xenopeltidae (xenopeltids).* Medium-sized (up to 1050 mm.) burrowing snakes; premaxillary teeth present; palatal teeth numerous; no vestiges of hind limb; Asia and tropical America; one or two genera, *Xenopeltis* and possibly *Loxocemus.*

*Family Acrochordidae\* (Oriental water snakes).* Medium-large (900 to 1800 mm.) marine or aquatic snakes; epidermal scales not overlapping; only labial scales enlarged on head; nostrils dorsal in position; supratemporal and quadrate united; no vestiges of hind limb; viviparous; southern Asia and East Indies east to the Solomon Islands; two genera, *Acrochordus* and *Chersydrus* (not given generic status by some).

*Family Colubridae (colubrids).* Small to large (0.2 to 4 m.) terrestrial, arboreal, fossorial, or aquatic snakes; pupil of eye round, vertical, or horizontal; teeth present in both jaws; aglyphous or opisthoglyphous; coronoid and postfrontal bones missing; no vestiges of hind limb; oviparous or viviparous; essentially worldwide; about 270 genera.

*Family Elaphidae (elaphids).* Small to large (0.35 to 6 m.) terrestrial or arboreal snakes; teeth present in both jaws; proteroglyphous; maxilla generally short; no vestiges of hind limb; tail not laterally compressed; North and South America, Asia, Africa, Australia, and the East Indies; about 41 genera.

*Family Hydrophidae (sea snakes).* Medium-large (up to 3 m.) marine snakes; teeth present in both jaws; proteroglyphous; no vestiges of hind limbs; tail laterally compressed as a swimming organ; tropical part of Indian and Pacific Oceans; about 15 genera.

*Family Viperidae (true vipers).* Small to moderately large (250 to 1800 mm.) terrestrial snakes; maxilla short and movable; sole-

---

*Regarded by some as a subfamily of the Colubridae.

noglyphous; body relatively stocky; tail short; no vestiges of hind limb; Europe, Asia, and Africa; 11 genera.

*Family Crotalidae (pit vipers).* Small to large (0.45 to 3.5 m.) terrestrial, arboreal, or aquatic snakes; loreal pits between nostril and eye; solenoglyphous; no vestiges of hind limb; North and South America, extreme southeastern Europe, Asia, and the East Indies; six genera.

SUBCLASS ARCHOSAURIA (ARCHOSAURIANS). Ruling Mesozoic reptiles developing specializations leading to a bipedal mode of locomotion; skull diapsid except in a few instances in which the supratemporal opening has secondarily closed; parietal foramen generally lacking; Triassic to Recent; five orders.

*Order Thecodontia (thecodonts).* Extinct ancestral carnivorous archosaurians with long, slender skulls; antorbital and palatal vacuities present; teeth thecodont; palatal teeth lacking; hind limbs larger than forelimbs, indicating bipedal tendency; Triassic.

*Order Crocodilia (crocodilians).* Large, degenerate survivors of Triassic archosaurians; skull elongate; nares terminal; a secondary palate present; teeth thecodont; locomotion quadripedal; toes webbed; pubis excluded from acetabulum; tail laterally compressed; Triassic to Recent; three living families (regarded by some as subfamilies only).

*Family Crocodylidae (crocodiles).* Snout slender, elongate, but not sharply demarked from rest of skull; supratemporal opening small; teeth strong and interlocking, with those of lower jaw fitting into pits in upper jaw; North and South America, Africa, Asia, the East Indies, and Australia; three genera.

*Family Alligatoridae (alligators and caiman).* Snout relatively broad and not sharply demarked from skull; supratemporal opening small or secondarily closed; teeth strong, with those of lower jaw fitting into pits inside those of upper jaw; North America, South America, and China; four genera.

*Family Gavialidae (gavials).* Snout slender, elongate, and sharply demarked from rest of skull; supratemporal opening large; teeth slender, numerous, and interlocking; pits for lower teeth lacking in upper jaw; Asia; one genus and one species, *Gavialis gangeticus.*

*Order Pterosauria (flying reptiles).* Extinct diapsid reptiles specialized for aerial life; bones light and pneumatic; antorbital openings present; orbits large; well developed thecodont teeth; forelimb adapted for flight with great elongation of fourth digit to form a wing covered with a membrane of skin; tail elongate; Jurassic and Cretaceous.

*Order Saurischia (reptile-like dinosaurs).* Extinct diapsid reptiles primarily adapted to a bipedal mode of life; teeth generally present throughout length of jaws; front limbs usually reduced in size; pelvis triradiate; Triassic to Cretaceous.

*Order Ornithischia (birdlike dinosaurs).* Extinct diapsid reptiles that were largely quadripedal; teeth either lacking or absent from front of jaws; front limbs not markedly reduced in size; pelvis tetraradiate because of a two-pronged pubis; Triassic to Cretaceous.

SUBCLASS SYNAPSIDA (MAMMAL-LIKE REPTILES). Extinct reptiles possessing a single, lateral temporal opening; skull fairly large and not flattened; a parietal foramen generally present; Carboniferous to early Jurassic; two orders.

Order Pelycosauria (pelycosaurs). Primitive synapsid reptiles; external nares lateral and widely separated; supratemporal present; no secondary palate; Carboniferous to Permian.

Order Therapsida (therapsids). Advanced synapsid reptiles; external nares dorsal and close to tip of snout; supratemporal absent; a secondary palate in advanced forms; Permian to early Jurassic.

## References Recommended

Barbour, T. 1926. Reptiles and Amphibians: Their Habits and Adaptations. Boston, Houghton Mifflin Co.

Bartholomew, G. A. 1966. A Field Study of Temperature Relations in the Galápagos Marine Iguana. Copeia 1966:241–250.

Bullock, T. H., and Cowles, R. B. 1952. Physiology of an Infrared Receptor: The Facial Pit of Pit Vipers. Science 115:541–543.

Camp, C. L. 1923. Classification of Lizards. Bull. Am. Mus. Nat. Hist. 48:239–481.

Carr, A. 1952. Handbook of Turtles. Ithaca, N.Y., Comstock Publishing Co.

Cowles, R. B., and Bogert, C. M. 1944. A Preliminary Study of the Thermal Requirements of Desert Reptiles. Bull. Am. Mus. Nat. Hist. 83:267–296.

Curran, C. H., and Kauffeld, C. 1937. Snakes and Their Ways. New York, Harper & Bros.

De Witt, C. B. 1967. Precision of Thermoregulation and its Relation to Environmental Factors in the Desert Iguana, Dipsosaurus dorsalis. Physiol. Zool. 40:49–66.

Ditmars, R. L. 1936. The Reptiles of North America. New York, Doubleday & Co., Inc.

Ditmars, R. L. 1939. A Field Book of North American Snakes. New York, Doubleday & Co., Inc.

Gadow, H. 1923. Amphibia and Reptiles. New York, St. Martin's Press, Inc.

Jackson, D. C. 1969. Bouyancy Control in the Freshwater Turtle, Pseudemys scripta elegans. Science 166:1649–1651.

Mayhew, W. W. 1961. Photoperiodic Response of Female Fringe-toed Lizards. Science 134:2104–2105.

Mayhew, W. W. 1962. Scaphiopus couchi in California's Colorado Desert. Herpetologica 18:153–161.

Mayhew, W. W. 1963. Temperature Preferences of Sceloporus orcutti. Herpetologica 18:217–233.

Mayhew, W. W. 1963. Biology of the Granite Spiny Lizard, Sceloporus orcutti. Am. Midland Nat. 69:310–327.

Mayhew, W. W. 1964. Photoperiodic Responses in Three Species of the Lizard Genus Uma. Herpetologica 20:95–113.

Oliver, J. A. 1955. The Natural History of North American Amphibians and Reptiles. Princeton, N.J., D. Van Nostrand Co., Inc.

Pope, C. H. 1937. Snakes Alive and How They Live. New York, The Viking Press, Inc.

Pope, C. H. 1939. Turtles of the United States and Canada. New York, Alfred A. Knopf.

Pope, C. H. 1955. The Reptile World. New York, Alfred A. Knopf.

Romer, A. S. 1956. Osteology of the Reptiles. Chicago, University of Chicago Press.

Schmidt, K. P., and Davis, D. D. 1941. Field Book of Snakes of the United States and Canada. New York, G. P. Putnam's Sons, Inc.

Schmidt, K. P., and Davis, D. D. 1953. A Check List of North American Amphibians and Reptiles. Chicago, University of Chicago Press.

Schmidt, K. P., and Inger, R. F. 1957. Living Reptiles of the World. Garden City, N.Y., Hanover House.

Schmidt-Nielson, K., Borut, A., Lee, P., and Crawford, E., Jr. 1963. Nasal Salt Excretion and the Possible Function of the Cloaca in Water Conservation. Science *142*: 1300–1301.

Smith, H. M. 1946. Handbook of Lizards. Ithaca, N.Y., Comstock Publishing Co.

Stebbins, R. C. 1954. Amphibians and Reptiles of Western North America. New York, McGraw-Hill Book Co., Inc.

Stebbins, R. C. 1963. Activity Changes in the Striped Plateau Lizard with Evidence on Influence of the Parietal Eye. Copeia *1963*:681–691.

Stebbins, R. C., and Barwick, R. E. 1968. Radiotelemetric Study of Thermoregulation in a Lace Monitor. Copeia *1968*:541–547.

Stebbins, R. C., and Eakin, R. M. 1958. The Role of the "Third Eye" in Reptilian Behavior. Am. Mus. Novitates *1870*:1–40.

Van Denburgh, J. 1922. The Reptiles of Western North America. 1. Lizards. 2. Snakes and Turtles. Occasional Papers Calif. Acad. Sci., No. 10.

Warren, J. W., and Proske, U. 1968. Infrared Receptors in the Facial Pits of the Australian Python, *Morelia spilotes*. Science *159*:439–441.

Wilhoft, D. C. 1964. Seasonal Changes in the Thyroid and Interrenal Glands of the Tropical Australian Skink, *Leilopisma rhomboidalis*. Gen. Comp. Endocr. *4* (No. 1):42–53.

Wright, A. H., and Wright, A. A. 1957. Handbook of Snakes of the United States and Canada. Ithaca, N.Y., Comstock Publishing Co.

*Chapter Five*

# BIRDS

## GENERAL CHARACTERS

Birds share with mammals the distinction of being the most recent vertebrates to inhabit the earth. The estimated 8700 living species are widely distributed over the world, ranging from the Arctic to the Antarctic and occurring at sea as well as on land. Even the most remote islands are visited by sea birds and many such small isolated land masses have their own endemic avifauna. Although birds have failed to adapt themselves to life underwater or underground as, for example, have whales and moles among the Mammalia, there are very few other vertebrate habitats into which they have not ventured.

As a class birds are strikingly uniform. The most obvious avian features are feathers and the possession of a horny bill. However, there are many other structural characters that readily distinguish them from other forms of animal life.

The lower vertebrates, considered in the preceding chapters, depend largely on the external environment as a source of body heat. Birds, however, are *endothermous*, producing their own heat. Furthermore they are referred to as *homoiothermous* because they can maintain a fairly high yet constant body temperature. This does not mean that a bird's temperature never fluctuates; it has been demonstrated in a number of species that there may be a daily range of several degrees. Furthermore, in hibernating poor-wills (*Phalaenoptilus nuttallii*) the body temperature may drop to as low as 45° F., and those species of hummingbirds that are subject to nocturnal torpidity have body temperatures which are greatly reduced at night as a means of decreasing energy output.

## Skeletal System

The skeletal system of the bird (Fig. 5.1) possesses many unique features. In general the bones are remarkably <u>light</u>, especially in species that fly. This is because the <u>larger bones contain pneumatic cavities</u> connected with the respiratory system. <u>Those of the skull, for the most part, are fused togethe</u>r so that most of the sutures are obliterated. The upper and lower mandibles are greatly elongated as a support for the bill. <u>Teeth are entirely lacking</u> in modern birds.

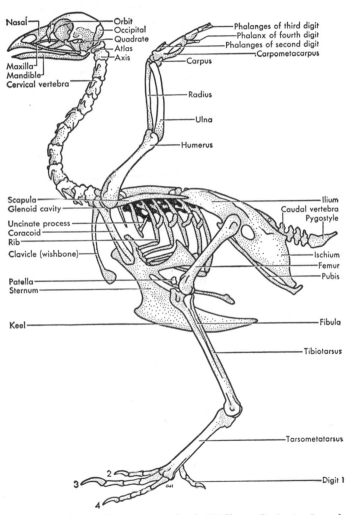

**Figure 5.1.** Skeleton of a domestic fowl. (Wallace, G. J.: An Introduction to Ornithology. After Hegner and Stiles: College Zoology. Copyright 1951. The Macmillan Co., 1955.)

Each lower mandible is composed of five bones and articulates with the skull by means of a movable quadrate. The orbits are very large and are separated from one another by a thin interorbital septum, thus pushing the braincase posteriorly. The skull articulates with the first cervical vertebra by means of a single occipital condyle.

The structure of the avian palate has long been one of the characters used in the diagnosis of larger taxonomic categories. It has been based on the shape of the prevomers (often called vomers but not to be confused with the mammalian vomer) and whether they are fused as well as the maxillopalatines are to one another. The *dromaeognathous* palate has the maxillopalatines separate and the prevomers extending far posteriorly and articulating with the posterior end of the palatines. It is found in ratite birds and kiwis. In the *schizognathous* type of palate the maxillopalatines are separate, but the prevomers are fused, pointed anteriorly, and either small or large. Widely separated orders such as the Charadriiformes, the Galliformes, the Gruiformes, and the Piciformes possess this type of palate. In the Ciconiiformes, Anseriformes, Falconiformes, and many others the palate is referred to as *desmognathous*, with the maxillopalatines meeting in the midline and the prevomers small, fused, or lacking. Passerine birds and swifts have an *aegithognathous* palate, with the maxillopalatines separate and the vomers fused, broad, and truncate anteriorly.

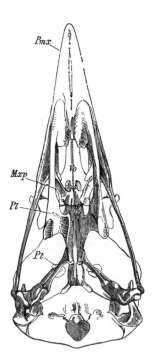

**FIGURE 5.2.** Ventral view of an aegithognathous palate of *Corvus: Pmx*, premaxillary; *mxp*, maxillopalatine; *pl*, palatine; *pt*, pterygoid; *vo*, vomer. (From Beddard, F. E.: The Structure and Classification of Birds. Longmans, Green, and Co., 1898; after Huxley.)

The vertebral column of the bird possesses many adaptations. The cervical vertebrae are more numerous and variable in number than in any other living group of vertebrates. They are also especially flexible because of the saddle-shaped articular surfaces on the ends of the centra, a condition referred to as *heterocoelous*. The anterior thoracic vertebrae are capable of slight movement, but those situated more posteriorly, as well as the lumbar, sacral, and anterior caudal vertebrae are fused with the pelvic girdle to form what is known as the *synsacrum.* The few free caudal vertebrae are compressed and the distal caudals are fused to form a single structure called the *pygostyle*, which terminates the short tail.

The avian ribs are flattened, and all but the first and last possess posterior projections referred to as *uncinate processes* (lacking in screamers) which overlap the next succeeding rib. The sternum or breast bone is very flat and broad so as to offer sufficient surface for the attachment of muscles used in flying. In all, except the *ratite* or flightless birds, the sternum is keeled or *carinate.* This long flat keel that extends down from the mid-ventral line of the sternum is an additional means of increasing the surface for muscle attachment.

The fusion of most of the trunk vertebrae with the pelvis and their firm connection with the very large sternum by means of the flattened ribs provide a very strong skeletal framework even though the bones themselves are relatively light. All of this is an advantage to a creature that must propel itself through the air.

The scapula of the bird is long and slender, while the coracoid is proportionately short and stout. The clavicles are fused together to form the *furcula* or wishbone.

The most conspicuous skeletal modifications of the forelimb are concerned with the wrist and hand. Only two carpal elements are found in the wrist, the *radiale* and the *ulnare*, which articulate, respectively, with the radius and ulna. Distal to the wrist is a composite bone called the *carpometacarpus* which represents some of the carpal elements of other vertebrates and the second, third, and fourth metacarpals. Four small bones which are remnants of three digits are attached to the carpometacarpus.

Although the hind limb is not as highly modified as the forelimb, it shows some interesting specializations. The fibula is proportionately small and partly fused with the tibia. Some of the tarsal elements are likewise fused with the distal end of the tibia. This composite bone is called the *tibiotarsus.*

The remainder of the tarsals are fused to the second, third, and fourth metatarsals forming a bone called the *tarsometatarsus*. A remnant of the first metatarsal is attached to this by means of ligaments. No more than four toes are found in birds, and the number is sometimes reduced to three and, in the ostrich, to two.

## MUSCULAR SYSTEM

The avian muscular system differs in many respects from that of most other land vertebrates. The muscles of the jaw and neck exhibit many specializations associated with food habits, use of the bill, and mobility of the neck. Since most of the trunk vertebrae are fused, there is a great reduction in the dorsal musculature. The abdominal muscles are also very poorly developed. On the other hand, the extrinsic muscles of the wing, especially the *pectoralis major*, which is a depressor and therefore concerned with the downstroke in flight, show enormous development in flying birds. The pectoralis major alone may constitute as much as one-fifth of the total body weight. It originates on the sternum and furcula and inserts on the underside of the humerus.

The *supracoracoideus*, which is concerned with the upstroke, also originates on the sternum, beneath the pectoralis major, and inserts on the upper side of the humerus. The *deltoid* muscle and *latissimus dorsi* have an action similar to the supracoracoideus. In hummingbirds, which have a very rapid wing stroke, the latter muscle is proportionally large. Rather closely associated with the deltoid is the *propatagialis*, which sends tendinous slips into the *patagium* or web of skin extending from the anterior part of the

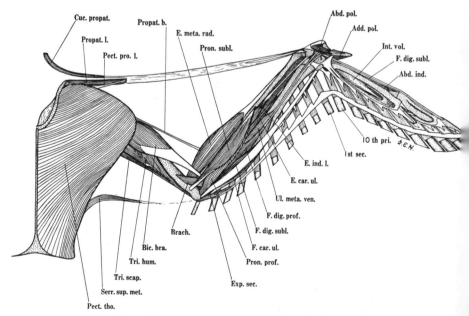

**FIGURE 5.3.** Ventral view of the wing muscles of the common crow (*Corvus brachyrhynchos*). (George, J. C., and A. J. Berger: Avian Myology. Academic Press, 1966; after Hudson and Lanzellotti.)

wing. One of these tensors is known as the *longus*. It extends from the head of the humerus along the anterior margin of the patagium to the wrist. Another is called the *brevis* and extends from the humerus to the proximal part of the forearm. A third tensor known as the "biceps slip" or *tensor accessorius* runs from the biceps muscle to the anterior edge of the patagium.

Despite the reduction in the number of skeletal elements in the avian pectoral limb, there are a number of intrinsic muscles present which are concerned with pronation, supination, and rotation of the wing during flight. The muscles of both the wings and the legs in birds, in general, tend to be concentrated close to the body and insert distally by long tendinous extensions. The avian shank consists largely of tendons surrounding the tarsometatarsus. Most important, especially in perching birds, are the flexors which enable the toes to grasp a perch. The tendons concerned with this action, such as the *flexor perforans* and the *flexor digitorum longus*, often behave more or less as a single unit.

## CIRCULATORY SYSTEM

The circulatory system of birds is about as advanced as that of mammals. There is a complete separation of venous and arterial blood and a four-chambered heart. The systemic aorta leaves the left ventricle and carries blood to the head and body by way of the right fourth aortic arch. Considerable variation occurs with respect to the carotid arteries. Generally the common carotids are paired. In bitterns, however, the two fuse together shortly after branching from the innominates and form a single stem. In other groups there may be a reduction in the size of either the right or left common carotid before they fuse, and in passerine birds only the left common carotid remains. There are two functional precaval veins and a complete postcaval. The former are formed by the union of the jugular and subclavian on each side. The postcaval receives blood from the limbs by way of the renal portals, which pass through the kidneys but do not break up into capillaries and are therefore not comparable to the renal portals of lower vertebrates. The avian erythrocytes are nucleate and larger than those of mammals.

## DIGESTIVE SYSTEM

The digestive system of birds shows many interesting modifications, some of which are associated with the absence of teeth in this group. Since lips are lacking, there are no labial glands in the mouth, nor is there an intermaxillary gland. However, sublingual glands are present. Amylase and ptyalin have both been found to occur in the saliva of fowl, although there is little indication that these enzymes play a part in the conversion of starch to sugar. In granivorous and carnivorous birds, a portion of the esophagus is enlarged into a sac-like pouch called the *crop* which is used for temporary storage of food. The crop is essentially lacking in digestive glands, although in pigeons and their relatives there are two glandlike structures present that are capable of producing a nourishing material called pigeon milk which is regurgitated by the parents and fed to the young. The action of these glands is stimulated by a hormone called *prolactin,* which is produced by the anterior lobe of the pituitary during the reproductive season.

The stomach of a bird consists of an anterior glandular portion called the *proventriculus* which secretes gastric juices and a thick-walled, muscular, posterior chamber known as the *gizzard.* The inner lining of the gizzard is horny and often corrugated. It is here that grit and small pebbles picked up by seed-eating birds play a part in grinding up the food. The small intestine is coiled or looped. Most

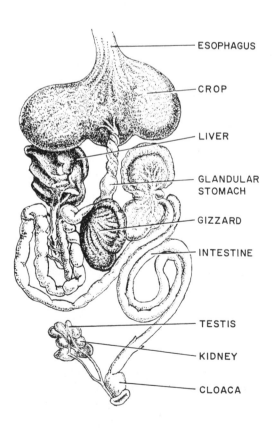

ESOPHAGUS

CROP

LIVER

GLANDULAR
STOMACH

GIZZARD

INTESTINE

TESTIS

KIDNEY

CLOACA

**Figure 5.4.** The digestive tract of a pigeon. (Welty, J. C.: The Life of Birds. W. B. Saunders Co., 1962; after Schimkewitsch and Stresemann.)

birds possess one or two colic caeca at the junction of the small and large intestines. The latter is short and straight and opens into the cloacal chamber.

## RESPIRATORY SYSTEM

The respiratory system of the bird is extremely efficient and, as a result, considerably more complicated than that of other air-breathing vertebrates (Fig. 5.6). As in the mammal the glottis is situated on the posterior floor of the pharynx and opens into the larynx or expanded upper part of the trachea. The larynx of the bird, however, is not a sound-producing organ, but serves to modulate tones that originate in the syrinx, which is at the lower end of the trachea where the latter bifurcates to form the right and left bronchi.

The expanded chamber of the syrinx is called the *tympanum* and is usually surrounded by both tracheal and bronchial rings. Extending up into the tympanum from the medial fusion of the bronchi is a bony ridge called the *pessulus* to which is attached a short vibratory membrane called the *semilunar membrane*. Additional membranes are present at the upper end of each bronchus where it joins the trachea. Sound results from air passing from the bronchi through the slits formed by these *tympaniform membranes* into the tympanum, where the vibratory semilunar membrane is situated. In songbirds all these structures are supplied with syringeal muscles, whose movements are responsible for the diversity of the sounds that are produced. There may be as many as nine pairs of syringeal muscles in some species. A few kinds of birds such as the ostrich and New World vultures lack a syrinx.

In some birds, such as swans and cranes, the trachea may be considerably longer than the neck with part of it extending back to the posterior end of the sternum. The lungs themselves are proportionately small and incapable of the great amount of expansion that is characteristic of mammalian lungs. The avian lungs, however, are connected with nine air sacs that are situated in various parts of the body. These are the unpaired interclavicular sac, the paired cervical, anterior thoracic, posterior thoracic, and abdominal sacs. The air sacs, which are not considered to have respiratory epithelium, serve essentially as reservoirs. Air passes through the bronchial circuit into the air sacs and then returns, generally by a separate set of bronchi, to the air capillaries in the lungs. Most investigators agree that on inspiration essentially pure air passes into the posterior air sacs. On the other hand there is evidence that some of the air that enters the anterior sacs has already passed through the lungs. On expiration air is forced from the sacs through the lungs. It has been suggested that

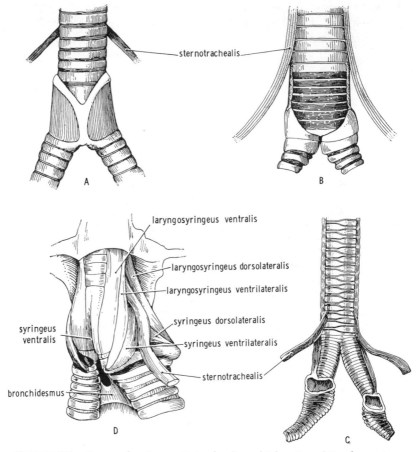

**FIGURE 5.5.** Types of syringes: *A*, tracheobronchial syrinx of *Neodrepanis coruscans* (after Amadon, 1951); *B*, tracheal syrinx of *Conopophaga aurita* (after Müller, 1878); *C*, bronchial syrinx of *Steatornis caripensis* (after Garrod, 1873c); *D*, tracheobronchial syrinx of the sunbird *Arachnothera longirostris* (after Köditz, 1925). (Van Tyne, J., and Berger, A. J.: Fundamentals of Ornithology. John Wiley & Sons, Inc., 1959.)

the anterior and posterior air sacs may alternate with one another in action. Despite the fact that there is still some confusion as to the exact mechanics of respiration in birds, there is no question but that there is essentially a steady stream of air passing through the air capillaries which insures an efficient exchange of gases.

Since birds lack a muscular diaphragm, respiration is effected through movement of the ribs and sternum. When in flight, breathing appears to be synchronized with the action of the wings.

Many birds possess air spaces in certain bones that are connected with the air sacs. The principal pneumatic bones are the humerus, the sternum, and the vertebrae, although in some species

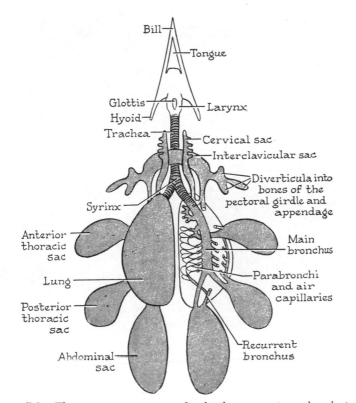

**FIGURE 5.6.** The respiratory organs of a bird as seen in a dorsal view. The course of the main bronchus through the lung to the air sacs, the major branches of the main bronchus, and the recurrent bronchi are shown on the right side. Minute parabronchi and air capillaries, a few of which are shown, interconnect the recurrent bronchi and the branches of the main bronchus. (Villee, C. A., Walker, W., and Smith, F.: General Zoology. 3rd Ed., W. B. Saunders Co., 1968.)

other bones may also have air spaces. A bird with an occluded trachea and a broken humerus may respire through the opening in the latter. These pneumatic bones appear to be of more common occurrence in large flying birds, although their physiological function is not clearly known.

There has been considerable speculation regarding the various possible functions performed by the avian air sacs in addition to aiding in respiration. Some of the functions suggested are to decrease the specific gravity of the body, to reduce friction between moving parts in flight, to assist in reducing body temperature, particularly during periods of activity, to facilitate spermatogenesis by reducing the temperature of the testes, to increase buoyancy in waterbirds, and to serve as pneumatic cushions to absorb shock in birds that dive into the water from the air. Not all of these suggested functions, however, have been satisfactorily proved.

**Figure 5.7.** Many birds pant on hot days. This increases evaporation and thereby reduces body temperature. These Heermann's gulls (*Larus heermanni*) nest on a desert island in the Sea of Cortez where the daily temperature at that season is over 100° F.

## Urogenital System

The urogenital system of birds is in many respects more closely allied to that of the reptiles than to that of mammals, except for the monotremes. The kidneys, as in all the amniotes, are of the metanephric type and paired. They are proportionately large, however, and irregularly lobed, fitting into depressions in the synsacrum. Each kidney has a ureter which opens directly into the cloacal chamber. The urine, therefore, mixes with fecal material. The only bird known to possess a bladder is the ostrich.

Recent studies on the supraorbital glands of certain birds, especially marine species, have shown that, as in some reptiles, these glands are used for the rapid excretion of salt from the blood. This accounts for the ability of marine species to ingest salt water without special modification of the kidney. In coastal areas birds such as gulls may often be seen with fluid dripping from their nostrils. This is really a concentrated salt solution. Such functional glands are not entirely restricted to marine birds. They have also been found to be functional in some kinds of waterbirds in the Great Plains area of North America, where the alkalinity of water may be quite high in ponds and lakes. Under these circumstances such structures have considerable survival value for some species. In a desert bird such as the ostrich, salt glands provide a means of conserving body water. By removing salt from the excretory system, more reabsorption of water may take place in the cloaca.

The testes are paired and remain in the upper part of the abdominal cavity. In most birds the ductus deferens on each side opens independently into the cloaca. In some, however, such as the ducks and geese, a single penis-like structure, similar to that of turtles and crocodiles and derived from the anteroventral wall of the cloaca, is present.

In most birds the right ovary and oviduct, although present in embryonic development, become vestigial so that only the left genital system is functional. Along the course of the oviduct are several glands which secrete membranes over the eggs, including layers of albumen, shell membranes, and a calcareous shell.

## NERVOUS SYSTEM

The central nervous system of birds shows a considerable advance over that of reptiles. The olfactory lobes of the brain are extremely small, as one might expect in a group that has a notably poor sense of smell. The cerebrum is large and covers the diencephalon and optic lobes. Its size, however, results from enlargement of the corpus striatum rather than the cerebral cortex. The latter is smooth. The optic lobes are exceptionally large. This seems to be correlated with the remarkable powers of sight that birds possess. The cerebellum is larger than in reptiles and is deeply fissured, although it is not as large as in mammals. Ventral to the cerebellum, the avian brain shows the beginning of the development of a pons. As do other amniotes, birds possess 12 cranial nerves.

## SENSE ORGANS

The nasal passages of birds show an advance over those of reptiles in that three shelves or conchae are present. In most birds there are external openings or nostrils leading into these passages from the outside, but in certain members of the Pelecaniformes these are closed. The position of the nostrils is usually lateral and proximal but in some of the Procellariformes the nostrils, which are in the form of tubes in members of this order, are dorsal in position. The kiwi of New Zealand is unique in having the nostrils nearly terminal on the bill. Generally the nostrils are separated internally from one another by a system, but in a few groups, such as the New World vultures as well as some of the Gruiformes, a partition is lacking. The olfactory epithelium in most birds is relatively restricted and confined to the surface of the uppermost or superior concha. This is correlated with the small size of the olfactory lobes of the brain and accounts for the relatively poor sense of smell that most birds reportedly have. Rather recently it has been demonstrated that the turkey vulture (*Cathartes aura*) is capable of detecting carrion by means of the sense of smell and that the portion of the brain concerned with this sense is markedly larger than in the related black vulture (*Coragyps atratus*). The kiwi is another exception; it has been rather clearly demonstrated that members of this species make use of their terminal nostrils in locating food while probing in the ground.

Taste buds are lacking on the tongues of most birds, although they are found on the lining of the mouth and pharynx. Jacobson's organ is rudimentary.

The eyes of birds are highly developed and quite large in proportion to body size. Accommodation is accomplished by action of the ciliary muscles which change the shape of the lens. An ossified sclerotic ring is present and surrounds the ciliary region. One of the unusual features of the bird's eye is the presence of a fan-shaped structure called the *pecten* which extends into the posterior chamber from the point at which the optic nerve emerges from the retina (Fig. 5.8). It has been suggested that the pecten may serve to provide nourishment to avascular parts of the eye, that its ridges may be an aid in perception by acting somewhat like a ruled grid, and that it may be an orienting mechanism which enables birds to direct their course in relation to the position of the sun or to star patterns. As in other groups of vertebrates the retina of diurnal birds predominates

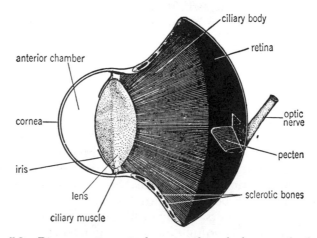

**FIGURE 5.8.** Diagrammatic sagittal section through the eye of a bird of prey. (Weichert, C. K.: Anatomy of the Chordates. McGraw-Hill Book Co., Inc., 1958.)

**FIGURE 5.9.** The oilbird or guacharo (*Steatornis caripensis*) of South America. (Photograph by Edward S. Ross.)

in cones, whereas rods are much more numerous in nocturnal species.

It appears likely that certain birds that frequent the deep recesses of caves where almost total darkness prevails emit series of sound pulses whose echoes enable them to orient themselves, much like some kinds of bats and marine mammals. The oilbird (*Steatornis caripensis*) of northern South America and the island of Trinidad, a species that is nocturnal and lives in caves, has been shown to rely on echo-location as a means of orientation while flying about at night. Likewise, certain species of cave-inhabiting swiftlets of the genus *Collocalia* are dependent upon acoustical orientation when flying about in the dark.

The ear in most birds lacks an external pinna. In the barn owl (*Tyto alba*), however, this structure is well developed. As in reptiles there is but a single bone, the *columella,* in the middle ear, which transmits vibrations from the tympanic membrane to the inner ear. A cochlea is present although it does not assume the complicated spiral form that is found in the higher mammals.

## SPECIAL CHARACTERS

### FEATHER STRUCTURE

Feathers are specialized structures that are confined entirely to members of the class Aves. Phylogenetically they are thought to have arisen from the same epidermal structures that gave rise to reptilian scales. Embryologically a feather starts as a dermal papilla with an overlying layer of epidermis. It is the latter, however, that gives rise to the final cornified structure, which of course, when fully formed, is nonliving like mammalian hair.

The main axis of a typical feather (Fig. 5.10) is called the *shaft*. The proximal portion of the shaft, called the *calamus*, is hollow and lacks any webbing. The remainder of the shaft, referred to as the *rachis*, is filled with pith and possesses a web or vane on either side. If the web is examined under a microscope it will be seen to be composed of *barbs*, which extend out laterally from the rachis, and *barbules*, which in turn extend out laterally from the barbs. Thus adjacent rows of barbules overlap one another. The tip and underside of each barbule possess small filaments called *barbicels* or *hamuli* which further assist in holding overlapping barbules together.

In some birds, a second feather that is complete in every respect grows out from the back of the shaft at the junction of the rachis and calamus. This accessory feather is called an *aftershaft*. In the emus it is about as long as the main shaft. It is well developed in gallinaceous birds, but in many other groups it is lacking or else represented only by a few tufts of down.

There are many modifications in the structure of certain types of feathers (Fig. 5.11). On plucking the contour or outer feathers of a bird, one will find small structures that superficially resemble hairs scattered over the body. These are called *filoplumes* and, if examined carefully, will be seen to consist of a slender shaft with a few barbs present at the tip. Anyone who has plucked ducks will be familiar with *down*. These are feathers in which the rachis is lacking and the slender barbs emerge from the end of the calamus. Although tiny barbules are present they lack barbicels to hold them together so as to form a web. The down feathers forming the natal plumage of young birds are called *neossoptiles* in contrast to *teleoptiles* or adult down.

A specialized type of feather found on the breast of heron-like birds is the *powder-down*. Structurally a powder-down is somewhat similar to a down feather but its barbs keep disintegrating into fine talc-like powder. The exact function or functions of powder-down are not known. It has been suggested that the powder may be used in preening to lubricate the feathers, that it may aid in insulating the body, and that it assists in warming the eggs during incubation. *Semiplumes* are feathers that do not have the barbs held together. Like filoplumes and down, the semiplumes are hidden beneath the contour feathers in adult birds. *Bristles* which represent only the feather shaft and which extend beyond the contour feathers are

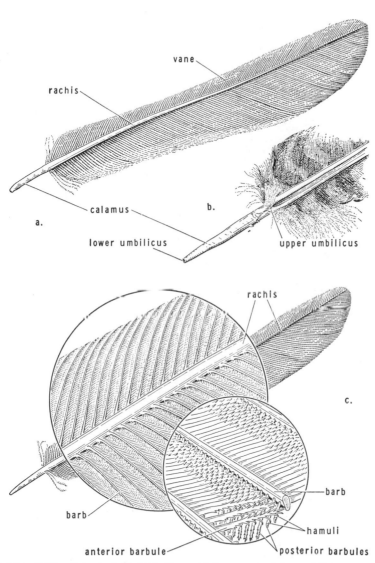

**FIGURE 5.10.** A typical flight feather and the nomenclature of its parts: *a*, general view; *b*, detail of the base of the feather; *c*, detail of the vane. (Van Tyne, J., and Berger, A. J.: Fundamentals of Ornithology. John Wiley & Sons, Inc., 1959.)

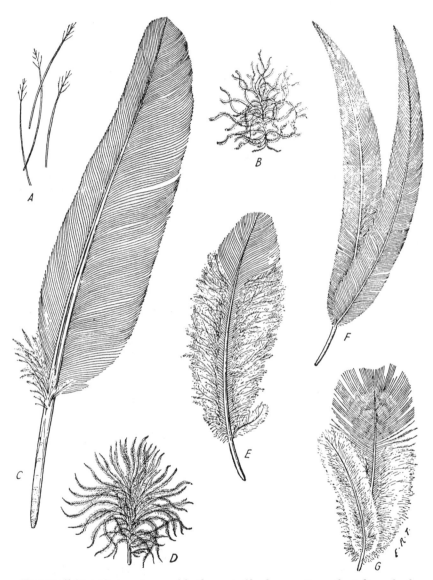

**FIGURE 5.11.** Various types of feathers: *A*, filoplumes; *B*, nestling down feather; *C*, primary wing feather of pigeon; *D*, permanent down feather; *E*, feather with free barbs; *F*, emu's feather with long aftershaft; *G*, contour feather of pheasant with aftershaft. (Partly after Thompson: The Biology of Birds, Sidgwick & Jackson, Ltd. In Young, J. Z.: The Life of Vertebrates. Oxford, Clarendon Press, 1950.)

FIGURE 5.12. A week-old peregrine falcon (*Falco peregrinus*) covered with down. (Photographed near Pt. Barrow, Alaska, by John Koranda.)

found in a number of groups of birds. Best known perhaps are the *rictal* bristles on either side of the head in caprimulgids and flycatchers and the bristles that cover the nostrils of some woodpeckers. Occasionally some of the latter feathers have a few remnants of the barbs near the base of the shaft rather than at the tip, as in a filoplume.

## COLORATION

Feather color may be produced by pigment granules, by the diffraction and reflection of light as a result of the feather structure, or by both pigment and structure.

As in other vertebrates the principal pigments are carotenoids and melanins. The former, frequently called lipochromes, are not soluble in water but may be dissolved by various fat solvents such as methanol, ether, or carbon disulfide. For convenience they are divided into two groups known as *zooerythrin*, or animal red, and *zooxanthin* or animal yellow. The melanins are soluble only in acids. *Eumelanin* granules vary from black to dark brown, and *phaeomelanin* granules may be almost colorless to reddish brown.

Pigment granules may occur in both the shaft and the vane and many of the feather colors that we are familiar with are the product of both carotenoids and melanins. The presence of spherical melanin granules near the tips of the contour feathers may produce the effect known as Newton's rings and result in iridescence. Some colors such as blue, most greens and violets are not the result of pigment but depend entirely on the feather structure. In bluebirds, for example, the blue feathers do not contain any blue pigment, but they absorb all but the blue rays of the spectrum. The latter are reflected. Greens are usually produced in the same manner except that the blue rays pass through a layer of yellow pigment. There are a few tropical birds that possess unusual pigments, such as the plantain-eaters of Africa. In this group a copper pigment called *turacoverdin* produces green while deep red is the result of *turacin.* In one species of plantain-eater, *Tauraco corythaix,* the yolk of the egg is vermilion in color instead of yellow as in most birds. Chemical analysis has shown that the pigments producing this unusually colored yolk are all carotenoids of which about three-fifths consists of a red pigment known as *astaxanthin.*

Although the color of various kinds of birds is under genetic control it may be modified by internal and external factors. As most aviculturists know, many species of birds that have red in their plumage tend to have the red replaced by yellow after several years in captivity. Even in the wild it is not unusual to see house finches (*Carpodacus mexicanus*) with yellowish or orange feathers on the head instead of red. In captive birds this change has been attributed by many to diet. Hormones also play a very important part in the control of feather color. In species where there is sexual dimorphism in color, the administration of estrogenic hormones in sufficient quantity to males prior to the onset of the molt may result in the assumption of female plumage when feather replacement occurs. Females may be similarly induced to assume male plumage by the administration of testosterone.

Oxidation and abrasion are external factors that effect changes in color to a lesser or greater degree in most species of birds. Carotenoids, especially, are subject to fading in sunlight, and feathers that are worn for a year may be quite different in color from those of the new plumage. In certain kinds of birds the bright nuptial plumage is a result of the wearing off of the dull tips of the feathers acquired the previous autumn.

## FEATHER ARRANGEMENT

Although a bird may superficially appear to have feathers evenly distributed all over the body, a cursory examination will prove that this is not generally the case. On parting the feathers or plucking them it will be seen that they are arranged in very definite tracts which are called *pterylae.* Between the pterylae there are extensive bare patches which are referred to as *apteria.* There are a very few exceptions, such as the penguins and kiwis, in which the feathers are found over most parts of the body.

In the study of *pterylosis,* or feathers and their arrangement, names have been applied to the various feather tracts (Fig. 5.13).

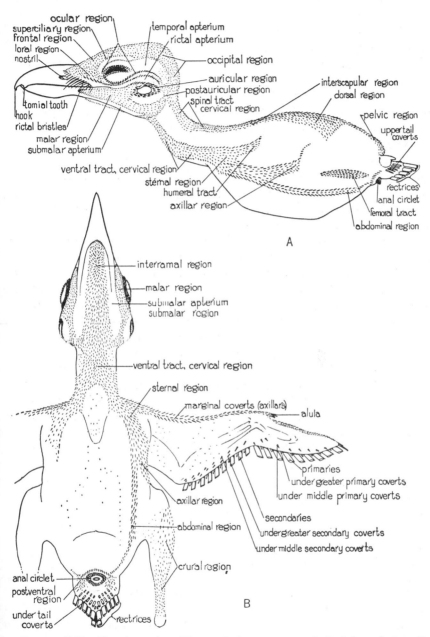

**FIGURE 5.13.** Pterylography of *Lanius ludovicianus gambeli: A*, lateral view; *B*, ventral view, natural size. (Miller, A. H.: University of California Publications in Zoology, Vol. 38, 1933.)

These tracts vary in shape and extent in different species, hence are of considerable importance in classification. The major tracts are listed and described as follows:

The *capital tract* covers the top, sides, and back of the head and is continuous with the next succeeding pteryla.

The *spinal tract* extends from the top of the neck, posteriorly along the back, to the base of the tail. It may be continuous or may be separated in the middle. Sometimes the spinal tract is split into two parts in the middle of the back enclosing an elliptical apterium, and occasionally it may fork into two branches anterior to the base of the tail.

The *ventral tract* begins between the rami of the lower mandible and extends down the ventral side of the neck, where it generally branches into two lateral tracts that pass along either side of the body and terminate in the vicinity of the vent. A portion of the apterium on the lower breast and abdomen becomes quite vascularized in some birds during the nesting period and constitutes what is called the *brood patch.* As the brood patch develops the skin becomes thicker and the feathers in the area are lost. This is believed to be an aid in incubation, since the skin that is in contact with the eggs receives more blood than the skin on other parts of the body.

The *humeral tracts* are a pair of pterylae that parallel one another as narrow bands extending posteriorly on either shoulder. The feathers on these tracts are called the *scapulars.*

The *caudal tracts* include the following: the *rectrices* which are generally the long, strong, flight feathers of the tail; the *upper tail coverts;* the *under tail coverts* which are often referred to as the crissum; the *anal circlet;* and, when present, the feathers of the *uropygial* or oil gland situated at the dorsal base of the tail.

The *alar tracts* include the various pterylae that are situated on the wing. The *remiges* are the strong flight feathers that grow from the posterior margin of the wing. They are divided into three groups. Those situated between the wrist and the tip are called *primaries.* Those between the wrist and the elbow are the *secondaries.* The innermost remiges, which appear to be a continuation of the secondaries in the elbow region, are referred to as *tertiaries.* On the so-called "thumb" of a bird's wing, which actually is believed to represent the remnant of the second finger, there are three remige-like feathers called the *alula* or bastard wing. Covering the upper and lower surfaces of the wing are series of feathers called coverts. The following groups comprise the upper wing coverts: the *greater primary coverts* overlap the bases of the primaries; the *median primary coverts* overlap the greater primary coverts; the *greater secondary coverts* overlap the bases of the secondaries; the *median secondary coverts* overlap the greater secondary coverts; anterior to the median secondary coverts are the *lesser coverts* or *cubitals* which usually consist of several rows of feathers; the *alula coverts* overlap the bases of the quills of the alula; the remaining small feathers of the upper surface which cover the anterior margin of the wing are called the *marginal coverts*. The coverts on the underside of the wing apart from the greater primary and secondary coverts are less distinct than those on the upper surface but have received names comparable to those of the upper surface. In addition to the remiges, the alula, and the coverts, there is a group of feathers in the axilla of the wing known as the *axillaries*.

The *femoral tract* extends along the outer surface of the thigh from near the knee joint to the body.

The *crural tract* constitutes the remaining feather tract on the leg. The shank is rather completely feathered.

## MOLT

Since feathers, once they are fully formed, are nonliving integumental structures they are constantly subject to deterioration as a result of oxidation and abrasion. This necessitates the periodic loss of the old feathers and their replacement by new ones. Such loss and replacement is referred to as molting. The process of molt is not haphazard but follows a very definite sequence within any one species. Certain feather tracts molt before others and even within single pterylae there is usually a definite sequence in which the feathers are shed and replaced. Furthermore, molting occurs at a very definite time or times during the year and is usually accomplished within a period of a few weeks.

As a rule all adult birds have a complete molt once a year. Exceptional instances, however, are known in which female hummingbirds that have had a prolonged nesting season have maintained the same plumage for two years. This, of course, is unusual. The annual molt, known as the postnuptial molt, normally occurs after the breeding season, which means that for North American species it takes place some time between June and October, depending on the habits of the birds involved. Some species have, in addition to the postnuptial molt, a prenuptial molt which precedes the breeding season. While this is generally only a partial molt involving certain feathers or feather tracts, there are a few species in which this is complete. The fact that the feathers of most birds last a year, often enabling their owners to travel thousands of miles as well as providing protection against sun, wind, rain, and other environmental factors, is a tribute to the durability of these seemingly delicate structures.

The physiological factors responsible for molting are far from known. It does appear that the thyroid gland is more active at such times and molting can be induced by large dosages of thyroxin. Nevertheless, the fact that thyroidectomized birds will still molt indicates that more than one factor is involved.

Although the various plumages and molts of immature birds might more appropriately be included in Chapter 13 (Growth and Development), it seems more convenient to include them here. The number of plumages that may be worn before the adult plumage is acquired varies in different species. Some birds of the year are indistinguishable from adults their first winter, while others may require up to five years to accomplish this.

*Natal Plumage.* There are a few species of birds that are entirely naked on hatching. Most kinds, however, have varying amounts of down on the body. This may consist of only a few tufts in the case of altricial species or a very thorough covering on the young of precocial birds (Fig. 5.14). The natal plumage is lost as a result of the *postnatal molt* and replaced by the following:

*Juvenal Plumage.* This is the plumage that is characteristic of most fledglings. The feathers are more substantial than their predecessors, although downy remnants of the natal plumage often adhere to the tips of the juvenal feathers for a while. In most passerine birds this plumage is only worn for a few weeks and is either completely

FIGURE 5.14. The protective color pattern of the natal plumage of a young killdeer (*Charadrius vociferus*) blends with the gravel background.

FIGURE 5.15. A young per-egrine falcon (*Falco peregrinus*) in the process of acquiring its ju-venal plumage. (Photograph by John Koranda.)

or partially lost by the *postjuvenal molt* and replaced by the following:

*First Winter Plumage.* This is acquired in the late summer or autumn and is retained either until the succeeding spring or for the next 12 months, depending upon the species. In some birds the first winter plumage is indistinguishable from the adult winter plumage. Generally, however, it is readily recognizable. In most species it is replaced, at least in part, as a result of the *first prenup-tial molt* by the following:

*First Nuptial Plumage.* This is the first breeding plumage, which may or may not resemble that of the adult. In some species this plumage is merely the result of wear on the part of the first winter plumage. More often it is a result of a partial replacement of the body plumage, although in a few kinds of birds the nuptial plumage is acquired by a complete molt involving all the feathers. In all birds this plumage is normally lost as a result of the *first postnup-tial molt* and is replaced by the following:

*Second Winter Plumage.* Except for those species that assume adult plumage the first year or those that require more than two years to attain this condition, this plumage is indistinguishable from the adult winter plumage. It is replaced the following spring by the *second nuptial plumage* as a result of the *second prenuptial molt.*

Apart from those species whose plumages are distinguishable from those of the adults after the second year, the succeeding molts and plumages are referred to as the *postnuptial molt,* the *winter plumage,* the *prenuptial molt,* and the *nuptial plumage.*

In quite a few North American birds the difference between the winter and nuptial plumage is quite marked. It is especially noticeable among some of the shorebirds, such as the dowitchers and knots, whose drab gray and white winter garb is replaced by brilliant rufous colors, or in the ptarmigans, whose white winter plumage is in marked contrast to the bright nuptial plumage. Although males and females of a number of kinds of birds are identical in coloration, in the majority of species the males are more brightly colored, especially in the nuptial plumage. An exception to this, however, is found among the phalaropes, in which most of the nesting duties are carried on by the male. Female phalaropes are more brilliantly colored than the males in spring and summer.

A rather unusual situation, as far as time of molt is concerned, occurs in the males of certain ducks. After the nesting season the postnuptial molt results in the males assuming a drab plumage, referred to as the *eclipse plumage*. While this molt is in progress the flight feathers of the wings are shed so rapidly that for a while the birds may be incapable of flying. The dull, female-like, eclipse plumage at the time of danger, therefore, makes the males rather inconspicuous. As soon as the flight feathers have grown in, however, the males then undergo a second molt of the body feathers which corresponds to the prenuptial molt of other species. Thus by autumn or early winter the males are in full breeding plumage. This appears to be correlated with early pairing that occurs in many kinds of ducks.

FIGURE 5.16. A male red phalarope (*Phalaropus fulicarius*) on its nest at Pt. Barrow, Alaska. (Photograph by John Koranda.)

## FUNCTIONS OF FEATHERS

Since birds are endotherms, one of the principal functions of feathers is the conservation of body heat. The layer of still air retained within these structures provides insulation against heat loss from within and the penetration of cold from without. The depth of this layer can be controlled by raising or lowering the feathers. On a cold day a resting bird will have its feathers fluffed out to increase the layer of insulation just as you or I would put on heavier clothing to accomplish the same end. When it is warm the feathers will be pressed tight against the body to reduce the depth of the insulating layer.

Feathers afford protection in a number of different ways. A small number of flightless birds as well as species with limited powers of flight depend primarily upon their ability to run in order to escape from danger. The great majority, however, rely upon flight under such circumstances. This of course is possible because of the development of powerful wing feathers. Feathers also afford direct protection against injury. The skin of a bird is relatively thin in contrast to that of many other kinds of vertebrates and would soon be injured by contact with twigs and branches if it were not covered with a layer of feathers. As those who have hunted waterfowl know, ducks and geese are almost impervious to small shot because of the dense covering of feathers on their bodies.

Feather coloration plays a very important part in the protection of many species. This is evident even in the downy young stage of many precocial birds. A visit to a nesting colony of gulls where there are newly hatched young or an attempt to locate young quail that have "frozen" in the grass will readily impress one with the protective value provided by the natal plumage. The concealing value of the color pattern of many caprimulgids, such as nighthawks and poor-wills, that sleep on the ground in the open during the day can hardly be disputed. Even when the exact location of a resting individual is known, the intricate pattern of gray, brown, black, and white so blends with the background that it often is difficult to distinguish the outline of a bird until one has approached to within a few feet of it. The white winter plumage of the ptarmigan seems clearly as adaptive as the winter coat of the varying hare.

Many birds appear to be rather conspicuously colored when examined as study skins in a museum, yet in real life their bold colors make them difficult to see. The broad black bands across the front of the killdeer for example tend to disrupt the outline of the bird when viewed from a distance. Similarly the various black and white patterns so characteristic of the backs of many kinds of North American woodpeckers make the birds more difficult to see against a background of bark than would a solid color. Such coloration is referred to as *disruptive.*

The principle of *countershading,* demonstrated by the late Abbott Thayer, refers to the fact that most birds, as well as many other kinds of animals, are lighter in color on the undersurface of the body than on the upperparts. The theory is that a light ventral surface counteracts the effect of shadow so that the outline of the bird does not stand out in relief against the background.

One may generalize as far as color is concerned with birds of various habitats. Most grass-inhabiting kinds tend to be streaked, those that live in undergrowth where it is well shaded are apt to be brown, while those that forage among leaves and branches in the sunlight, like many warblers, are frequently various shades of green or yellow. Less apparent to the average observer, but of considerable significance to the student of systematics, are the racial trends within many species as far as color is concerned. Populations in arid regions are often considerably paler than populations of the same species in areas where the humidity and rainfall are greater and the vegetation more luxuriant.

It must not be inferred, however, that avian coloration is always protective. Obviously the brilliant plumage of certain birds makes them very conspicuous. In the majority of birds the bright colors are characteristic of the males. Sexual dimorphism of this sort is believed to aid in sex recognition within the species and is, therefore, referred to as *epigamic* coloration. Sometimes males perform elaborate dis-

FIGURE 5.17. When at rest birds spend much of their time preening or caring for their feathers, as this white ibis (*Eudocimus albus*), a tropical American species, is doing.

plays during the period of courtship. There are, of course, many birds in which the color of the sexes is identical, and sex recognition appears to depend upon differences in the behavior patterns of males and females.

Another important function performed by feathers in waterbirds is to aid in increasing buoyancy (Fig. 5.18). The undersurface of the body of most swimming birds is densely covered with feathers between which there are pockets of air. These birds, therefore, rest on their own life rafts.

*Flight.* It is obvious that if it were not for feathers, birds would be unable to fly. Many other factors, however, are essential to this mode of locomotion. The avian body is streamlined in shape and proportionally light due to the structure of the skeletal system and to the presence of numerous air chambers in various parts of the body. The pectoral musculature which provides the driving force for the wings is strongly developed and the respiratory system has attained a high degree of efficiency, functioning both for the rapid exchange of gases and as a cooling system.

The mechanics of flight is a complicated subject that relates to aerodynamics and, as such, has been the object of considerable study in recent years. The same principles of lift, drag, tip vortex, pressure distribution, and aspect ratio that are used in aviation apply to bird flight. The wing of a bird and of a plane are, in certain respects, comparable. Both are streamlined so as to reduce resistance to the air and both possess camber, with the dorsal surface convex so that the

FIGURE 5.18. Even the natal down of waterfowl, such as the young mallards (*Anas platyrhynchos*) shown above, serves to increase buoyancy in the water.

pressure from beneath exceeds the pressure from above. The inner half of the bird's wing, however, is primarily concerned with lift while the outer half, from the wrist to the tip, must serve as the propelling force, unlike the comparable part of a plane's wing. While the propeller of a plane swings in a complete circle, the distal part of a bird's wing is limited to swinging through a semicircle. The outer wing, nevertheless, is very versatile; not only is it capable of producing a forward drive but it may go in reverse, function as a helicopter blade for vertical lift, or, in gliding, perform the same function as the inner part of the wing.

Studies on the motion of the wing in flight, by means of high speed motion pictures, show that the general movement is down and forward in the downstroke and up and backward in the upstroke. Furthermore, in the upstroke the wing is partly folded so as to reduce resistance to the air. On alighting, birds make use of flaps just as a plane does on landing. In the former this is accomplished by sharply increasing the angle of the wing so that the back part is directed downward. This increases lift temporarily at a reduced speed and terminates at the stalling point when the feet have contacted the landing point.

Birds that soar or glide must make use of rising currents of air (Fig. 5.19). On land, wind rises when it is deflected by objects such as hills or mountains. Air also rises as a result of being heated close to the ground. This produces thermals, which are used so much by

**FIGURE 5.19.** Gulls frequently make use of updrafts in gliding and soaring. (Photograph by Cecil Tose.)

many of the large birds of prey. Oceanic gliders such as albatrosses make similar use of updrafts which are believed to be the result of wave motion on the surface of the ocean.

## BILLS

The bill or beak of a bird is a curious modification of the upper and lower jaws which serves many purposes, including, among others, the procuring of food, defense, nest building, and the preening of feathers. It, therefore, varies greatly in different groups of birds, depending upon their habits. Overlying the bony framework of the upper and lower mandible is a horny sheath that is called the *rhamphotheca*. Since the sheath of each mandible is derived embryologically from several separate plates which later fuse together, it has been suggested that these structures may have been derived from the plates that surround the mouths and nostrils of reptiles. These would include such scales as the rostral, the internasals, nasals, upper labials of the upper jaw, and the mental and lower labials of the lower jaw. In certain groups of birds, notably the kiwis, ostriches, tinamous, albatrosses, and petrels, sutures showing the lines of fusion remain in the adult bird. One structure that appears at the tip of the upper mandible prior to hatching and is lost following this stage is a small sharp protuberance called the *egg-tooth*. This egg-tooth assists the young bird in breaking through the shell.

Various names are applied to different parts of the bills of birds. The central ridge of the uppermost part of the upper mandible, extending from the base to the tip, is called the *culmen*. The cutting edges of the upper and lower mandibles, respectively, are called the *maxillary* and *mandibular tomia*. The tomia may be smooth as in sparrows, they may be "toothed" or notched to varying degrees as in vireos, shrikes, and falcons, or they may be serrate as in some hummingbirds, barbets, and mergansers. In ducks, geese, swans, and flamingos there are numerous thin plates or *lamellae* just inside the tomia which serve as strainers. Sometimes the basal portion of the upper mandible is soft and fleshy as in hawks and parrots and is called the *cere*. In all North American birds the nostrils are situated on the proximal part of the upper mandible. In some species each nostril is covered, at least in part, by a fleshy or horny fold known as the *operculum*. There may or may not be an internal septum separating the two nostrils. In the former case the nostrils are referred to as *imperforate* and in the latter as *perforate*. The median ridge formed by the fusion of the mandibular rami anteriorly in the lower jaw is called the *gonys*.

The shape of the bill (Fig. 5.20) reflects in large measure the

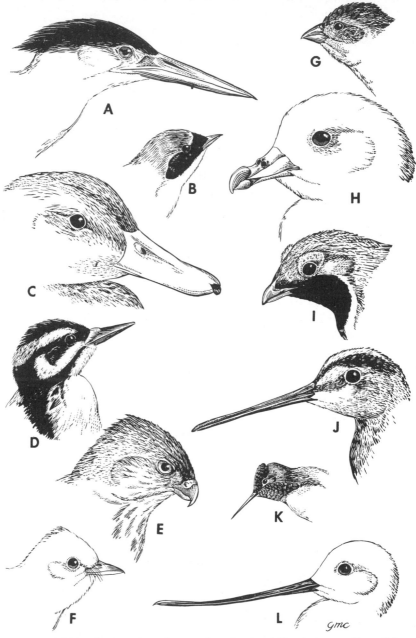

**FIGURE 5.20.** Some selected examples of avian bills: A, heron; B, warbler; C, duck; D, woodpecker; E, hawk; F, flycatcher; G, finch; H, fulmar; I, grouse; J, snipe; K, hummingbird; L, avocet.

habits of the species. In seed-eating species such as finches the bill is usually stout and conical in shape, and tapers abruptly. This facilitates not only the gathering but the shelling of seeds. In cross-bills, which are also finches, the tips of the mandibles cross each other so as to enable the birds to pry seeds from cones. The bills of birds that live upon flesh are hooked at the end to aid in tearing up their food into pieces sufficiently small to swallow. Birds such as herons that capture fish have long spear-shaped bills that can readily be thrust into water after an unsuspecting victim. Woodpeckers, likewise, have strong chisel-like bills that are capable of cutting wood and also of penetrating bark to secure insects. The lamellated bills of ducks are obviously useful in straining food from the water. Hummingbirds that secure nectar from deep in the corollas of flowers have long tubular bills. Similarly the bills of many shorebirds that probe deep in mud or sand to secure food are long and slender. In birds like warblers that pick insects off foliage the bill is slender and pointed like a small pair of forceps. On the other hand, birds such as swifts, most caprimulgids, and flycatchers that capture insects on the wing, have dorsoventrally compressed bills that are very broad at the base.

Special mention should be made of members of the Pelecani-formes to which the pelicans, cormorants, frigate-birds, and their relatives belong. Most members of this order have a pouch or gular sac under the chin. In the pelicans this sac is used to store fish temporarily and assists in the process of swallowing. Furthermore, in the feeding of the young, food is regurgitated into this sac. In other members of this group the gular sac appears to be more significant in the sexual display. During courtship the male frigate-bird inflates this pouch until it presents the appearance of a balloon.

## LEGS AND FEET

There are a few kinds of birds in which the tarsus (tarsometatar-sus) and feet are completely feathered as, for example, most of the owls and ptarmigans. There are a few others, for example the rough-legged hawks, that have the tarsus but not the feet feathered. The great majority of birds, however, have featherless tarsi and feet that are covered with horny scales (Fig. 5.21). Sometimes these scales are *imbricate*, that is, they overlap one another in an orderly manner on the anterior surface of the tarsus. This type of tarsus is referred to as *scutellate* and is characteristic of most sparrows and finches. In other groups, such as thrushes, the horny covering presents a smooth appearance without any evidence of separate scales. This is referred to as a *booted* tarsus. In geese and many shorebirds the covering on the tarsus is broken up into numerous, small, irregular scales that tend to be polygonal in shape. This is called a *reticulated* tarsus.

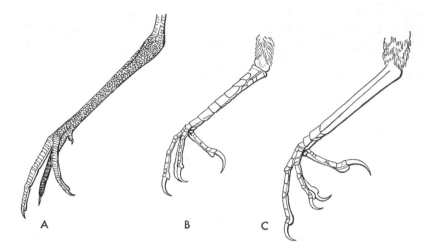

**FIGURE 5.21.** Examples of the three major types of horny sheaths on the avian tarsometatarsus: *A*, reticulate (plover); *B*, scutellate (flycatcher); *C*, booted (robin).

The feet of birds, like their bills, reflect the habits of the species (Fig. 5.22). In passerine or perching birds there are normally three toes in front with the hallux situated behind. In most of the woodpeckers the fourth toe is reversed so that there are two toes in front and two behind. This yoke-toed condition is referred to as *zygodactylous*. However, in the three-toed woodpeckers the hallux is completely missing. Some of the swifts have *palmprodactylous* feet in which all four toes are directed forward. Presumably this is an aid in clinging to vertical surfaces. In certain groups, such as the kingfishers, the outer and middle toes are partly joined together, a condition referred to as *syndactylous*. The *talonid* feet of raptorial birds have toes which are unusually strong and widely separable and the outer one is capable of being reversed in some species.

Birds that use their feet for swimming or wading usually have the toes joined together, at least to some extent, by webbing, or else the toes are lobed. Both methods serve to increase the surface of the foot. Some birds, such as the pelicans, have all four toes joined together by webs that extend out to the tips of the digits and are therefore *totipalmate*. Others, such as ducks and geese, have the three front toes fully webbed with the hallux separate, a condition referred to as *palmate*. The feet in a number of shorebirds and herons have the three front toes joined by webbing that only extends partway to the tips of the digits and are called *semipalmate*. The toes of grebes and coots possess broad, flat lobes which function like webs when these birds are swimming. The hind toe of diving ducks possesses a similar structure. In many members of the grouse family the sides of the toes are fringed or *pectinated*.

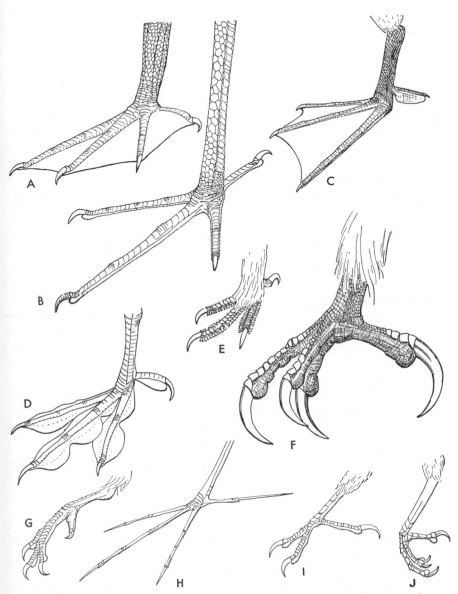

**FIGURE 5.22.** Some examples of avian feet: *A*, cormorant; *B*, heron; *C*, diving duck; *D*, coot; *E*, spruce grouse; *F*, eagle; *G*, kingfisher; *H*, jacana; *I*, three-toed woodpecker; *J*, finch.

Even the nails of birds show a limited amount of variation in different groups. In most birds they tend to be laterally compressed, curved, and pointed, a condition well illustrated by the nails of most perching birds. In swifts the curvature is more marked and probably is an aid in enabling these birds to cling to vertical surfaces. In hawks and owls, size is accentuated since the claws are used to capture and hold prey. Sometimes the curvature is so reduced that the nails are nearly straight, as in certain of the rail-like birds. The nails of grebes, instead of being compressed, are flattened, somewhat resembling the condition found in man. In certain seemingly unrelated groups of birds the inner edge of the nail on the middle toe possesses a series of teeth or serrations. This pectination or *comb claw* is found in herons, nighthawks and their relatives, barn owls, and cormorants. Although various suggestions have been offered to account for it, the true function of the comb claw is still not known.

## CLASSIFICATION OF BIRDS

### CLASS AVES (BIRDS)

Warm-blooded vertebrates with skull possessing only one condyle; a single bone in the middle ear; front limbs basically modified for flight; body covered with feathers; scales present on feet, which usually retain the reptilian formula of 2-3-4-5 phalanges; Jurassic to Recent; two subclasses and six extinct and 27 living orders.

SUBCLASS ARCHAEORNITHES (ANCESTRAL BIRDS). Extinct birds possessing teeth in both jaws and a long, feathered tail; front limbs with three clawed fingers; vertebrae separate to sacrum and amphicoelous; Jurassic; one order, Archaeopterygiformes, one family, Archaeopterygidae, and two genera, *Archaeopteryx* and *Archaeornis*.

SUBCLASS NEORNITHES (TRUE BIRDS). Extinct and living birds with well developed sternum which usually possesses a keel; tail not elongate; vertebrae amphicoelous or heterocoelus; Cretaceous to Recent; four extinct and 27 living orders and about 165 living families.

*Order Hesperornithiformes (Western Birds).* Extinct, flightless birds possessing teeth and lacking a keeled sternum; vertebrae heterocoelous; wings degenerate; Cretaceous; three families.

*Order Ichthyornithiformes (Fish Birds).* Extinct birds lacking teeth and possessing a keeled sternum; vertebrae amphicoelous; wings well developed; Cretaceous; two families.

*Order Sphenisciformes (Penguins).* Medium to large (400 to 1200 mm.)* flightless marine birds; wings modified as flippers; apteria lacking; body densely covered with feathers, those on wings

---

*Measurements given in this classification represent total length.

being scalelike; continental coasts of the southern part of the Southern Hemisphere, subantarctic islands and the Galápagos Islands;
one family.*

*Family Spheniscidae (Penguins).* Characters and distribution,
those of the order; 17 species.

**Order Struthioniformes (Ostrich).** Very large (up to 1800 mm.), flightless, ratite birds; body plumage soft; thighs bare; toes reduced to
two; Africa and southern Asia; one family.

*Family Struthionidae (Ostrich).* Characters and distribution,
those of the order; one living species, *Struthio camelus.*

**Order Rheiformes (Rheas).** Large (900 to 1300 mm.), flightless, ratite
birds; body plumage soft; thighs feathered; toes reduced to three;
South America; one family.

*Family Rheidae (Rheas).* Characters and distribution, those of
the order; two genera and two species, *Rhea americana* and
*Pterocnemia pennata.*

**Order Casuariiformes (Cassowaries and Emus).** Very large (1300 to
2000 mm.), flightless, ratite birds; plumage coarse and hairlike;
thighs feathered; toes reduced to three; Australia, Tasmania, and
New Guinea; two families.

*Family Casuariidae (Cassowaries).* Bill strong and somewhat laterally compressed; a large casque on head; wattles on neck;
the New Guinea and northern Australian region; one genus,
*Casuarius,* and three species.

*Family Dromiceidae (Emus).* Bill stout and somewhat dorsoventrally compressed; head lacking any casque or wattles; Australia and Tasmania; one genus and two species, *Dromiceius
n. hollandiae* and *D. diemenianus.*

**Order Aepyornithiformes (Elephant Birds).** Very large (height, up to
3 m.), extinct, flightless, ratite birds with a heavy body (weight,
estimated up to 450 kg.); wings reduced; legs very heavy; Africa
and Madagascar; Eocene to (possibly?) Recent; one family, 11
species.

**Order Dinornithiformes (Moas).** Gigantic (height, up to more than 4
m.), extinct, flightless, ratite birds; wings reduced; New Zealand;
Pleistocene to Recent; two families and 22 species.

**Order Apterygiformes (Kiwis).** Medium-sized (500 to 800 mm.),
flightless, ratite birds with coarse, hairlike feathers; wing and tail
feathers lacking; wings vestigial; bill long, slender, and slightly
curved; nostrils nearly terminal; four toes; New Zealand; one
family.

*Family Apterygidae (Kiwis).* Characters and distribution, those
of the order; one genus, *Apteryx,* and three species.

**Order Tinamiformes (Tinamous).** Moderately small to medium-sized
(200 to 530 mm.) terrestrial birds superficially resembling certain
gallinaceous species; palate dromaeognathous, as in ratites; ster-

*Only living families, genera, and species are listed unless otherwise indicated.

num carinate; powder-down feathers present; Mexico south to South America; one family.

*Family Tinamidae (Tinamous).* Characters and distribution, those of the order; about 45 species.

**Order Gaviiformes (Loons).** Medium-large (600 to 950 mm.) diving birds with first three toes fully webbed; bill fairly long, straight, and pointed; plumage dense; 11 functional primaries; an aftershaft present; legs far posterior and encased within body down to tarsus; tarsus reticulate and laterally compressed; holarctic; one family.

*Family Gaviidae (Loons).* Characters and distribution, those of the order; one genus, *Gavia,* and four species.

**Order Podicipediformes (Grebes).** Medium-sized (220 to 650 mm.) diving birds with lobed rather than webbed toes; bill moderately short, slender, and usually pointed; plumage soft, dense, and shiny; 11 functional primaries, the outer three or four emarginate; an aftershaft present; tarsus scutellate and laterally compressed; nearly worldwide; one family.

*Family Podicipedidae (Grebes).* Characters and distribution, those of the order; four genera and 18 species.

**Order Procellariiformes (Albatrosses, Shearwaters, and Petrels).** Very small to large (130 to 1350 mm.) marine birds possessing nostrils consisting of raised tubes; sutures evident in rhamphotheca; bill hooked; wings mostly long and narrow; 10 functional primaries; a small aftershaft present; front toes fully webbed; hind toe rudimentary or lacking; tarsus reticulate; oceans of the world; four families.

*Family Diomedeidae (Albatrosses).* Large (700 to 1350 mm.) procellariiforms with a wingspread up to 3.5 m.; tubular nostrils lateral; wings very long and narrow; southern oceans and the north Pacific Ocean; two genera and 13 species.

*Family Procellariidae (Shearwaters and Fulmars).* Small to medium-sized (280 to 900 mm.) procellariiforms; nasal tubes dorsal and relatively long; wings very long and narrow; oceans of the world; 13 genera and 53 species.

*Family Hydrobatidae (Storm Petrels).* Very small to small (130 to 250 mm.) procellariiforms with a single, dorsally situated nasal tube which is divided internally; wings long and narrow; oceans of the world; nine genera and 22 species.

*Family Pelecanoididae (Diving Petrels).* Small (165 to 250 mm.), auklike procellariiforms; tubular nostrils dorsal and opening upward; wings short and used in diving; southern oceans; one genus, *Pelecanoides,* and five species.

**Order Pelecaniformes (Pelicans and Allies).** Medium to large (400 to 1650 mm.) aquatic or marine birds; nostrils generally very small or absent; a gular pouch present; tarsus reticulate; feet totipalmate; cosmopolitan; six families.

*Family Phaëthontidae (Tropic Birds).* Medium-sized (400 to 500 mm., exclusive of elongate tail feathers) marine birds; bill strong, slightly curved, and pointed; nostrils large for the order; gular pouch very small; central tail feathers greatly elongated; tropical seas; one genus, *Phaëthon*, and three species.

*Family Pelecanidae (Pelicans).* Large (1250 to 1650 mm.) marine or freshwater birds; bill straight, long, and flattened, with a terminal hook; external nostrils absent; gular pouch very large; tail short; nearly worldwide; one genus, *Pelecanus*, and six species.

*Family Sulidae (Boobies and Gannets).* Medium-large (700 to 1000 mm.) marine birds; bill long, conical, and pointed; external nostrils absent; gular pouch small; tail of medium length; tropical and temperate seas except the north Pacific Ocean; two genera and nine species.

*Family Phalacrocoracidae (Cormorants).* Medium to medium-large (450 to 1000 mm.) aquatic or marine birds; bill of medium length, cylindrical, and hooked; nostrils vestigial; gular pouch small and usually colored; tail moderately long; nearly worldwide; three genera and 31 species.

*Family Anhingidae (Anhingas).* Medium-large (800 to 900 mm.) freshwater birds; bill long, slender, and sharply pointed; nostrils vestigial; gular pouch small; head and neck nearly of same diameter; tail relatively long; tropical and subtropical parts of the world; one genus, *Anhinga*, and about four species.

*Family Fregatidae (Frigate Birds).* Medium-large (750 to 1000 mm.) marine birds; bill long, cylindrical, and hooked; nostrils small; gular pouch capable of great distention in male; tail deeply forked; tropical and subtropical seas; one genus, *Fregata*, and five species.

**Order Ciconiiformes (Herons, Storks, and Allies).** Small to large (280 to 1450 mm.) wading birds; bill, neck, and legs relatively long; lower part of tibiotarsus bare; four toes with anterior three at least slightly webbed; worldwide; seven families.

*Family Ardeidae (Heronlike Birds).* Small to large (280 to 1450 mm.) birds with long, straight, pointed bill; lores naked; middle toenail pectinate; possessing powder-down; worldwide; 31 genera and 62 species.

*Family Cochleariidae (Boatbill Heron).* Medium-sized (500 to 525 mm.) heron with a bill that is about two-thirds as broad as it is long; eyes relatively large; throat bare with a distensible gular pouch; middle toenail pectinate; four pairs of powder-down patches present; Mexico to Brazil; one genus and one species, *Cochlearius cochlearius.*

*Family Balaenicipitidae (Shoebill Stork).* Large (1150 mm.) stork-like bird with massive, flattened bill; middle toenail pectinate; pair of powder-down patches on rump; north-central Africa; one genus and one species. *Balaeniceps rex.*

*Family Scopidae (Hammerhead).* Medium-sized (500 mm.),

storklike bird with a long, straight, laterally compressed bill; powder-down patches lacking; middle toenail pectinate; neck extended in flight; southwestern Asia, Africa, and Madagascar; one genus and one species, *Scopus umbretta.*

Family *Ciconiidae (Storks).* Medium to large (750 to 1525 mm.) birds with long legs and long neck; bill large and long; powder-down patches lacking; middle toenail not pectinate; toes relatively short; North and South America, Europe, Asia, Africa, and Australia; 11 genera and 17 species.

Family *Threskiornithidae (Ibises and Spoonbills).* Medium to large (480 to 1050 mm.) birds with long, decurved, or spatulate bill; grooves extending from nostrils to tip of bill; powder-down patches lacking; middle toenail slightly scalloped; subtropical and tropical parts of North and South America, Eurasia, Africa, and Australia; 20 genera and 28 species.

Family *Phoenicopteridae (Flamingos).* Large (900 to 1200 mm.) birds with pinkish plumage; bill thick, lamellate, and abruptly bent downward in the middle; legs very long; toes joined by webs; South America, the Caribbean, Eurasia, and Africa; three genera and six species.

Order *Anseriformes (Waterfowl).* Moderately small to very large (290 to 1525 mm.) aquatic or semiaquatic birds; nostrils perforate; 11 primaries with the first reduced in size; oil gland feathered; down present on both pterylae and apteria; toes at least partly webbed; cosmopolitan; two families.

Family *Anhimidae (Screamers).* Moderately large (700 to 900 mm.), long-legged birds with a crest or horny spike on head; bill gallinaceous in shape; wings possessing two sharp spurs; toes partly webbed; South America; two genera, *Anhima* and *Chauna,* and three species.

Family *Anatidae (Ducks, Geese, and Swans).* Moderately small to very large (290 to 1525 mm.) birds with a lamellate bill that is usually broad and flat; legs short or of medium length; toes webbed; cosmopolitan; five subfamilies and about 145 species.

Order *Falconiformes (Diurnal Birds of Prey).* Small to very large (150 to 1500 mm.) flesh-eating birds; bill hooked; base of bill possessing a fleshy cere through which the nostrils open; feet powerful, with an opposable hind toe; worldwide except Antarctica; five families.

Family *Cathartidae (American Vultures).* Moderately large to very large (635 to 1320 mm.) carrion eaters with a bare head colored black, yellow, or red; bill relatively weak compared with other members of the order; claws weakly hooked; syrinx lacking; North and South America; six species.

Family *Sagittariidae (Secretary Bird).* Very large (up to 1500 mm.) terrestrial bird of prey; a long crest of feathers on nape; legs very long and slender; toes short and partly webbed; Africa; one species, *Sagittarius serpentarius.*

*Family Accipitridae (Old World Vultures, Hawks, and Eagles).* Moderately small to large (280 to 1150 mm.) birds of prey; bill strongly hooked; wings generally broad and rounded; legs of medium length; toes possessing strongly hooked nails; worldwide except Antarctica; 205 species.

*Family Pandionidae (Osprey).* Moderately large (up to 600 mm.) hawk with rather pointed wings; outer toe reversible as in owls; short spine present on pads of feet; essentially worldwide but associated with coastal lines, rivers, and lakes; one species, *Pandion haliaetus.*

*Family Falconidae (Falcons and Caracaras).* Small to medium-sized (150 to 635 mm.) birds of prey with long, pointed wings; bill usually possessing a tooth or notch; legs of medium length in the falcons and proportionately long in the caracaras; toes possessing powerful, hooked claws; essentially worldwide except Antarctica; 58 species.

**Order Galliformes (Fowl-like Birds).** Small to very large (125 to 2335 mm.) terrestrial birds with short, downcurved bill; head often possessing colored, bare areas or wattle-like structures; wings short and rounded, mostly capable of rapid flight for a limited distance; legs and feet strong with toes curved; tail very short to extremely long; essentially worldwide, being absent from Antarctica and Polynesia; seven families.

*Family Megapodiidae (Mound Builders).* Medium-sized (250 to 650 mm.), ground-inhabiting, gallinaceous birds with a relatively small head; feet proportionately large; hind toe large and on same level as other toes; Philippines south to Australia and east to parts of Polynesia; 10 species.

*Family Cracidae (Curassows, Guans, and Chachalacas).* Medium to fairly large (500 to 1000 mm.) arboreal, gallinaceous birds; tail long and flat; hind toe large and on same level as other toes; southwestern United States to Argentina; 38 species.

*Family Tetraonidae (Grouse).* Medium-sized to moderately large (300 to 900 mm.) gallinaceous birds possessing a tarsus that is at least partly feathered; inflatable sacs or erectile feathers usually present on neck; hind toe smaller than front toes and elevated; temperate and north temperate parts of the Northern Hemisphere; 18 species.

*Family Phasianidae (Pheasants, Quails, and Partridges).* Small to very large (125 to 2335 mm.) gallinaceous birds usually possessing a tarsus that lacks feathers; inflatable sacs absent from neck; tail very short to very long; hind toe slightly elevated and smaller than front toes; spurs often present on back of tarsus of male; nearly worldwide in tropical and temperate parts of the world except the Pacific islands; 165 species.

*Family Numididae (Guinea Fowls).* Medium-sized (430 to 750 mm.), ground-dwelling, gallinaceous birds with head and neck largely bare; tail very short; hind toe short and somewhat elevated; Africa and Madagascar; seven species.

*Family Meleagrididae (Turkeys).* Moderately large (900 to 1200 mm.) terrestrial, gallinaceous birds with bare head and neck possessing red and blue carunculated skin; neck proportionately long; feet large with hind toe slightly elevated; North and Central America; two species.

*Family Opisthocomidae (Hoatzin).* Medium-sized (600 mm.) gallinaceous bird possessing a long, loose crest of feathers; head proportionately small; legs of medium length with long, strong toes; young possessing two functional claws at the tip of the wing for several weeks after hatching; northern South America; one species, *Opisthocomus hoatzin.*

**Order Gruiformes (Cranes, Rails, and Their Allies).** Very small to very large (112 to 1525 mm.), mainly aquatic birds; neck relatively long; wings rounded; tail short; legs ranging from medium length to long; worldwide except Antarctica; 12 families.

*Family Mesitornithidae (Mesites).* Moderately small (250 to 275 mm.) terrestrial birds with short, rounded wings; clavicles greatly reduced in size; five patches of powder-down present; tail moderately long and rounded; Madagascar; three species.

*Family Turnicidae (Bustardquails).* Very small (155 to 200 mm.) terrestrial, quail-like birds lacking a hind toe; wings short; tail very short; southern parts of the Eastern Hemisphere from southern Europe south through Africa and Madagascar and east through Asia and Australia to the Solomon Islands; 15 species.

*Family Pedionomidae (Plains Wanderer).* Small (up to 170 mm.), quail-like bird resembling one of the Turnicidae, but possessing a well developed hind toe; coloration cryptic; wings short; tail very short; central deserts of Australia; one species, *Pedionomus torquatus.*

*Family Gruidae (Cranes).* Large to very large (785 to 1525 mm.), long-necked, long-legged birds; head possessing bare patches or ornamental plumes; wings large; tail short; hind toe small and elevated; North America, Europe, Asia, Australia, and Africa; 14 species.

*Family Aramidae (Limpkin).* Medium-large (up to 635 mm.) marsh bird possessing a long, laterally compressed bill; wings relatively small and rounded; tibia partly bare; legs and toes long; hind toes long and not elevated; southern United States to South America; one species, *Aramus guarauna.*

*Family Psophiidae (Trumpeters).* Medium-sized (430 to 535 mm.) birds with long legs and neck but fowl-like bill; feathers velvet-like on head and neck; tail short but tail coverts long; feet of moderate size; northeastern South America; three species.

*Family Rallidae (Rail-like Birds).* Small to medium-sized (140 to 500 mm.) birds usually associated with marshes; bill strong, long to short; wings short and rounded; toes long; worldwide; 132 species.

*Family Heliornithidae (Sun Grebes or Finfoots).* Medium-sized (300 to 625 mm.) marsh birds with elongate body; neck and bill moderately long; tail long; legs short; toes lobed as in grebes; southern Mexico to South America, tropical Africa, and southeastern Asia; three species.

*Family Rhynochetidae (Kagu).* Medium-sized, terrestrial, forest-dwelling bird with a large, crested head; plumage grayish; bill and feet orange-red; legs long; New Caledonia; one species, *Rhinochetos jubatus.*

*Family Eurypygidae (Sunbittern).* Medium-sized (up to 450 mm.) wading bird with soft, short feathers on neck and bold color pattern on wings; bill elongate and straight; tail long; legs and toes long; hind toe slightly elevated; southern Mexico to northern South America; one species, *Eurypyga helias.*

*Family Cariamidae (Seriamas).* Fairly large (750 to 900 mm.) terrestrial birds with very long legs and an erectile crest on the head; bill short, broad, and decurved; neck moderately long; wings short; tail long; toes short; central South America; two species, *Cariama cristata* and *Chunga burmeisteri.*

*Family Otididae (Bustards).* Medium to large (370 to 1320 mm.), heavy-bodied, cursorial birds possessing broad wings and lacking a hind toe; bill short, strong, and flattened, neck long; tail short; legs long; toes short and thick; southern Europe and Asia, Australia, and Africa; 23 species.

**Order Charadriiformes (Shorebirds, Gulls, and Auks).** Small to fairly large (125 to 750 mm.) birds primarily of seacoast, lake, and marsh; bill short to long, straight, decurved, or recurved; little sexual dimorphism; wings long; spinal tract forked between shoulders; feathers possessing an aftershaft; oil gland tufted; worldwide; 16 families.

*Family Jacanidae (Jacanas).* Small to medium-sized (165 to 535 mm.) marsh birds with toes and claws greatly elongated; bill straight and of medium length; a frontal shield present in most species; sharp spur on bend of wings; tail short; legs long; tropical parts of America, Africa, Asia, and northern Australia; seven species.

*Family Rostratulidae (Painted Snipes).* Small (190 to 240 mm.) marsh birds with large eyes, a long bill, and cryptic coloration; neck short; wings broad; tail short; southern South America, Africa, southern Asia, and Australia; two species, *Rostratula benghalensis* and *Nycticryphes semi-collaris.*

*Family Haematopodidae (Oystercatchers).* Medium-sized (380 to 500 mm.), stocky-bodied shorebirds; bill red, long, laterally compressed, and truncate terminally; coloration, black or a combination of black or brown and white; legs and feet stout and pink; hind toe missing; temperate and tropical seacoasts of the world as well as some inland waters in Europe and Asia; six species.

*Family Charadriidae (Plovers).* Small to medium-sized (150 to 400 mm.), stocky shorebirds with relatively short bill that is slightly enlarged near the tip; head and eyes proportionately large; neck short; hind toe rudimentary or lacking; worldwide; 63 species.

*Family Scolopacidae (Sandpipers and Their Allies).* Small to medium-sized (125 to 600 mm.) wading birds with thin long bill; bill straight, decurved, or slightly upcurved; neck, medium length to long; toes moderately long; worldwide; 82 species.

*Family Recurvirostridae (Stilts and Avocets).* Medium-sized (300 to 480 mm.) shorebirds with extremely long legs; head proportionately small; bill very long, slender, straight, and upcurved or decurved; neck long; hind toe vestigial or lacking; North and South America, central and southern Eurasia, Africa, Madagascar, Australia, New Zealand, and the Pacific islands; seven species.

*Family Phalaropodidae (Phalaropes).* Moderately small (190 to 250 mm.) aquatic birds with long, slender bill; toes lobed and semipalmate; arctic and subarctic parts of North America and Eurasia during breeding season, migrating to equatorial areas to winter; three species.

*Family Dromadidae (Crab Plover).* Medium-sized (up to 380 mm.) wading bird with long, strong, pointed bill that is laterally compressed; plumage white, black, and brown; legs long; toes partly webbed; hind toe large; middle toenail pectinate; coasts of east Africa, Madagascar, and southwestern Asia bordering the Indian Ocean; one species, *Dromas ardeola.*

*Family Burhinidae (Thick-Knees).* Medium-sized (350 to 520 mm.), cursorial, bustard-like birds with enlarged tibiotarsal joints; bill short and ploverlike; head and eyes proportionately large; feet partly webbed; hind toe lacking; temperate and tropical parts of the world; nine species.

*Family Glareolidae (Pratincoles and Coursers).* Medium-small (150 to 250 mm.), gregarious, insect-eating charadriiforms representing two diverse subfamilies; coursers are cursorial, possess long legs with only three toes, short wings and tail, and long pointed downcurved bill; pratincoles have short legs with four toes, long wings and long forked tail, and short bill; Europe, Africa, Asia, and Australia; 17 species.

*Family Thinocoridae (Seedsnipes).* Medium-small (170 to 280 mm.) charadriiforms with long, pointed wings and short legs; bill short and conical for eating seeds and buds; rostrals covered by an operculum; hind toe elevated; higher Andes south to the tip of South America; four species.

*Family Chionididae (Sheathbills).* Medium-sized (350 to 430 mm.), white-bodied waders with a horny sheath at base of

bill; bill short and stout; wattles present on face; wings long
and possessing a spur at the bend; legs short; feet large with
hind toe elevated; islands and coast of the antarctic and subantarctic; two species, *Chionis alba* and *C. minor.*

*Family Stercorariidae (skuas and jaegers).* Medium-sized (430
to 600 mm.), predatory, gull-like birds; a fleshy cere present
at base of bill; bill strongly hooked; wings long; legs short;
front toes webbed; hind toe small and slightly elevated; high
latitudes of Northern and Southern Hemispheres; four species.

*Family Laridae (gulls and terns).* Moderately small to medium-
sized (200 to 760 mm.) waterbirds with webbed feet; bill
slender to heavy and pointed to hooked; plumage (adult) a
combination of black or gray and white, or all white; bill and
feet usually brightly colored; legs short to medium length; hind
toe small or vestigial; essentially worldwide where there is
available water; 82 species.

*Family Rynchopidae (skimmers).* Medium-sized (up to 500
mm.) ternlike bird with the lower mandible considerably longer than the upper; mandibles laterally compressed; knifelike
wings long and pointed; legs short; feet slightly webbed; east
coast of North America, coasts and larger rivers of Central and
South America, Africa, and southern and southeastern Asia;
three species.

*Family Alcidae (Auks and Their Allies).* Small to large (165 to
760 mm.) diving birds, somewhat resembling penguins; head
large; bill variable in shape; neck short; body heavy and
stout; legs situated posteriorly on body; toes webbed; hind
toe vestigial or absent; coasts of northern Pacific, northern
Atlantic, and Arctic Oceans; 22 species.

**Order Columbiformes (Sandgrouse, Dodos, Pigeons, and Doves).**
Small to large (150 to 1200 mm.) land birds with rather plump
body and loosely attached feathers; bill possessing a cere in most
species; nearly worldwide except high latitudes in the Northern
and Southern Hemispheres; three families.

*Family Pteroclidae (Sandgrouse).* Medium-small (225 to 400
mm.), partridge-like birds with dovelike head; bill short and
lacking a cere; tail coverts elongate; legs short; tarsi feathered; Europe, southern Asia, and Africa; 16 species.

*Family Raphidae (Dodos and Solitaire).* Large (probably up to
1200 mm.), flightless, pigeon-like birds; head large; bill large
and hooked; wings very small; legs and feet strongly developed; the Mascarene Islands in the Indian Ocean; three
species, all extinct by the end of the eighteenth century.

*Family Columbidae (Pigeons and Doves).* Small to fairly large
(150 to 840 mm.), stout-bodied birds with small head; bill
possessing a fleshy cere and usually slightly enlarged distally; distribution, that of the order; 289 species.

**Order Psittaciformes (Parrots and Their Allies).** Very small to fairly
large (95 to 1000 mm.) birds, possessing a short, deep, hooked

bill; plumage usually brightly colored with various shades of green, yellow, red, or blue; bill possessing a cere; feet strong with toes zygodactylous; tropical parts of the world and much of the Southern Hemisphere; one family.

Family Psittacidae (Parrots and Their Allies). Characters and distribution, that of the order; 315 species.

Order Cuculiformes (Cuckoos and Their Allies). Moderately small to medium-sized (160 to 700 mm.) birds that are mostly arboreal; upper mandible not movable; bill lacking a cere; feet zygodactylous; nearly worldwide except high latitudes; two families.

Family Musophagidae (Touracos). Medium-sized (375 to 700 mm.) cuculiform birds with long tail; bill short, stout, and serrate; wings short and, in most species, possessing a patch of crimson; Africa; 20 species.

Family Cuculidae (Cuckoos, Roadrunners, and Anis). Moderately small to medium-large (160 to 700 mm.), cuculiform birds with downcurved bill; wings rather long; tail usually long with feathers graduated; nearly worldwide; 127 species.

Order Strigiformes (Owls). Small to fairly large (130 to 700 mm.) nocturnal birds of prey; head large; bill hooked and possessing a fleshy cere at base; eyes directed forward; plumage soft and cryptic, mostly various shades of brown and gray with varying amounts of white; feet taloned with the outer toe reversible; essentially worldwide; two families.

Family Tytonidae (Barn Owls). Medium-sized (300 to 530 mm.) owls with a long heart-shaped face; eyes relatively small; bill proportionately long; wings long and rounded; legs long and completely feathered; middle claw pectinate; worldwide except New Zealand and some oceanic islands; 11 species.

Family Strigidae (Typical Owls). Small to fairly large (130 to 700 mm.) owls with round head; eyes very large; bill relatively short and stout; wings broad and rounded; legs short to medium length; middle claw not pectinate; essentially worldwide; 123 species.

Order Caprimulgiformes (Goatsuckers and Their Allies). Medium-sized (190 to 535 mm.), largely nocturnal birds with very short legs and small feet; bill very small but mouth very large; plumage cryptic; wings long and pointed; nearly worldwide; five families.

Family Steatornithidae (Oilbird). A medium-sized (up to 480 mm.), cave-inhabiting, vegetarian caprimulgid; bill short; rictal bristles long; wings long; legs very short; toes long; northern South America and the Island of Trinidad; one species, Steatornis caripensis.

Family Podargidae (Frogmouths). Medium to large (215 to 535 mm.) caprimulgids with large, broad, flat, hooked bill; wings relatively small for members of the order; middle toe elongated; eastern Asia and Australia to the Solomon Islands; 12 species.

Family Nyctibiidae (Potoos). Moderately large (400 to 500 mm.)

caprimulgids that perch upright on stumps and branches; bill small and narrow; rictal bristles lacking; wings and tail long; legs short but toes long; middle toenail not pectinate; Mexico to central South America; five species.

*Family Aegothelidae (Owlet Frogmouths).* Small (190 to 300 mm.) caprimulgids with large head and rictal bristles possessing barbs; flank feathers elongate; legs and feet small; middle toenail not pectinate; Australia-New Guinea region; eight species.

*Family Caprimulgidae (Nightjars).* Small (190 to 290 mm.) caprimulgids with very small feet and a pectinated middle toenail; head large; bill small; rictal bristles present; wings moderately long; temperate parts of North and South America, Africa, and Eurasia except northern latitudes; 67 species.

**Order Apodiformes (Swifts and Hummingbirds).** Extremely small to medium-small (60 to 330 mm.) birds with long, pointed wings and capable of very rapid flight; humeri proportionately short and thick; legs very short; nearly worldwide except extreme northern and southern latitudes and many oceanic islands; three families.

*Family Apodidae (Swifts).* Small to medium-small (90 to 230 mm.) swifts with small bill and a wide gape; feet very small; hallux reversible; distribution, that of the order; 76 species.

*Family Hemiprocnidae (Crested Swifts).* Small to medium-small (165 to 330 mm.) swifts possessing a crest on the head and a patch of silky feathers on the flank; eyes very large; feet small; hallux not reversible; southeastern Asia and adjacent islands; three species.

*Family Trochilidae (Hummingbirds).* Very small to medium-small (60 to 215 mm.) birds with long, slender, tubular bill; tongue very elongate; wings long and slender; wingbeat very rapid; legs very short; feet small and weak; temperate and tropical parts of North and South America; 319 species.

**Order Coliiformes (Mousebirds).** Medium-small (300 mm.), dull-colored birds with long tail and red feet and legs; bill short and stout; wings short and rounded; outer toe reversible; central and southern parts of Africa; one family.

*Family Coliidae (Mousebirds).* Characters and range, those of the order; one genus, *Colius*, and six species.

**Order Trogoniformes (Trogons).** Medium-sized (250 to 350 mm.) birds with small legs and feathered tarsi; plumage generally colorful; skin very thin with feathers loosely attached; tail long and truncate; bill short, broad, and flat; toes zygodactylous with the inner or second toe posterior in position; front toes syndactylous; tropical and subtropical parts of North and South America, Africa, Asia, and from the Philippine Islands south to Sumatra and Java; one family.

*Family Trogonidae (Trogons).* Characters and range, those of the order; 34 species.

**Order Coraciiformes (Kingfishers and Their Allies).** Small to large

(90 to 1600 mm.) tropical and subtropical birds with three front toes syndactylous in all except one family; bill usually large in proportion to size; plumage generally brightly colored; worldwide except northern parts of the Northern Hemisphere, but most abundant in the tropics; nine families.

*Family Alcedinidae (Kingfishers).* Small to medium-sized (100 to 450 mm.) birds with large head and long, strong bill; neck short; body compact; legs very short; essentially worldwide; 84 species.

*Family Todidae (Todies).* Small (90 to 115 mm.) birds with long, flattened bill possessing serrations along cutting edge; plumage colored bright green on upper parts of body; throat red; legs slender; toes relatively long; Greater Antilles; one genus, *Todus,* and five species.

*Family Momotidae (Motmots).* Medium-sized (170 to 500 mm.) birds with brownish green plumage; bill moderately long, broad, and decurved with serrations along edges; tail generally long with central feathers racquet-tipped; legs short; New World tropics; eight species.

*Family Meropidae (Bee-Eaters).* Medium-sized (150 to 350 mm.) birds with long, pointed, laterally compressed bill; color generally predominately green with a black stripe on each side of the head; wings long and pointed; middle tail feathers elongated in most species; temperate and tropical parts of the Old World and Australia; 24 species.

*Family Coraciidae (Rollers).* Medium-sized (240 to 450 mm.) birds with the two inner front toes united basally; plumage usually colorful except in ground rollers; bill strong and decurved; neck short; wings and tail long; outer front toe free; central and southern Europe and Asia, Africa, northern Australia, and east to the Solomon Islands; 17 species.

*Family Leptosomatidae (Cuckoo-Rollers).* A medium-sized (up to 450 mm.) bird with a short crest and possessing patches of powder-down; bill stout and decurved; sexes unlike in color of plumage; tail long and truncated; legs very short; Madagascar and adjacent islands; one species, *Leptosomus discolor.*

*Family Upupidae (Hoopoe).* A medium-sized (up to 300 mm.) crested bird with a long slender bill; general color pinkish brown; feathers of crest tipped with black; prominent black and white bands on wings and tail; central and southern Europe, Asia, Africa, and Madagascar; one species, *Upupa epops.*

*Family Phoeniculidae (Woodhoopoes).* Medium-sized (220 to 380 mm.) birds with long slender bill; head lacking a crest; plumage possessing a metallic gloss; tail long and pointed; forested parts of Africa; six species.

*Family Bucerotidae (Hornbills).* Medium to large (380 to 1600 mm.) birds with very large curved bill that usually possesses a casque on the culmen; eyelashes conspicuous; feathers rather coarse and loosely webbed; wings strong; tail long; Africa, Asia, Malaysia, the Philippines, and east to the Solomon Islands; 45 species.

*Order Piciformes (Woodpeckers and Their Allies).*   Small to medium-large (80 to 600 mm.) arboreal birds with zygodactylous feet; bill small to extremely large; legs and feet usually strong; outer toe reversible in most species; temperate and tropical parts of the world except New Zealand, the Australian region, and Madagascar; six families.

*Family Galbulidae (Jacamars).*   Medium-small (125 to 300 mm.), slender birds with long, thin pointed bill; plumage usually metallic green above; tail long and graduated; southern Mexico south to Brazil; 15 species.

*Family Bucconidae (Puffbirds).*   Medium-small (140 to 320 mm.), stocky birds with large head and rather heavy, broad bill; neck short; tail of medium length and either truncate or rounded; southern Mexico south to Paraguay; 30 species.

*Family Capitonidae (Barbets).*   Rather small (90 to 315 mm.), heavy-bodied, woodpecker-like birds with proportionately heavy bill; prominent bristles at base of bill; plumage usually bright; wings short and rounded; tail short to medium; tropics of Asia, Africa, and the Americas; 72 species.

*Family Indicatoridae (Honeyguides).*   Rather small (110 to 200 mm.), dull-colored birds possessing only nine primaries; bill short; wings long and pointed; Africa, the western Himalayas, Malaya, Sumatra, and Borneo; 11 species.

*Family Ramphastidae (Toucans).*   Medium to fairly large (300 to 600 mm.) birds with very large bill; edges of bill serrate; skin around eye, bare; tongue long and narrow; wings short and rounded; New World tropics, southern Mexico to Argentina; 37 species.

*Family Picidae (Woodpeckers).*   Small to medium-large (90 to 560 mm.) birds with straight, pointed bill; head proportionately large and neck slender; plumage rather harsh except in the wrynecks; rectrices stiff and pointed except in the piculets and wrynecks; legs short; feet strong; worldwide except Madagascar, New Zealand, the Australian region, and oceanic islands; 210 species.

*Order Passeriformes (Perching Birds).*   A very large order of small to medium-large (75 to 1015 mm.) birds containing about three-fifths of the known living species; bill variable in shape; wings ranging from short and rounded to long and pointed; tail very short to very long; feet possessing four functional, unwebbed toes all on the same plane; inner or outer front toes never reversible; worldwide; four suborders and 64 families recognized here.

SUBORDER EURYLAIMI (BROADBILLS).   Suboscine Passeriformes possessing one pair of syringeal muscles; 15 cervical vertebrae; deep plantar tendons (flexor digitorum longus and flexor hallucis longus) separate; hallux weak; one family.

*Family Eurylaimidae (Broadbills).* Medium-small (125 to 275 mm.) passerine birds with brightly colored plumage; bill flat, broad, and hooked at the tip; syrinx possessing a single pair of muscles; front toes syndactylous; tropical parts of Asia and Africa; 14 species.

SUBORDER TYRANNI (OVENBIRDS, TYRANT FLYCATCHERS, AND ALLIES). Suboscine Passeriformes possessing zero to two pairs of syringeal muscles; 14 cervical vertebrae; deep plantar tendons (flexor digitorum longus and flexor hallucis longus) united; hallux strong; 13 families.

*Family Dendrocolaptidae (Woodcreepers).* Medium-small (150 to 370 mm.) passerines with rounded nostrils; bill strong and laterally compressed; syrinx possessing two pairs of muscles; tail feathers spiny-tipped; front toes syndactylous; Mexico to northern Argentina; 48 species.

*Family Furnariidae (Ovenbirds).* Medium-small (120 to 280 mm.) passerines with elongated nostrils; bill slender; syrinx possessing two pairs of muscles; tail feathers not spiny-tipped; front toes only slightly syndactylous; southern Mexico to southern South America; 215 species.

*Family Formicariidae (Antbirds).* Small to medium-sized (95 to 370 mm.), dull-colored passerine birds with loosely webbed plumage; bill strong and hooked at the tip; wings short and rounded; front toes slightly syndactylous; southern Mexico south to Argentina; 222 species.

*Family Conopophagidae (Antpipits).* Small (100 to 140 mm.), stocky, dull-colored passerine birds with loosely webbed plumage; head proportionately large; bill broad, flat, and hooked at tip; wings short and rounded; legs long and feet strong; Central America south to Brazil; 26 species.

*Family Rhinocryptidae (Tapaculos).* Small to medium-small (115 to 250 mm.), stocky, terrestrial, passerine birds; plumage soft and loosely webbed; bill sharply pointed; wings small and rounded; legs, feet, and claws strong; Central America south to southern South America; 26 species.

*Family Cotingidae (Cotingas).* Small to medium-large (90 to 460 mm.), stocky, forest-dwelling passerines; plumage very conspicuously colored in a number of species; caruncles, lappets, or erectile crests present on the head or neck of some species; legs short; feet large; southwestern United States south to southern South America and Jamaica; 90 species.

*Family Pipridae (Manakins).* Small (80 to 160 mm.), stocky, conspicuously colored, forest-dwelling passerines; bill short, broad, and slightly hooked; legs short; two front toes partly joined together; southern Mexico south to Paraguay; 59 species.

*Family Tyrannidae (Tyrant Flycatchers).* Small to medium-sized (75 to 405 mm.), insectivorous, passerine birds; head moderately large; bill usually broad, flat, and slightly hooked; rictal bristles well developed as a rule; legs and feet usually weak; North and South America; 365 species.

*Family Oxyruncidae (Sharpbill).* A medium-small (up to 175 mm.), olive-green passerine with a scarlet crest; underparts yellowish white, barred or spotted; bill moderately long, straight, and tapering to a sharp point; nostrils elongated and covered by an operculum; wings and tail fairly long; legs short; Costa Rica south to Paraguay; one species, *Oxyruncus cristatus.*

*Family Phytotomidae (Plantcutters).* Moderately small (165 to 175 mm.), stocky, finchlike birds; a crest present on the head; bill short, stout, conical, and serrate along the edges; wings short and pointed; tail fairly long; legs short; South America; three species.

*Family Pittidae (Pittas).* Medium-small (150 to 280 mm.), plump birds with extremely short tail; head large; bill strong and slightly curved; wings short and rounded; legs long and feet large; south-central Africa, southern Asia, Australia, and east to the Solomon Islands; 23 species.

*Family Acanthisittidae (New Zealand Wrens).* Very small (75 to 100 mm.), dull-colored, wrenlike birds with extremely short tail; bill straight and slender; wings short; legs long; toes long and slender; New Zealand; four species.

*Family Philepittidae (Asities and False Sunbirds).* Small (100 to 165 mm.), soft-plumaged, plump-bodied, arboreal birds; bill fairly long, slender, and decurved; bare area around eye of male; wings rounded; tail short and rounded; legs and feet stout; Madagascar; four species.

SUBORDER MENURAE (LYREBIRDS AND SCRUB-BIRDS). Suboscine birds possessing two to three pairs of syringeal muscles; 14 cervical vertebrae; deep plantar tendons (flexor digitorum longus and flexor hallucis longus) united; hallux strong; two families.

*Family Menuridae (Lyrebirds).* Very large (760 to 1015 mm.) passerine birds with elaborate lyre-shaped tail plumes in the male; bill elongated and pointed; neck long; three pairs of syringeal muscles; 16 rectrices with outer pair curved and banded in the male; legs and feet large; southeastern Australia; two species.

*Family Atrichornithidae (Scrub-Birds).* Medium-small (165 to 230 mm.), terrestrial passerines with brown, vermiculated plumage; bill fairly large and pointed; two pairs of syringeal muscles; wings small; tail long and graduated; legs and feet large; Australia; two species.

SUBORDER PASSERES (SONGBIRDS). Oscine birds with five to seven pairs of syringeal muscles; 14 cervical vertebrae; deep plantar tendons (flexor digitorum longus and flexor hallucis longus) united; hallux strong; 48 families recognized here, essentially following Wetmore's classification (1960) with a few modifications.

*Family Alaudidae (Larks).* Moderately small (120 to 230 mm.) terrestrial, passerine birds with cryptic coloration; bill variable in shape; syrinx possessing five pairs of muscles; wings long and pointed; tarsus scaled on posterior surface; hind claw long and straight; nearly worldwide except central and southern South America, the Pacific islands, and Antarctica; 75 species.

*Family Hirundinidae (Swallows).* Small or moderately small (95 to 230 mm.) passerine birds with long, pointed wings; bill small and triangular; plumage very compact; legs very short; feet small; almost worldwide except New Zealand and the polar regions; 75 species.

*Family Dicruridae (Drongos).* Moderately small to medium-large (175 to 635 mm.), arboreal, passerine birds; plumage usually black; bill strong, hooked at the tip, and possessing a notch; eyes generally red; prominent bristles over nostrils; wings long; legs short; feet strong; Africa, Asia, Australia, and east to the Solomon Islands; 20 species.

*Family Oriolidae (Old World Orioles).* Medium-sized (175 to 305 mm.), tropical or subtropical, arboreal, passerine birds that are predominately yellow and black in color; bill strong and pointed; wings long; legs short; Africa and Eurasia; 26 species.

*Family Corvidae (Crows, Jays, and Magpies).* Medium to large (175 to 700 mm.) passerine birds with strong bill; nostrils rounded; bill lacking a notch; 10 primaries present in wing; tarsus scutellate; legs and feet strong; nearly worldwide except New Zealand, Antarctica, and most oceanic islands; 100 species.

*Family Cracticidae (Bell Magpies and Australian Butcherbirds).* Medium-large (260 to 585 mm.), corvid-like passerine birds; bill large and stout; plumage black and white or gray; legs and feet strong; Australia, Tasmania, New Guinea, and adjacent islands; 10 species.

*Family Grallinidae (Magpie Larks).* Medium-sized (190 to 500 mm.), corvid-like passerines with plumage black and white or gray to brown; neck short; nest, a deep mud bowl lined with fibers or feathers; Australia and New Guinea; four species.

*Family Ptilonorhynchidae (Bowerbirds).* Medium-sized (230 to 370 mm.) birds resembling members of the Paradisaeidae, but possessing a hind toe that is shorter than the middle front toe; wings short to medium-sized; legs moderately short; feet strong; Australia and New Guinea; 18 species.

*Family Paradisaeidae (Birds of Paradise).* Moderately small to large (140 to 1015 mm.) passerine birds in which the male possesses a specialized plumage for display; wings of medium size and rounded; legs short; feet strong; Australia, New Guinea, and adjacent islands; 43 species.

*Family Paridae (Titmice).* Small (75 to 200 mm.), arboreal, thickplumaged birds; bill short, stout, and pointed; wings rounded; 10 primaries present; legs short but strong; North America, Europe, Asia, and Africa; 65 species.

*Family Sittidae (Nuthatches).* Small (95 to 190 mm.), largely arboreal birds usually gray or blue-gray above (except the Australian nuthatches) with a dark line through the eye; bill slender; wings pointed; tail short and truncate to medium; legs short and strong; toes long; North America, Europe, Asia, Australia, and New Guinea; 22 species.

*Family Hyposittidae (Coral-billed Nuthatches).* Small (up to 125 mm.), creeper-like, greenish blue bird; bill short and reddish; tail fairly long; feet syndactylous; Madagascar; one species, *Hypositta corallirostris.*

*Family Certhiidae (Creepers).* Small (120 to 175 mm.) birds capable of creeping over vertical surfaces; bill laterally compressed and slender; plumage usually streaked, barred, or spotted on back; tail either short and soft or long with stiff spines at tip of rectrices; toes long; claws long and sharp; North America, Europe, Asia, Africa, Australia, and New Guinea; 17 species.

*Family Timaliidae (Babblers and Their Allies).* Small to medium-large (90 to 405 mm.) insectivorous birds with loose, fluffy plumage; wings short and rounded; 10 primaries present; legs and feet relatively large and strong; Europe, Asia, Australia, Africa, Madagascar, and western North America; 282 species.

*Family Campephagidae (Cuckoo-Shrikes and Minivets).* Small to medium-sized (125 to 310 mm.), arboreal, passerine birds with a decurved bill that is notched and hooked terminally; plumage soft with feathers loosely attached; nostrils partly hidden by bristles; tail usually graduated; legs short; Africa, Asia, Australia, and east to Samoa; 71 species.

*Family Pycnonotidae (Bulbuls).* Medium-small (140 to 285 mm.), usually dull-plumaged birds with a patch of hairlike feathers on the nape; bill short or of medium length and slightly curved; rictal bristles usually well developed; neck short; wings rounded; tail moderately long; legs short; Asia, Africa, Madagascar, and adjacent islands; 120 species.

*Family Irenidae (Leafbirds).* Medium-small (120 to 240 mm.), brightly colored, arboreal birds resembling bulbuls; bill moderately long, curved, and slightly hooked; a patch of hairlike feathers often present on nape; tail coverts sometimes elongated; Asia; 14 species.

*Family Cinclidae (Dippers).* Moderately small (140 to 190 mm.), stocky, solitary, aquatic birds; bill straight, slender, and laterally compressed; plumage dense; oil gland very large; tail short; legs and feet strong; Europe, Asia, and North and South America; one genus, *Cinclus,* and five species.

*Family Troglodytidae (Wrens).* Small to medium-small (95 to 220 mm.), brownish or grayish passerine birds with slender,

pointed bill; wings short and rounded; tail short to long; legs
and feet strong; front toes somewhat syndactylous; North and
South America, Europe, Asia, and northern Africa; 59 species.

*Family Mimidae (Thrashers and Mockingbirds).* Medium-sized
(200 to 300 mm.) passerine birds with slender body and long
tail; bill strong and usually fairly long; small rictal bristles
present; wings short and rounded; tail moderately long;
middle and outer front toes partly syndactylous; North and
South America and West Indies; 31 species.

*Family Turdidae (Thrushes).* Small to medium-sized (115 to 330
mm.), stocky songbirds usually possessing a booted tarsus;
bill of medium length; spots generally present in juvenal
plumage; 10 primaries present; worldwide except New Zealand, certain oceanic islands, and the polar areas; about 300
species.

*Family Zelodoniidae (Wren-Thrush).* A small (up to 120 mm.),
soft-plumaged, wrenlike bird resembling some of the
thrushes; bill short and somewhat flattened; neck relatively
short; wings rounded and with tenth primary greatly reduced
in size; tail short; legs long; Central America; one species,
*Zelodonia coronata.*

*Family Sylviidae (Old World Warblers).* Small to medium-small
(90 to 290 mm.), mostly dull-colored birds; bill slender; rictal
bristles present or absent; spots absent in juvenal plumage;
10 primaries present; legs short or of medium length; nearly
worldwide; four subfamilies (accorded family rank by some
ornithologists); 398 species.

*Family Muscicapidae (Old World Flycatchers).* Small to medium-small (90 to 230 mm.), thrushlike birds; bill broad, flat,
and possessing a subterminal notch; rictal bristles present;
spots generally present in juvenal plumage; 10 primaries present; tarsus scutellate; Europe, Asia, Africa, Australia, New
Zealand, and most of the Pacific islands; 328 species.

*Family Prunellidae (Accentors).* Fairly small (125 to 175 mm.),
sparrow-like birds; bill pointed and thrushlike; plumage rather plain, sometimes streaked or spotted; spots present in
juvenal plumage; 10 primaries present; tarsus reticulate;
North Africa, Europe, and Asia; 12 species.

*Family Motacillidae (Wagtails and Pipits).* Medium-small (125
to 220 mm.), slender-bodied, ground-dwelling birds that walk
rather than hop; bill slender and pointed; wings pointed;
nine primaries present; toes long; essentially worldwide; 54
species.

*Family Bombycillidae (Waxwings).* Medium-small (160 to 190
mm.), moderately plump birds with fairly broad bill; plumage
soft and silky; wings pointed; 10 primaries present; Northern
Hemisphere; two subfamilies, two genera (*Bombycilla* and
*Hypocolius*), and four species.

*Family Ptilogonatidae (Silky Flycatchers).* Medium-small (185 to 250 mm.), slender-bodied, silky-plumaged birds; bill small and broad; a crest on the head; wings short; 10 primaries present; tail long; legs short; southwestern United States to Panama; four species.

*Family Dulidae (Palmchat).* A medium-small (175 mm.), olive-brown bird; bill rather heavy and laterally compressed; wings rounded; 10 primaries present; legs and toes stout; West Indies; one species, *Dulus dominicus.*

*Family Artamidae (Wood-Swallows).* Medium-small (145 to 205 mm.), stout-bodied birds with long, pointed wings; bill moderately stout, long, and curved downward; plumage soft; legs short and stout; India and southeastern Asia to the Philippines, Australia, and the Fiji Islands; 10 species.

*Family Vangidae (Vanga-Shrikes).* Small to medium-sized (125 to 310 mm.), stout-bodied birds usually colored metallic black above; bill strong, hooked, and toothed; tail moderately long; legs and feet short and stout; Madagascar; 12 species.

*Family Laniidae (Shrikes).* Moderately small to medium-sized (160 to 370 mm.) birds with stocky body and broad head; bill strong, hooked, and notched at the tip; usually gray or brown above with a black mask on the head; 10 primaries present; tail long; legs and feet strong; Europe, Asia, Africa, and North America; 72 species.

*Family Prionopidae (Wood-Shrikes).* Medium-small (190 to 255 mm.), shrikelike birds patterned in black and white; a crest usually present on head; bill stout and hooked; wattle present around eyes; 10 primaries present; Africa; 13 species.

*Family Callaeidae (Wattlebirds).* Medium-sized (250 to 535 mm.) birds possessing a pair of orange or blue wattles at the gape; bill strong but variable in length; wings rounded with first primaries very long; 10 primaries present; tail long; legs and feet large and strong; New Zealand; three species.

*Family Sturnidae (Starlings).* Generally medium-sized (175 to 430 mm.) birds with straight or downcurved bill; plumage silky, usually with a metallic sheen; 10 primaries present; legs and feet strong; Europe, Asia, Africa, Australia, and some of the Pacific islands; 106 species.

*Family Meliphagidae (Honeyeaters).* Small to medium-sized (100 to 355 mm.) birds with slender, downcurved bill; tongue long, extensible, and brush-tipped; wings long and pointed; 10 primaries present; legs strong; South Africa, Australia, New Zealand, parts of Indonesia, and some of the Pacific islands; 160 species.

*Family Nectariniidae (Sunbirds).* Small to medium-sized (95 to 255 mm.), brightly colored birds; bill long and curved with serrations along the edges; tongue long and tubular; wings short and rounded; 10 primaries present; legs short with tarsus scutellate; toes short; Africa, Madagascar, Asia, Indonesia, Australia, and some of the Pacific islands; 104 species.

*Family Dicaeidae (Flowerpeckers).* Small (75 to 190 mm.), stocky birds with short tail; bill variable but usually possessing serrations along terminal third; tongue long and tubular; neck short; wings moderately long; tenth primary reduced in size and vestigial in some species; legs short; Asia, Australia, and east to the Solomon Islands; 54 species.

*Family Zosteropidae (White-Eyes).* Small (100 to 140 mm.), yellowish green birds with a white eye ring; bill slender and pointed; tongue brush-tipped; wings pointed; tenth primary vestigial or absent; tail truncate; legs short; Africa, Madagascar, Asia, Australia, New Zealand, and east to the Philippines; 80 species.

*Family Vireonidae (Vireos).* Small (100 to 185 mm.), olive-green or gray-brown birds; bill of medium length, rather thick, hooked at the tip, and possessing a subterminal notch; plumage never streaked or spotted; tenth primary small or vestigial; short but strong legs and feet; North and South America and the West Indies; 42 species.

*Family Drepanididae (Hawaiian Honeycreepers).* Moderately small (115 to 220 mm.) birds showing great variability in color; bill ranging from long, slender, and sickle-shaped to short, thick, and hooked; wings pointed; nine functional primaries; feet strong; Hawaiian Islands; 22 species.

*Family Parulidae (Wood Warblers and Bananaquits).* Small (110 to 185 mm.), often colorful arboreal birds; bill usually slender and pointed; rictal bristles present; nine primaries present; feet small but strong; North and South America and the West Indies; about 119 species.

*Family Icteridae (Troupials).* Moderately small to medium-large (170 to 545 mm.) birds with strong, conical bill; plumage often black or black in combination with other colors; wings generally long and pointed; nine primaries present; feet strong; North and South America and the West Indies; 94 species.

*Family Tersinidae (Swallow-Tanager).* A fairly small (up to 160 mm.), turquoise-blue bird with a black face and chin; bill short, conical, somewhat flattened, and with a slight hook at the tip; wings long; nine primaries; legs short; South America north to Panama; one species, *Tersina viridis.*

*Family Thraupidae (Tanagers).* Small to medium-sized (75 to 305 mm.), brightly colored birds mostly with short, rounded wings; bill rather conical, generally slightly hooked, and possessing a notch in cutting edge; rictal bristles present; nine primaries present; legs short; North and South America and the West Indies; 222 species.

*Family Catamblyrhynchidae (Plush-capped Finch).* A fairly small (up to 150 mm.) bird with erect, golden, plushlike feathers in the crown; body plumage blue-gray above, chestnut below; bill short, thick, and slightly hooked; wings short and rounded; nine primaries; hind claw proportionately large; Andes of South America; one species, *Catamblyrhyn-chus diadema.*

*Family Ploceidae (Old World Seedeaters).* Small to medium-large (75 to 650 mm.) passerine birds; bill conical with exposed portion of upper mandible more than twice the length of the gonys; tenth primary either reduced, vestigial, or absent; tarsus never longer than middle toe with claw; essentially worldwide; three subfamilies (Carduelinae, Estril-dinae, and Ploceinae) and about 375 species.

*Family Fringillidae (New World Seedeaters).* Small or medium-small (95 to 375 mm.) passerine birds; bill short, thick, and conical with the exposed portion of the upper mandible less than twice the length of the gonys; rictal bristles usually present; nine functional primaries; tarsus longer than middle toe with claw; essentially worldwide except Madagascar, Australia, and Oceania; three subfamilies (Richmondeninae, Geospizinae, and Fringillinae) and nearly 700 species.

## References Recommended

Allen, G. M. 1925. Birds and Their Attributes. Boston, Marshall Jones Co.

Austin, O. L., Jr. 1961. Birds of the World. New York, Golden Press.

Aymar, G. C. 1936. Bird Flight. New York, Dodd, Mead & Co.

Beddard, F. E. 1898. The Structure and Classification of Birds. London, Longmans, Green, and Co.

Brush, A. H., and Allen, K. 1963. Astaxanthin in Cedar Waxwings. Science 142:47-48.

Chamberlain, F. W. 1943. Atlas of Avian Anatomy. Michigan State College, Agric. Exp. Station.

Collias, N. E. 1960. An Ecological and Functional Classification of Animal Sounds. *In* Lanyon, W. E., and Tavolga, W. N., Eds. Animal Sounds and Animal Communication. AIBS Symposium, Publ. 7:368-391.

Cooch, F. G. 1964. A Preliminary Study of the Survival Value of a Functional Salt Gland in Prairie Anatidae. Auk 81:380-393.

Dwight, J., Jr. 1900. The Sequence of Plumages and Moults of the Passerine Birds of New York. Annals N.Y. Acad. Sci. 13:73-360.

Frings, H., Anthony, A., and Schein, M. W. 1958. Salt Excretion of Nasal Gland of Laysan and Black-footed Albatrosses. Science 128:1572.

George, J. C. 1966. Avian Myology. New York, Academic Press.

Goldsmith, T. H. 1965. The Red-yolked Egg of the Touraco, *Tauraco corythaix.* Postilla, No. 91.

Gunston, D. 1960. Story of Bird Anting. Part I. Audubon Mag. 62:268-269, 297-298.

Hudson, G. E. 1937. Studies on the Muscles of the Pelvic Appendage in Birds. Am. Midland Nat. 18:1-108.

Humphrey, P. S., and Parkes, K. C. 1959. An Approach to the Study of Molts and Plumages. Auk 76:1-31.

Irving, L. 1955. Nocturnal Decline in the Temperature of Birds in Cold Weather. Condor 57:362-365.

Jaeger, E. C. 1949. Further Observations on the Hibernation of the Poorwill. Condor 51:105-109.

Macdonald, J. D., Goodwin, D., and Adler, H. E. 1962. Bird Behavior. New York, Sterling Publishing Co., Inc.

Marler, P. 1957. Specific Distinctiveness in the Communication Signals of Birds. Behavior 11:13-39.

Marshall, J. T., Jr. 1955. Hibernation in Captive Goatsuckers. Condor 57:129-134.

Medway, Lord. 1959. Echo-location Among *Collocalia*. Nature *184*:1352-1353.

Miller, A. H. 1963. Desert Adaptations in Birds. Proc. XIII Internat. Ornith. Cong., pp. 666-674.

Pearson, O. P. 1950. The Metabolism of Hummingbirds. Condor *52*:145-152.

Peterson, R. T. 1947. A Field Guide to the Birds. Boston, Houghton Mifflin Co.

Peterson, R. T. 1961. A Field Guide to Western Birds. Boston, Houghton Mifflin Co.

Pettingill, O. S., Jr. 1946. A Laboratory and Field Manual of Ornithology. Minneapolis, Burgess Publishing Co.

Schmidt-Nielsen, K. 1959. Salt Glands. Scientific American *200*:109-116.

Schufeldt, R. W. 1890. The Myology of the Raven (*Corvus corax sinuatus*). London, Macmillan and Co.

Selander, R. K., and Kuich, L. L. 1963. Hormonal Control and Development of the Incubation Patch in Icterids, with Notes on Behavior of Cowbirds. Condor 65:73-90.

Stager, K. E. 1964. The Role of Olfaction in Food Location by the Turkey Vulture (*Cathartes aura*). Contrib. in Science, Los Angeles County Museum, No. 81.

Storer, J. H. 1948. The Flight of Birds Analyzed Through Slow-motion Photography. Cranbrook Institute of Science, Bull. 28.

Sturkey, P. D. 1954. Avian Physiology. Ithaca, N.Y., Comstock Publishing Associates.

Test, F. H. 1942. The Nature of the Red, Yellow, and Orange Pigments in Woodpeckers of the Genus *Colaptes*. Univ. Calif. Publ. Zool. 46:371-390.

Thayer, G. H. 1909. Concealing Coloration in the Animal Kingdom. New York, The Macmillan Co.

Thomason, A. L. 1927. Birds: An Introduction to Ornithology. New York, Henry Holt and Co., Inc.

Van Tyne, J., and Berger, A. J. 1959. Fundamentals of Ornithology. New York, John Wiley & Sons, Inc.

Wagner, H. O. 1955. The Molt of Hummingbirds. Auk 72:286-291.

Wallace, G. J. 1955. An Introduction to Ornithology. New York, The Macmillan Co.

Welty, J. C. 1962. The Life of Birds. Philadelphia, W. B. Saunders Co.

Wetmore, A. 1960. A Classification for the Birds of the World. Smithsonian Misc. Coll. *139* (No. 11):1-37.

Young, J. Z. 1955. The Life of Vertebrates. Oxford, Clarendon Press.

# Chapter Six

# MAMMALS

## GENERAL CHARACTERS

Mammals have risen rapidly from their reptilian origin back in the Mesozoic and have expanded into nearly every available niche and habitat on the earth. They are found in the ocean, along the shores, in lakes and rivers, underground, above ground, in trees, and some have even taken to the air. They range from the polar regions to the tropics and, on most continental areas, exceed all other terrestrial vertebrates in individual numbers. The presently recognized number of Recent species is 4060. Some of these, however, may ultimately be shown to be geographic variants, so that the real number may be somewhat less than this.

As a group mammals possess many structural characters that readily distinguish them from other living vertebrates. The most diagnostic mammalian feature is the presence of mammary glands, which provide nourishment for the young. Other glands, such as sebaceous and sweat glands, are commonly found on parts of the body. Hair is present during some period of life, although it may be reduced or completely absent in the adult stage of certain specialized forms such as whales. Mammals, like birds, are endotherms, since they possess internal thermoregulating mechanisms which control body temperature.

## SKELETAL SYSTEM

In the mammalian skeletal system there is a greater ossification as well as a reduction in the number of bony elements, both dermal and replacement, than that found in lower forms. This is especially true of the skull (Fig. 6.1). Bones such as the prefrontals, postfrontals, postorbitals, and quadratojugals are lacking. In many mammals the four occipital bones are fused together. Various degrees of fusion occur in the sphenoidal area. The presphenoid, orbitosphenoids, basisphenoid, and alisphenoids may be separate or all fused into a single bone. In some mammals the petrosal and tympanic bones fuse with the squamosal to form the temporal bone, which therefore is both dermal and replacement. The mammalian cranium is relatively large to accommodate the proportionately enlarged brain. A hard palate is present which is responsible for the posterior position of the internal choanae. The skull articulates with the first cervical vertebra by means of two condyles, each of which is on one of the exoccipital bones. In the middle ear the *stapes*, which is derived from the columella of reptiles, is supplemented by two other auditory ossicles: the *incus*, derived from the quadrate, and the *malleus*, derived from the articulare. This means that the lower jaw articulates directly with the squamosal of the skull without the intervention of

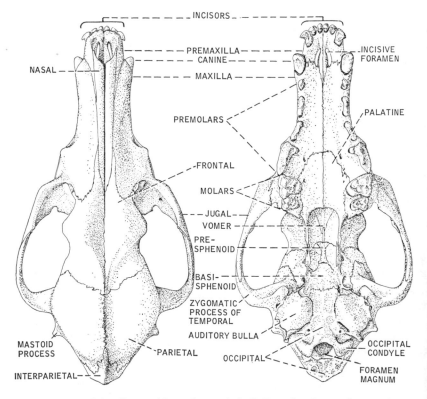

**FIGURE 6.1.**    Coyote (*Canis latrans*) skull. Dorsal and ventral view.

the quadrate. In the upper jaw, teeth are borne only by the premaxillaries and maxillaries. The lower jaw is composed of the paired dentary bones.

The vertebrae generally have the ends of the centra flattened or acelous, a condition referred to as *amphiplatyan*. The cervical or neck vertebrae are seven in number in all but four kinds of mammals. The exceptions are the manatee (*Trichechus*) and the two-toed sloth (*Choloepus*), each of which possesses six; the anteater (*Tamandua*), which has eight; and the three-toed sloth (*Bradypus*), which has nine. It is interesting to note that all these exceptions are inhabitants of tropical America. The cervical vertebrae are generally capable of considerable flexibility. In cetaceans, however, which lack cervical flexibility, these vertebrae exhibit various degrees of fusion.

The thoracic or rib-bearing vertebrae show relatively little flexibility. They vary in number according to the number of pairs of ribs present. The latter may range from nine to 24. Behind the thoracic are the lumbar vertebrae, which are stout and fairly flexible. Although the number is variable, it is most frequently between five and seven. Caudal to the lumbar are the sacral vertebrae, which are fused together to form the sacrum. Most mammals have three sacral vertebrae but in some kinds there may be four or five. The caudal vertebrae are extremely variable in number, depending upon the length of the tail.

The mammalian rib typically possesses two heads. One of these, the *capitulum*, articulates at a point of junction between adjacent centra, and the other, the *tuberculum*, articulates with the transverse process of a thoracic vertebra. The anterior, or so-called *true ribs*, connect directly with the sternum, while those situated more posteriorly either connect with the costal cartilage of the last true rib or are free terminally. The sternum usually consists of a series of bony elements in a linear arrangement.

Although mammals basically possess four pentadactyl limbs, they have been subject to considerable modification in many specialized groups, as will be noted later in this chapter. Reduction has most frequently occurred in the number of toes, but in cetaceans and sirenians all external evidence of the hind limb has disappeared. In all the living mammals above the monotremes the coracoid of the shoulder girdle has been reduced to a process on the scapula. The interclavicle is also lacking in marsupial and placental mammals. The clavicles are well developed in some mammals but reduced or absent in others. Except in specialized groups, such as cetaceans and sirenians, the pelvic girdle consists of three bony elements on each side which generally are fused to form the *innominate* bone. The iliac portion of the girdle joins the sacrum on either side.

## Muscular System

It is not possible to give a brief description of the mammalian muscular system, since it varies so greatly in different specialized forms. A few distinctive features, however, may be pointed out. The metameric arrangement of the trunk muscles, so evident in lower vertebrates, is largely gone or obscured (Fig. 6.2). With the development of powers of locomotion by use of the limbs, we find that the extrinsic muscles assoicated with these structures cover much of the trunk musculature that remains. This is especially true in the thoracic region.

The branchial muscles, which in fishlike vertebrates are concerned primarily with movement of the gill arches, develop into muscles of the face, neck, and shoulder in mammals. Such important muscles as the masseter, temporal, digastric, mylohyoid, trapezius, sternomastoid, and cleidomastoid represent derivatives of the original branchial musculature and are called, therefore, *branchiomeric muscles*. Like their forerunners in fishes they are innervated by cranial nerves.

Dermal or integumentary muscles reach their greatest development in mammals. These include muscles of facial expression, muscles of the eyelids, nose, and lips, muscles capable of twitching or moving the skin and of erecting the hair, and sphincter muscles controlling body apertures.

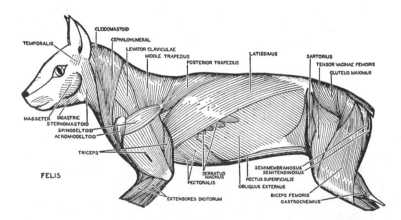

FIGURE 6.2.  Superficial lateral trunk muscles in a mammal (*Felis*). The metamerism of the lateral trunk muscles, which is such a striking feature of the lower vertebrates, is reduced and modified in reptiles and almost disappears in adult mammals. (Courtesy of Neal and Rand: Chordate Anatomy. McGraw-Hill Book Co., Inc. 1950.)

## CIRCULATORY SYSTEM

The mammalian circulatory system possesses many advanced features, including a four-chambered heart composed of two atrial and two ventricular chambers. As a result of complete interatrial and interventricular septa, there is a complete separation of venous and arterial blood as in birds. The right atrium is separated from the right ventricle by a *tricuspid* valve, whereas the left atrium is separated from the left ventricle by a *bicuspid* or *mitral valve.* The systemic aorta is derived in part from the left fourth aortic arch. The right fourth aortic arch becomes the right subclavian artery. There may be one or two precaval veins. There is a single postcaval vein whose embryological development is very complicated. All the caval veins enter the right atrium directly, as the sinus venosus is absorbed into the wall of this chamber of the heart in embryonic life. There is no renal portal system, although the hepatic portal venous system very much resembles that of other vertebrates. The mammalian erythrocytes are enucleate.

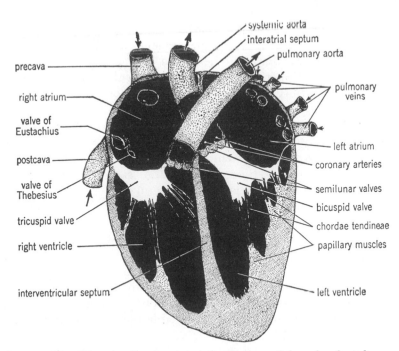

FIGURE 6.3. Diagram showing internal structure of four-chambered mammalian heart, ventral view. (Weichert, C. K.: Anatomy of the Chordates. McGraw-Hill Book Co., Inc., 1958.)

## DIGESTIVE SYSTEM

There are a number of unique features that characterize the mammalian digestive system. Movable lips are present in all except certain groups such as monotremes and cetaceans. Teeth are generally present and are so specialized in many species that they are considered separately under Special Characters.

*Oral glands*, primarily concerned with the secretion of mucus, are present in all classes of vertebrates. They are most highly developed, however, in terrestrial forms, since they serve to keep the mouth moist, bathe the taste buds, and assist in the swallowing of food. In mammals several glands whose ducts lead into the oral cavity, principally the parotid, the submaxillary, and the sublingual glands, are specialized as salivary glands. In many species they are capable of producing an enzyme called *salivary amylase*. This product, when present in the saliva, is activated by chloride ions and serves to initiate the conversion of starch to sugar as it is mixed with food in the mouth.

In most mammals, except whales, the tongue is highly developed and capable of considerable movement, in addition to extension and retraction, as a result of the presence of a number of intrinsic muscles. On its upper surface there are numerous papillae of several types, some of which are associated with taste buds. The esophagus is easily distinguished from the stomach, lacks glands, and varies in relation to the length of the neck. Mammal stomachs exhibit a wide variety of shapes and forms which are correlated with food habits. They range from relatively simple, sac-like structures to those which are composed of a series of chambers each of which may have a separate function. Vampire bats (*Desmodus rotundus*), whose food consists of fresh blood which they lap from their victims within a fairly short period of time, have a large sac-shaped stomach for storage. An interesting specialization is seen in the grasshopper mice (*Onychomys*), which are small rodents living in western North America. These mammals live largely on insects and therefore must process considerable chitin through their digestive systems. To accomplish this they have developed a stomach in which the digestive glands are concentrated in the fundus, which is a pocket that secretes gastric juices into the stomach through a small canal. The cardiac and pyloric parts of the stomach through which food must pass are lined with a cornified epithelium that is capable of withstanding the abrasive effect of the chitin.

The most complicated stomachs are found in ruminants, cetaceans, and sirenians. They may be composed of several chambers. In ruminants there are four parts. The first is a temporary storage chamber called the *rumen*. Forage is chewed hastily and goes to this compartment, where it becomes moistened and churned into a mass.

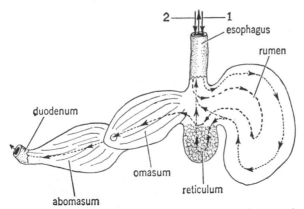

2———1
esophagus
rumen
duodenum
omasum
reticulum
abomasum

**FIGURE 6.4.** Diagram of stomach of ruminant, showing functional relationships: *1* denotes initial swallowing; *2* indicates path taken by food upon reswallowing. (Weichert, C. K.: Anatomy of the Chordates. McGraw-Hill Book Co., Inc., 1958. Modified from Kingsley: Comparative Anatomy of Vertebrates. The Blakiston Division, McGraw-Hill Book Co., Inc., with permission.)

From here it passes into the second stomach called the *reticulum*. The latter contains numerous small compartments and packs the food into small masses called cuds. These cuds are then regurgitated when the animal is resting, and the vegetation is thoroughly chewed. It is then swallowed a second time and passes back into the third stomach, which is the *omasum* or *psalterium*. Here it is subject to further churning as a result of peristaltic action and passes back to the fourth chamber or *abomasum*. The latter possesses digestive glands within its walls, and the secretion from these glands is mixed with the food, which then goes into the *duodenum* or anterior part of the small intestine.

The small intestine is proportionately long and coiled in most mammals. In domestic cattle, for example, it may measure 165 feet in length. There is a blind pocket or caecum at the junction of the colon and small intestine which is generally small in carnivorous species but quite long in many herbivores. A cloaca is absent in mammals above the monotremes.

## RESPIRATORY SYSTEM

The respiratory system of the mammal is much less complicated than that of the bird. In front of the slit-like glottis on the floor of the pharynx is an elevated cartilaginous flap known as the *epiglottis*. Air going through the glottis enters the larynx and then passes into the trachea. The latter is supported by cartilaginous rings that are usually incomplete on the upper side. From the trachea air passes into paired primary bronchi, then into secondary bronchi which divide into smaller and smaller bronchioles, finally terminating in tiny *alveoli* or blind pockets in which there is an exchange of gases.

The most rapid respiratory rate in mammals is probably to be found among some of the insectivores. Long-tailed shrews have been found to respire 850 times per minute. These same animals may have a pulse rate of 800 beats per minute.

In some aquatic and marine mammals there are certain adaptive modifications of parts of the respiratory system. These frequently involve the development of flaps and valves for closing the external nares. In the toothed whales the epiglottis and part of the larynx are elongated into a tube which extends upward into the nasopharynx, where it may be closed tightly by muscles. In sirenians the lungs are elongated and have a relatively large capacity in proportion to body size.

Recent studies on seals have shown that they have developed some remarkable adaptations which enable them to dive to great

FIGURE 6.5.   The valves of the nostrils of a manatee just opening as it surfaces to breathe. (Courtesy of Steinhart Aquarium, San Francisco.)

depths without suffering from oxygen deprivation. This is accomplished in part by a marked reduction in the pulse rate as soon as the animal is submerged. It has been found that the pulse of an elephant seal will drop from 85 to 12 beats per minute as soon as its muzzle goes beneath the water. This also explains why seals can remain submerged for long periods of time. Experiments performed at McMurdo Sound in the Antarctic have shown that Weddell seals (with depth gauges attached) may dive to depths of nearly 2000 feet and remain submerged for 25 to 35 minutes. When moving about under the ice these animals regularly cruise at depths of 125 to over 200 feet and can rise to surface at a blowhole in a matter of a few seconds without any sign of the "bends."

The deepest-diving mammals are cetaceans. Sperm whales are capable of diving to depths of over 3000 feet, which entails the ability to withstand enormous pressure and to hold the breath for at least an hour. Observations made on bottlenose dolphins show that the lungs suffer complete alveolar collapse at depths of about 200 feet; this prevents any exchange of gases in deep dives and avoids nitrogen narcosis. It appears likely that this alveolar collapse occurs in all deep-diving cetaceans. Cetacean lungs are no larger in propor-

FIGURE 6.6. Two migrating gray whales (*Eschrichtius gibbosus*), one of which is spouting. The spout is largely the result of condensation of water vapor from the lungs.

tion to body size than those of land mammals, but the oxygen-carrying capacity of the blood is considerably greater as is the tolerance to carbon dioxide. Furthermore, in a dive the rate of heart beat is reduced to about 10 per minute and much of the blood is automatically shut off from the skin, body musculature, and tail region to ensure a rich supply to the brain and heart. In deep-diving mammals there is also a large amount of *myoglobin* in the muscle tissue which, like hemoglobin, serves to store oxygen. Myoglobin is responsible in part for the dark red color of the muscles.

## UROGENITAL SYSTEM

Both ovaries are generally functional in the mammal. Paired oviducts or fallopian tubes are present. In the egg-laying monotremes these ducts open separately into the cloaca, although it has been stated that only the left one is functional. In marsupials and placentals the lower part of the fallopian tube is enlarged into a *uterus,* in which embryonic development of the young takes place. In marsupials the uteri remain separate but in the placentals there are various degrees of fusion from the duplex uterus to a simple one (Fig. 6.7). The mammalian embryo, like that of the reptile and bird, is protected by a fetal membrane called the *amnion.*

The mammalian testes are either situated far posteriorly in the body or else they are outside the body cavity in a sac called the *scrotum.* In certain species the testes descend into the scrotal sac only during the breeding season. The male possesses a single penis, which in the monotremes is located on the floor of the cloaca. In all higher forms, however, it is either in a sheath that opens to the outside or else is external.

Adult mammals, like other amniotes, possess a metanephric type of kidney. They also have a urinary bladder. In mammals, as in other vertebrates, the kidneys are concerned not only with the elimination of nitrogenous waste in the form of urea resulting from protein

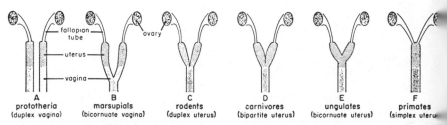

**FIGURE 6.7.** Fusion of the oviducts in mammals, in successive stages *A* to *F* (From Smith, H. M.: Evolution of the Chordate Structure. Holt, Reinhart and Winston, Inc., 1960.)

metabolism, but also with water balance. The latter problem varies greatly and is largely dependent on the environment and the habits of a species. Marine mammals must contend with the problem of salt ingestion with their food, while desert species are sometimes confronted with the absence of exogenous water. The latter condition is common within the range of many rodents on the deserts of western North America, much of Australia, the Sahara, and parts of Asia where only air-dried seeds may be eaten. Water conservation is accomplished mainly by the kidneys secreting urine that is concentrated to an extreme degree.

## NERVOUS SYSTEM

The mammalian nervous system is more highly developed than that of other vertebrates. In many species the cerebral hemispheres, derived from the telencephalon, have convolutions on the surface so that there are ridges or *gyri* and depressions or *sulci*. The outer layer or cortex of the cerebrum is composed of gray matter. The right and left cerebral hemispheres are connected with one another by a broad white commissure called the *corpus callosum*. In mammals the olfactory lobes are small compared with those of the lower vertebrates (Fig. 6.8).

The diencephalon consists of a dorsal *epithalamus*, a lateral *thalamus*, and a ventral *hypothalamus*. A pineal gland is present on the roof of the diencephalon, but it shows no eyelike structures. The thalamus is an important relay center. The hypothalamus is very important in mammals and consists of four parts. These are the *infundibulum*, forming the stalk and posterior lobe of the *pituitary*;

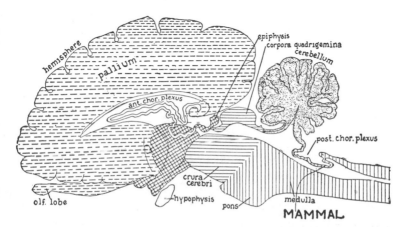

FIGURE 6.8.  Sagittal diagram of the brain. (After Edinger.)

the *optic chiasma,* where the right and left optic nerves cross en route to the brain; the *tuber cinereum,* which is believed to be the parasympathetic center; and the *mammillary bodies,* which integrate olfactory impulses. The hypothalamus controls a great many mammalian functions including blood pressure, sleep, water content, fat and carbohydrate metabolism, body temperature, and possibly rhythmic activities such as molt, migration, and pituitary secretion.

The midbrain in mammals is of less importance than in lower vertebrates, in which it is really the brain center. It is divided into four prominences called the *corpora quadrigemina.* The two superior lobes are concerned with sight, whereas the two inferior lobes are probably concerned with hearing. The *cerebellum,* which is the center of control of body movement, is most highly developed in mammals. Its surface is convoluted and it is divided into a number of lobes. Beneath the cerebellum is a typical mammalian structure called the *pons.* There is a slight indication of the pons appearing in some birds, but in mammals this relay center is a conspicuous feature of the ventral metencephalon.

There is a marked tendency toward a shortening of the spinal cord in mammals. In only a few species does it extend as far back as the sacrum. The central nervous system is surrounded by three protective layers or *meninges.* The innermost, which is in contact with the nervous system itself, is called the *pia mater.* Outside of this is a second meninx called the *arachnoidea.* Between the two is a space called the *subarachnoid space,* which is filled with cerebrospinal fluid. The outermost layer is the *dura mater.* It is separated from the arachnoidea by the *subdural space,* which contains a small amount of fluid.

*[handwritten margin notes: dura mater / subd. space / arachnoidea / suba. space / pia / cortex]*

As do other amniotes, mammals possess 12 cranial nerves. The spinal nerves have dorsal and ventral roots which unite to form a main trunk before emerging intervertebrally. Complicated plexuses, resulting from the intermixing of fibers from the ventral branches of spinal nerves, are found in mammals. These are differentiated into cervical, brachial, lumbar, and sacral plexuses.

## SENSE ORGANS

Although the olfactory lobes are not as large as in many lower vertebrates, the sense of smell is highly developed in many mammals. It is not only employed to detect members of the same species, but also enemies and food. The efficiency of this sensory system has been effected in mammals by the development of the nasal conchae into elaborate scroll-like structures (Fig. 6.9) which greatly increases the surface available for olfactory epithelium. There are some mam-

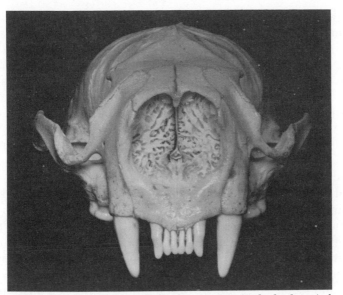

**FIGURE 6.9.** Front view of the skull of a sea otter (*Enhydra lutris*) showing the highly developed scroll-like structure of the nasal conchae.

mals, however, such as the whales, in which the olfactory organs are believed to be essentially nonfunctional.

The mammalian eye, in general, shows no special modification. Some species are adapted to nocturnal vision, others to diurnal vision, still others to vision under water. In strictly fossorial types such as moles the eye may be of little importance.

The ear, however, shows several advances over the ears of other vertebrates. The cochlea, which is essentially straight in reptiles and birds, is generally coiled to accommodate its increase in length. The middle ear contains three bony ossicles which transmit vibrations from the tympanic membrane to the inner ear. An external auditory canal is present, and in the great majority of mammals there is a well developed pinna which aids in funneling sounds into the auditory canal.

The auditory apparatus of some mammals shows remarkable specialization. Many bats, cetaceans, and pinnipeds depend largely upon the echoes of sounds that they themselves produce to detect the presence of objects in their environment as they move about. Back in the nineteenth century a scientist named Spallanzani proved, by a series of experiments, that bats with their vision occluded could fly about a room strung with wires yet avoid hitting these obstacles. They could not do this, however, with their ears plugged or their mouths sealed. By means of proper instrumentation, Dr. Donald R. Griffen and Dr. Robert Galambos in the early 1940's demonstrated that bats navigate by means of *echo-location*. Using a high frequency microphone, an amplifier, and an oscilloscope, they proved that small, insectivorous bats produce sounds of high frequency with great rapidity as they fly. These sounds are reflected back from nearby objects as echoes, thereby apprising the animals not only of the presence of the objects but also of their size and shape. The use of sounds of very high frequency (often in excess of 100,000 cycles) and, therefore, of very small wave length is essential to detect little objects such as flying insects.

Not long after this discovery it was found that many kinds of cetaceans produce underwater sounds with an even wider range of frequency than bats. Some sounds with frequencies as high as 300,000 cycles per second have been detected, although many cetacean utterances are much lower than those of bats. These sounds can be detected by the use of a hydrophone. Following the discovery of underwater cetacean sounds, it remained to be demonstrated how such pulses could be of use to the organisms producing them. An ear adapted to underwater hearing must have some basic modification from the type of ear used by a land animal. Sound not only travels approximately four times as rapidly through water as air, but it is passing through a medium of approximately the same density as the body, which is not true in air. In air, sound waves reach the inner ear through the auditory canal, the tympanic membrane, and the auditory ossicles. In water, because of the similarity in density of the mammalian body and the water, sound waves may be conducted right through the body. They therefore could pass from one ear to another rather than be received independently. This then presents a problem, since independent reception, by each of the auditory organs, is essential for accurate orientation. Cetaceans have solved this by having the tympanic bone, in which the cochlea is located, suspended from the skull by ligaments and surrounded by a cavity filled either with air or foam. Sound waves, therefore, can only be transmitted to the cochlea by specially modified auditory ossicles. The sound itself is produced in nasal sacs just inside the blowhole, rather than in a larynx as in terrestrial mammals. Many kinds of toothed cetaceans have an enlarged forehead called the "melon" containing oil-filled chambers; this is thought to serve as a means of beaming the sounds outward from the nasal sacs which are directly behind it. Recent experiments also indicate that sound reception from the water to the middle and inner ear may be by way of the lower jaw. The mandibles of porpoises, dolphins, and other toothed cetaceans are hollow and each is composed of an outer covering of

FIGURE 6.10.   The Amazon dolphin (*Inia*) lives in turbid South American rivers, where visibility is poor. It is believed to locate food and orient itself to a large extent by means of echo-location. (Courtesy of Steinhart Aquarium, San Francisco.)

very thin bone around an oil-filled cavity, which would favor sound transmission. The back of the mandible is very close to the middle ear.

Not only is directional hearing determined in whales, porpoises, and dolphins, as in man, by means of differences in the intensity of sounds received in each ear, but it is believed that they make use of frequency modulation. Most of their own underwater sounds show a great range in frequency. Consequently if both ears are not the same distance away from the object from which a sound is reflected, each will receive pulses of different frequency at the same time.

Experiments on porpoises and dolphins in captivity have shown that, when blindfolded, they are able to avoid obstacles and locate small objects with great accuracy. They may even distinguish between pieces of food and capsules of the same size and weight under such circumstances.

More recently it has been demonstrated that a number of kinds of pinnipeds make use of echo-location when swimming underwater. They too can, in total darkness, not only locate objects in the water with great rapidity, but may distinguish between objects of different composition but of the same size and shape. Seals and sea lions have fairly good eyesight, which they also make use of in clear water when the light is good and there is no need for using their sonar mechanism.

## ENDOCRINE GLANDS

The endocrine glands reach their peak of development in mammals. As in other vertebrates the most important of these is the pituitary, which has a dual embryonic origin. The posterior lobe develops from the floor of the brain, whereas the anterior lobe, the pars intermedia, and the pars tuberalis originate from the roof of the stomodaeum, just in front of the oral plate. A number of different hormones are produced by the various parts of this composite gland. Some hormones regulate the activity of other endocrine glands such as the thyroid, the adrenals, and the gonads. These are the *thyrotrophic,* the *adrenocorticotrophic,* and *gonadotrophic hormones.* Others are concerned with the regulation of such body activities as growth, fat, and carbohydrate metabolism and urine secretion.

The *pineal body* of mammals is small and represents the basal portion of the most posterior of the median, eyelike structures, sometimes called the epiphysis. Although it tends to decrease in size with age its function, if any, in mammals has not yet been determined.

The *thyroid gland,* which originates from the floor of the embryonic foregut, is large in mammals and consists of two lobes situated on either side of the trachea just below the larynx and connected together by an isthmus. Its product is a hormone called *thyroxin,* which is rich in iodine. In certain areas where the food and water is extremely low in iodine content, human beings develop abnormally large thyroids, a condition known as goiter. Thyroxin is important in regulating metabolism.

The *parathyroids* of most mammals are either embedded in or very closely associated with the thyroid gland. They are essential for life, since they play a major role in calcium metabolism. The *thymus gland,* like the parathyroids and the thyroid, is derived embryologically from the foregut. In mammals this gland migrates during the course of its development posteriorly to a position close to the anterior part of the heart. For many years it was known that the thymus was much larger in young than in adult mammals but its function was unknown. Recently it has been found that this gland is concerned with conditioning the lymphatic system to form antibodies and reject graft tissues. Mice, rats, and hamsters that are thymectomized within a week of birth exhibit a greatly reduced ability to synthesize antibodies and will successfully accept skin grafts from other individuals.

The endocrine activity of the mammalian ovaries and testes is considered in some detail in Chapter 12.

## SPECIAL CHARACTERS

A great many specializations of mammals are associated with the integumentary system. The mammary glands, skin glands, hair, antlers, horns, nails, claws, and hoofs are all basically a part of the integument or accessories to it. The integument, consisting principally of the skin, is composed of a thin outer layer, the *epidermis*, and a deeper and generally much thicker layer called the *dermis*. The former is derived embryologically from ectoderm whereas the latter arises from mesenchyme, which is mostly mesodermal.

Since the integumentary system is the outermost part of the body, it serves for protection in many ways. Because of its somewhat impenetrable nature the skin protects more delicate underlying structures from injury and serves as a barrier against infection. Hair assists in providing greater security for the body. The reduction in hair in most terrestrial mammals is usually compensated for by a thickening of the skin, as in the pachyderms. Sometimes, in addition to structures such as nails and hoofs, the skin will form horny, keratinized scales as on the tails of many rodents and some marsupials. The entire body or major parts may be shielded by such plates, as in pangolins and the many armadillo-like mammals.

A most important function of the mammalian integumentary system is to assist in thermoregulation. As will be seen, certain glands function for this purpose. The skin also plays a part in protecting the body from high temperature and solar radiation; mammals exposed to intense sun in desert areas tend to have pigmented skin which is thicker than that of their counterparts in temperate regions. The mammalian integumentary specializations follow.

### Mammary Glands

The mammary or milk glands are of primary importance to mammals and are responsible for the common name applied to this class of vertebrates. In certain respects these structures are allied to sweat and sebaceous glands, and all may have developed from a common ancestral type of epidermal structure. Embryologically, the mammary glands arise from epidermal thickenings that extend as a line down either side of the body from the axilla to the groin. At certain points along these *milk lines* or ridges the future glands arise. They appear in the development of both sexes but normally become functional only in the female and are under hormonal control.

The position of the glands varies greatly in different species. They may be pectoral, abdominal, or inguinal. In most mammals the outlets of these glands are the nipples or *mammae*. Only in the

monotremes do we find nipples lacking. In this group the ducts open
on the flat surface of the skin and the young lick the milk that exudes
onto the tufts of hair on the ventral surface of the mother. In the
marsupials and placentals the ducts from the mammary glands open
into the elevated nipples. In some groups, such as primates, the
ducts open directly at the tip of the nipple, whereas in others, such
as artiodactyls, there is a false nipple, with the ducts opening at the
bottom of a common milk canal.

The number of mammae varies considerably, depending upon
the species. They are always paired but the number of pairs may
range from one to 13 and is roughly correlated with the number of
young that the female bears. Mammals that usually have one or two
young have a single pair of mammae. This is true of primates, cetaceans, and certain bats. Other species that are more prolific have a
greater number of mammae to accommodate the young.

As has already been stated, the milk exudes from the glands in
the monotremes. In most mammals, however, it is sucked out by the
young. An exception to this is seen in the cetaceans, in which special
muscles are capable of forcibly ejecting the milk from the nipple into
the mouth of the young. This adaptation prevents loss of this nourishing material in this group, whose life is spent in the water and
whose lips, which are essential for sucking, have disappeared.

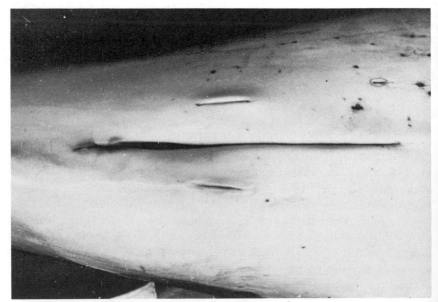

**FIGURE 6.11.** In cetaceans the nipples of the female when not in use are retracted within slits situated on either side of the vent, as seen on the undersurface of
this female harbor porpoise (*Phocoena phocoena*).

The *milk* or product of the mammary gland is the sole food of the young for the first part of postnatal life. In addition to water, it contains butterfat, lactose, albumin, and various salts. The proportion of these substances varies to a considerable degree in different kinds of mammals and is reflected in the growth rate of the young. Milk that is high in albumin content accelerates growth. In the human the albumin content is low and growth is correspondingly slow, whereas young guinea pigs feeding on milk rich in albumin double their weight in a few days. In most marine mammals the fat content of milk is also very high, sometimes amounting to 50 per cent. This accounts for the ability of fur seal pups to grow even though they may nurse their mothers only once every eight days. It also accounts for the rapid increase in body weight in certain other seals that, although weaned at two to four weeks of age, may quadruple their weight in this time.

## SKIN GLANDS

Apart from the mammary glands the other major mammalian skin or integumentary glands are the sweat glands, sebaceous glands, and scent glands. Sweat or *sudoriferous glands* are of common occurrence in mammals. They not only serve to eliminate waste substances from the body but greatly assist in thermoregulation. The evaporation of sweat from the skin serves to cool the body. Although in man sweat glands are widely distributed over the skin, this is not true for most mammalian species. The presence of heavy, dense fur in many kinds of mammals would make it highly disadvantageous if sweat glands were universally distributed over the body. In some species of rodents and certain carnivores, these glands are confined to the soles of the feet or the skin between the toes. The sweat glands of the wandering shrew (*Sorex vagrans*) are small in size, few in number, and restricted to the ventral surface of the body. Some bats seem to lack sudoriferous glands while others have them localized in the facial area (Fig. 6.12). These glands are reported to be completely lacking in whales, in which, of course, there can be no evaporation from the surface of the skin.

*Sebaceous glands* are usually associated with hair follicles, and their excretion serves to lubricate not only the hair but also the skin.

Many different kinds of *scent glands* are to be found among mammals. Some of these appear to be slight modifications of the sebaceous or sudoriferous types, whereas others are highly specialized. Scent glands serve a number of different purposes. Some are for defense against enemies or to warn other members of the same species away from a selected territory. Others function to attract

FIGURE 6.12.  Skin glands are found on the head in many kinds of bats. In the pallid bat (*Antrozous pallidus*) such glands occur on either side of the muzzle. A secretion, possessing a distinct odor, is exuded from the dark, porelike openings.

members of the same species or, frequently, individuals of the opposite sex. These glands are most highly developed in mammals that have a keen sense of smell. In forms that have the olfactory apparatus greatly reduced, such as whales, scent glands may be entirely lacking.

Probably the best known glands belonging to this category, among North American mammals, are those possessed by skunks. In these animals the secretion of the perianal glands may be forcibly expelled through papillae situated on either side of the anus. As all who have had unfortunate experiences with these mustelids know, this fluid can be directed at a suspected enemy with great accuracy and serves as a powerful defense weapon. The potent ingredient which produces the penetrating odor and which also may produce temporary or even permanent blindness if it contacts the eye is known as *methyl mercaptan*. It is interesting to note that the conspicuous black and white color so characteristic of the various species of American skunks is also found in certain of the Old World civets of the family Viverridae that are capable of expelling a similar type of scent for defense.

Other members of the family Mustelidae possess *anal glands* which may exude a fairly strong secretion when the animals are excited but, in all except the skunks, these glands appear primarily to serve other purposes. They may function to attract other members of the same species or members of the opposite sex. Scent from the anal glands of a weasel if rubbed on a trap will frequently lure other weasels to the trap. Martens, which are members of the weasel family, also possess *abdominal skin glands.* Both wild and captive individuals have been observed to drag their bodies over logs and other objects seemingly in an effort to deposit the secretion of the gland. Glands of a similar nature have also been found on the wolverine and badger.

In various members of the Canidae there are musk glands on the upper side of the tail near the base. The guard hairs on these areas are coarse and manelike, and underfur is lacking. The length of these glands varies both with the individual and the species. In the gray fox it may be from 3 to 4½ inches in length. Such glands appear to serve as a means of communication.

Peccaries exude a very strong odor from *dorsal glands* situated along the back. The glandular product is rubbed off on branches and twigs along trails and is often noticeable to members of the human species. Although the odor becomes more intense when the peccaries are excited, it is unlikely that it functions for defense. A small dorsal gland, situated between the shoulders, is also found on kangaroo rats. The exact function is not known, but it has been suggested that it may be of importance in the recognition of individuals or property and also may serve to waterproof the hair.

A number of American shrews possess skin glands on the sides of the body. These are thought to function in correlation with the reproductive cycle. Somewhat similar specialized sudoriferous glands are found on the head, chin, wrists, and ventral parts of the body in a number of genera of American moles. These glands appear to be most active during the breeding season. Moles also possess perineal glands.

*Ventral glands,* located on the surface of the abdomen, are found in wood rats. In the dusky-footed wood rat (*Neotoma fuscipes*) these glands are most highly developed in adult males during the breeding season. It has been suggested that they may play a part in the establishment and maintenance of territory as well as the recognition of places within the territory.

The pronghorn antelope (*Antilocapra americana*) has 11 scent glands, located as follows: one at the base of each ear, one on the lower back, one on each side of the rump, one behind each hock, and one interdigital gland on each foot. Some of these undoubtedly serve for recognition and sexual attraction. It has been noted, however, that when the hair on the rump is raised or lowered a strong musky odor is given off, suggesting the possibility that the glandular secretion there may serve for protection against various noxious insects.

Members of the deer family possess a number of skin glands, some of which are of considerable importance to taxonomists. Anterior to the eye is the sac-like *preorbital gland,* which fits into a pit in the underlying lacrimal bone. Cowan (1936) is of the opinion that this gland is functionless in American deer, although it is considerably larger in the mule deer (*Odocoileus hemionus*) than in the white-tailed deer (*Odocoileus virginianus*). In certain genera of antelope (Bovidae), however, the preorbital glands become enlarged during the breeding season and can actually be everted, producing a peculiar "four-eyed" effect. The *tarsal gland* is situated on the median side of the tarsal joint and consists of thickened areas of skin containing enlarged sebaceous and sudoriferous glands. It is covered by tufts of long, coarse hair, and the oily secretion is reported to have an ammoniacal odor. This gland is most active, at least in the male, during the breeding season. Studies made on the tarsal glands of black-tailed deer (*Odocoileus hemionus columbianus*) have shown that members of a herd regularly smell each other's tarsal gland during both day and night. It appears to be important in recognition. If a strange individual approaches the herd, smelling of this gland is the first stage in aggressive behavior. Subsequently the intruder may be chased. Fawns are said to recognize the mothers by the odor of the tarsal gland. Many members of the deer family possess *metatarsal glands,* although these are lacking in the moose (*Alces*) (Fig. 6.13). This gland is elongate in shape and situated on the outside of the metatarsus. A dark, horny ridge runs down the center of the

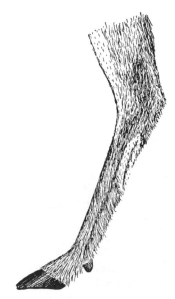

**FIGURE 6.13.** The elongate metatarsal gland on the outer side of the lower hind limb in the mule deer (*Odocoileus hemionus*) is surrounded by a tuft of long hairs.

gland and is devoid of hair although the rest of the gland is marked by stiff, long hairs. The metatarsal gland varies considerably in size in different subspecies of mule deer and is situated entirely or largely above the midpoint of the shank, whereas in the white-tailed deer it is below the midpoint of the shank and small in size. It has been suggested that the musky-smelling secretion aids the animals to locate various regular resting spots. *Interdigital glands* are present in many artiodactyls, and in American deer are found on all four feet. These glands, situated between the two main digits, are active the year around and may aid individuals in tracking other members of the same species or even in retracking themselves.

Many kinds of bats possess skin glands, which are frequently located on the head. In some species the odor produced by these glands is very strong. In the pallid bat (*Antrozous pallidus*) it faintly resembles the scent of a skunk. As is true of many mammalian skin glands, the secretion is exuded from the glandular pores in greater quantity when the animals are frightened, although this does not necessarily mean that it is a defense mechanism. Perhaps it aids individuals in locating their roosts, since the latter take on the strong odor so characteristic of the occupants.

*Preputial glands* are found in many kinds of mammals. The most notable perhaps are those of the beaver. Large sacs containing a secretion known as *castoreum* lie beneath the skin on either side of the penis and open by ducts into the prepuce. During the breeding season these glands become very much enlarged and the secretion, which is deposited at various places, probably assists the beavers in the securing of mates. Glands of a similar nature are found in many other rodents and also in members of the dog family. Male coyotes, foxes, and wolves regularly deposit their urine, which is mixed with the products of the preputial glands, on various scent posts which are visited periodically by other members of the same species.

## HAIR

Next to mammary glands, the most diagnostic mammalian feature is hair. There are a few groups of mammals in which this integumentary structure is greatly restricted, as, for example, in the armadillos and the whales. In some of the toothed whales it is reported present only in embryonic life. Most members of this class, however, have a well developed covering of hair on the body.

From the developmental standpoint, hair is an ectodermal structure derived from the malpighian layer of the epidermis. Structurally, on cross sectioning a hair, it is seen to be composed of three principal layers. The inner core, called the *medulla*, consists of

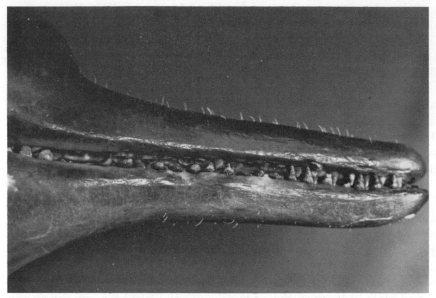

**FIGURE 6.14.** Some cetaceans have a few hairs on the head, even as adults. Here we see a profile of an Amazon dolphin (*Inia geoffrensis*) showing the presence of hair on the snout and lower jaw. (Photograph by Jacqueline Schonewald.)

cuboidal cells which are arranged loosely and may or may not contain pigment granules. In some hairs the medulla may be entirely lacking, whereas in others it may occupy most of the shaft. Outside the medulla is the *cortical layer*. The cells of the cortex are fusiform in shape, packed tightly together, and may or may not contain pigment. The relative extent of this layer varies with different species. The outer layer of the shaft is composed of the *cuticle*. The cuticle consists of flattened, scalelike cells which are generally unpigmented. These scales vary greatly in shape and arrangement in different species of mammals. Generally, however, they are divided into two groups, imbricate and coronal (Fig. 6.15). The *imbricate scales* are arranged so that they overlap one another, as do shingles. *Coronal scales* surround the shaft and fit into one another like stacked thimbles or glasses. A hair in cross section may be circular or flattened. Straight hairs have circular shafts, while curly hairs have flattened shafts; the degree of flattening is directly correlated with the amount of curliness.

Generally speaking hairs are roughly divided into two categories, guard hairs and underhairs or underfur. The *guard hairs* are the larger coarse hairs that are most apparent on the outer surface of the fur, whereas the hairs that constitute the *underfur* are usually much finer, shorter, and are not very apparent until the fur is parted. In most fur-bearing mammals the number of guard hairs is generally

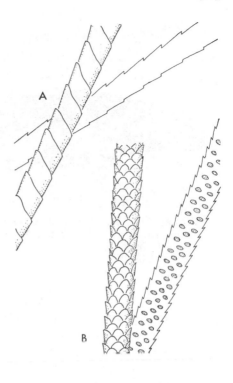

FIGURE 6.15. Examples of mammalian hair shafts illustrating (A) coronal scales in the red bat (*Lasiurus borealis*) and (B) imbricate scales in the pocket gopher (*Thomomys bottae*). (Redrawn from Hausman.)

exceeded by the underfur. Not all mammals possess underfur. In some of the Pinnipedia the underfur is greatly reduced or even entirely lost. Likewise, in some of the Insectivora the fur is so fine throughout that one cannot satisfactorily distinguish guard hairs from underfur.

Since hair, once it is formed, is a nonliving, keratinized structure, it is constantly subject to wear, just as the pigment within it is subject to fading. This necessitates periodic replacement, which is accomplished by molt of the old hair and growth of new hair. Such molts occur annually in some species, generally in the fall of the year. In other mammals there may be two replacements of hair each year, in which instance there is usually a spring and autumn molt. In a few species molts are of irregular occurrence. The process of molt, although it may occur within a few weeks, is a gradual one, and the old hairs are not lost until the new hairs have nearly attained their full growth. The manner in which the molt takes place generally follows a very definite pattern within any given species. A rather unusual type of molt occurs in the elephant seals (*Mirounga*). In these pinnipeds the old hair along with a layer of epidermis is molted in patches that may be several inches in diameter. By the time this occurs the new hair has already erupted from the surface of the skin beneath.

**FIGURE 6.16.** The adult and subadult males of the northern elephant seal (*Mirounga angustirostris*) of the eastern Pacific Ocean come to offshore islands to molt during midsummer.

Hair serves many functions for the mammal, but its prime purpose is the conservation of body heat. As a body insulator, the hair prevents undue loss of energy in this group of warm-blooded vertebrates, performing much the same function as do feathers for birds. Mammals living in cold regions usually have heavier coats of fur than do those in warm climates. In some marine forms, such as cetaceans, long, heavy fur would obviously be a detriment. The reduction or loss of this insulating layer is compensated for by the presence of a thick layer of fat or blubber. In pachyderms, such as the elephants, in which there is reduction in the amount of hair, we find that the skin has become greatly thickened.

In addition to preserving body heat, thus serving in part as a thermoregulator, hair performs many other functions, one of the principal of these being protection. The stiff coarse hairs which we refer to as *quills* in the porcupine are easily recognizable as serving for defense in these relatively slow-moving animals. This is likewise true of the specialized hairs of the Old World hedgehogs. Less obvious, but probably of equal importance in the survival of many species, is the color of the hair. In many mammals we find that the body color, which is due largely to the presence, absence, or distribution of pigment in the hair shaft, tends to blend with the environmental background. Mammals living in dark, forested areas are frequently much darker than those living in pale, desert regions. Even within single species this tendency toward a direct correlation between the color of the animal and its environment is often very noticeable. Perhaps the best illustration of color adaptation to environmental background is shown by *variable* animals, forms that have a dark summer pelage and a winter coat that is essentially white.

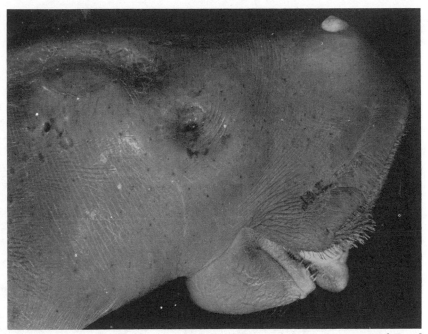

**FIGURE 6.17.** In the dugong (*Dugong dugong*) of the western tropical Pacific Ocean and the Indian Ocean, body hair is essentially reduced to stiff bristles that surround the mouth.

Such animals as the varying hare and the weasels exhibit this variableness in regions where there is snow in winter (Fig. 6.18).

This tendency for the color of many mammals to blend with that of the environment would seem to be definitely beneficial from the standpoint of making the animal less conspicuous. In the case of predatory species the individual has a better chance to approach its prey unnoticed. As far as the species preyed upon are concerned, protective color would seem to give them a great chance to elude their enemies.

Some mammals use hair as a means of signaling or warning other members of the species of danger. The white rump patch of the pronghorn antelope serves such a purpose. The hairs on the rump may be elevated rapidly by means of small muscles so as to produce a conspicuous white flash that can be seen for a considerable distance and serve as a warning when necessary.

Certain arboreal mammals, such as squirrels, have the hairs so arranged on the tail that they serve essentially as a rudder when the animal is leaping from branch to branch in a tree. Such a tail, with long hairs extending laterally, is also very important as a balancing organ in these species.

FIGURE 6.18. *Top,* Captive varying hare (*Lepus americanus*) photographed October 5 shortly after the onset of the autumnal molt with the brown summer pelage still quite prominent. *Bottom,* The same individual 17 days later with the white winter pelage rapidly replacing the summer pelage.

There may be specialized hairs on various parts of the body that are closely associated with sensory nerves and serve as tactile organs to acquaint an animal with the presence of objects in its environment. The moustachial whiskers or *vibrissae* come into this category and would seem to be of considerable importance, especially to the animals living in holes or burrows, where there is little or no light. Here the sense of touch is very important. Tactile hairs are often present on other parts of the head and on the legs of certain mammals.

## TEETH

Although teeth are found among fishes, amphibians, and reptiles, and are known to have been present in ancestral birds, they are most highly specialized in mammals as a group. It is true that these structures have been lost in certain forms, such as baleen whales and anteaters, but in most mammals teeth play an important part in everyday life, assisting in the securing of food, the chewing of food, and in many instances serving as weapons of self-defense.

A typical tooth consists of a thin but extremely hard outer covering referred to as the *enamel*, which overlies a thicker but softer layer called the dentine. The center of the tooth contains a cavity in which there are nerves and blood vessels. The enamel, however, may be greatly modified in certain species as a result of infolding, reduction, or even complete elimination.

Most mammals are *diphyodont;* that is, they have two sets of teeth, unlike many of the lower vertebrates in which tooth replace-

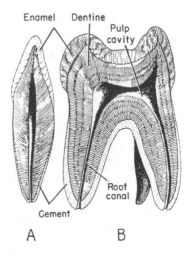

**FIGURE 6.19.** Sections through a mammalian incisor, *A*, and molar, *B*. (Romer, A. S.: *The Vertebrate Body.* W. B. Saunders Co., 1970; after Weber.)

ment may be an almost continuous process throughout life. The first set is referred to as the *deciduous* or *lacteal dentition,* commonly called the milk teeth. In some species the milk teeth are present only in embryonic life, being resorbed prior to birth. In many species the deciduous teeth are present only for a few days or weeks after birth, whereas in others they may be retained for greater periods of time. In moles the deciduous teeth remain throughout life and the permanent dentition is suppressed. Sometimes the milk teeth are highly specialized. For example, in certain bats these little teeth are recurved so as to form hooklets that enable the young to hold on more firmly to the mother's nipple.

In most mammals the first set of teeth is succeeded by the *permanent dentition,* which is not replaced in the event of loss or injury. These teeth are generally divided into four groups which are designated, as we proceed from anterior to posterior on the jaw, as the incisors, canines, premolars, and molars. The number of teeth present in each group varies with different kinds of mammals except for the canines, no more than one of which occurs on each side of each jaw. In expressing the number of teeth for a given species, mammalogists frequently make use of a *dental formula* which represents the number and type of teeth present on one side of the head. The dental formula for a coyote (*Canis latrans*), for example, is $I\frac{3}{3}$, $C\frac{1}{1}$, $Pm\frac{4}{4}$, $M\frac{2}{3}$ or 21 teeth. By multiplying this number by two, we arrive at the total number of teeth possessed by this species, which is normally 42. In the usual dental formula the initials used here to indicate the types of teeth are left out.

The *incisor teeth,* situated as they are in the front of the mouth, are used principally for cutting, grasping, and gnawing. They are, therefore, most highly developed in herbivorous animals. In most of the larger herbivores they have become flat and bladelike. Sometimes, as we find in the Cervidae and Bovidae, the upper incisors are lacking, and their function is taken over by the lips and tongue so that the cutting of vegetation is performed by the lower incisors.

The most remarkable incisor teeth are found in the proboscideans, which are now confined to the Old World. The large tusks of the elephants as well as those of the extinct mastodons and mammoths represent the second upper incisors.

Although many mammals possess six pairs of incisors the number has been reduced to two pairs in the rodents. In this group these teeth are highly specialized and chisel-like in shape. Furthermore, the incisors of rodents continue growing throughout the life of the individual. To produce the chisel shape the rodent incisors possess enamel only on the anterior surface. This edge wears down more slowly than the back edge. Wear, therefore, is essential to keep the incisors of rodents from becoming too long. In some rodents, such as certain microtines, harvest mice, and some genera of pocket gophers, the anterior surface of the incisors is grooved as a result of a longitudinal fold in the enamel. Lagomorphs differ from rodents in the possession of a second pair of upper incisors which are very small and situated directly behind the first pair. Furthermore, the crowns of the incisors of lagomorphs are completely encased in enamel.

Bats exhibit considerable variation in the number of incisors present. No more than two such teeth are found on each side of the upper jaw, although the full number of three in the lower jaw occurs in many species. In some forms, however, the upper incisors are completely lacking and, as in certain American bats of the genus *Molossus,* the lower incisors may be reduced to a single tooth on each side. Perhaps the greatest specialization of the incisors occurs among the vampires (*Desmodus*) in which the single upper incisor on each side is shaped like a miniature sharp blade for bloodletting. The incisors of most carnivores are small and insignificant.

Unlike the incisors the *canines* appear to be most highly developed in mammals that subsist on animal matter. In the carnivores these teeth are relatively large and pointed. They serve principally for the capturing and killing of prey and for tearing flesh. The upper canines of the walrus have become enlarged to form tusks which enable these large animals to pull themselves up onto the ice and also to dislodge shellfish from the floor of the ocean. In most bats the canines are elongate and probably enable insectivorous forms more readily to grasp their prey. The upper canines of the vampires, however, are similar to the upper incisors in shape, having a blade-like shape for piercing the skin of their victims.

In the artiodactyls the canine teeth may be elongate and tusklike as in some of the pigs, or they may be lacking or rudimentary in the upper jaw and incisiform in shape in the lower jaw as in most of the deer family. Canines are completely lacking in rodents and lagomorphs. In these two groups there is a large space on either side of each jaw, between the incisors and molariform teeth, called the *diastema.* The most unusual canine development is to be seen in the narwhal (*Monodon monoceros*) of the Arctic. In this cetacean the left upper canine of the male has developed into a great tusk that may grow to nine feet. It spirals to the left and is grooved helically and there are no other functional teeth in the mouth. Only rarely does the female have a tusk.

The *premolars* and *molars* exhibit great diversity in form in various groups of mammals, depending on the food habits of the species involved. The premolars are anterior to the molars on each jaw and are preceded in appearance by deciduous teeth. The molars do not have deciduous predecessors. While the premolars sometimes

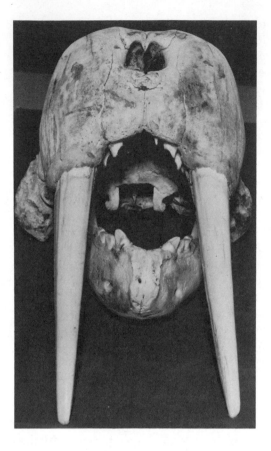

FIGURE 6.20. In the walrus (*Odobenus rosmarus*) the canine teeth are enlarged to form tusks in members of both sexes. (Photograph by John Koranda.)

differ from the molars in shape, they are often so similar that both may be included under the term *molariform* teeth.

Because of the diversity of the shape of the molariform teeth in different mammalian groups, a number of terms have been used to describe the basic differences (Fig. 6.21). Teeth with low crowns, such as those of squirrels, are referred to as *brachyodont*, while those with high crowns, such as the molariform teeth of microtines, are called *hypsodont*. Sometimes the surface of the crowns are *bunodont;* that is, they possess rounded cusps, as in members of the genus *Peromyscus*. With wear the surface of these teeth tends to become flattened. Such teeth are best adapted for crushing and bruising food. In many herbivores the enamel on the crown is folded so that with wear the surface presents a series of crescentic ridges, as in deer. This condition is referred to as *selenodont*. Teeth in which the enamel ridges run crosswise are called *lophodont*. Selenodont and lophodont teeth are best suited for cutting and slicing vegetation. Many carnivorous kinds of mammals have the crowns laterally compressed into sharp, bladelike cutting edges. This type of denti-

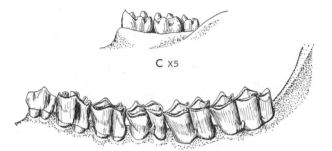

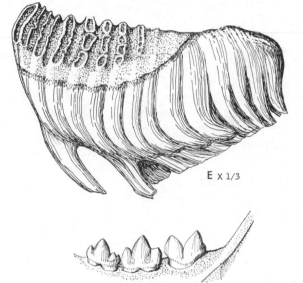

**FIGURE 6.21.** Selected examples of mammalian molariform teeth: *A*, brachyodont teeth of a ground squirrel (*Citellus*); *B*, hypsodont teeth of a wood rat (*Neotoma*); *C*, bunodont teeth of a white-footed mouse (*Peromyscus*); *D*, selenodont teeth of a deer (*Odocoileus*); *E*, lophodont teeth of an elephant (*Elephas*); *F*, secodont teeth of a mountain lion (*Felis*). Approximate size indicated.

tion is called *secodont*. In carnivores, specialized cutting teeth known as *carnassial* teeth consist of the last premolar in each upper jaw and the first molar in each lower jaw. Although the basic number of premolars and molars in mammals is thought to be four and three, respectively, on each side of each jaw, many species exhibit a marked reduction in the number of these teeth.

*Heterodont* dentition, or the specialization of the teeth into incisors, canines, premolars, and molars, is regarded as a mammalian characteristic. As has already been pointed out, however, certain forms have lost their teeth and others, such as the porpoises and dolphins, have reverted to essentially a *homodont* condition, in which the teeth appear as more or less uniform pegs in each jaw. Functional teeth are lacking in the baleen whales and their place is taken by long plates of a horny material called *baleen* that hangs in rows from the roof of the mouth. Each plate is fringed on its inner edge and the collective action of this series of strainers serves to screen masses of planktonic organisms from the sea water for food.

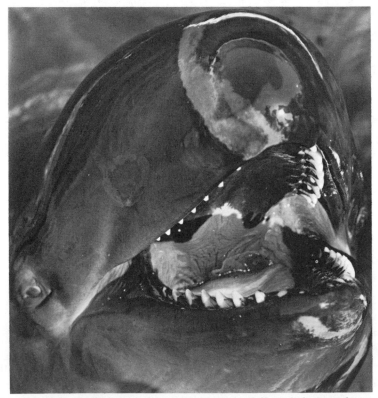

**FIGURE 6.22.** Looking into the mouth of a false killer whale (*Pseudorca crassidens*) one sees the uniform homodont dentition characteristic of most toothed cetaceans. (Courtesy of Marineland of the Pacific, Los Angeles, California.)

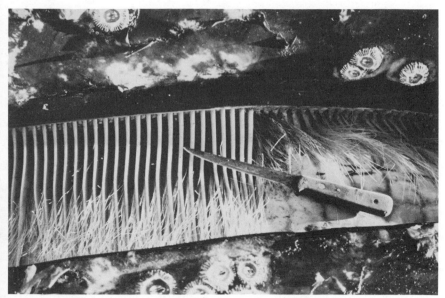

FIGURE 6.23. A lateral view into the partly opened mouth of a gray whale (*Eschrichtius gibbosus*) showing the plates of baleen hanging from the palate. Some of those on the right side have been cut off.

## ANTLERS AND HORNS

A common expression among sportsmen is: "Don't shoot until you see its horns." A well informed farmer might shudder if he were to overhear a deer hunter use these words because he would know that in all probability the only horned animals on his property are cattle, sheep, or goats. Actually the hunter does not mean horns but rather antlers.

Both antlers and horns are outgrowths found on the heads of artiodactyls, but they differ from one another basically. Antlers are confined to the family Cervidae (although lacking in the Asiatic genera *Moschus* and *Hydropotes*), are normally present only in males, except in the genus *Rangifer*, and are shed annually. Horns of the so-called "hollow" type are found among members of the family Bovidae and, in a peculiarly modified form, in the family Antilocapridae. They commonly occur on females as well as males, although when present in both sexes, they are usually larger in the males. Furthermore, if any shedding occurs only the outer sheath is involved.

When one considers that antlers are generally grown in a matter of a few months, even those of the Alaskan moose which may measure more than 6 feet from tip to tip, one cannot help but regard these structures as among the most unique to be found in mammals.

The antler is made of bone and during its period of growth is covered by the *velvet,* consisting of skin on which there are short, fine hairs. The bone comprising the inner portion or core of the antler is spongy while that on the outer surface beneath the velvet is quite compact. The growth of the antler each year in the spring begins at the top of the *pedicel,* an outgrowth of the frontal bone. In deer this pedicel appears as a protuberance on each frontal bone when the young males are about three months old. By the first winter these bulges have grown to a length of 1 or 2 inches and represent not only the pedicels, which are permanent, but the beginning of the first set of antlers, which will have attained their full growth by the following autumn. In yearling bucks these antlers are usually single prongs or *spikes,* although they may on occasion be forked or even have three points each. When the antlers are mature in the late summer or early autumn the supply of blood to the bony structure as well as the velvet is reduced and ultimately shut off. The velvet is then shed, partly as a result of the buck's rubbing its antlers against branches and objects, and the hard underlying bone is exposed. The antlers, which are important weapons of defense used by the bucks when fighting each other for possession of females during the rutting season, become weakened above the pedicel by the end of the winter. The first antlers in deer, therefore, require about a year and one half from the original outset of growth until they are finally shed. This is not true of all members of the family. Moose, for example, may develop small growths, referred to as *buttons,* during the first summer and autumn, but they are shed the first winter. Irrespective of these variations during the early period of life, all the North American cervids require only a few months to grow their antlers after the first set is shed. Growth in most deer begins about March and is completed by the end of August. Following this the velvet is shed and the antlers are kept until the following January or February. There is considerable variation in these times, of course, depending on the individual, the locality, and the species.

There are several decidedly different styles of antlers possessed by various species of cervids in North America (Fig. 6.24). Best known, perhaps, is that of the white-tailed deer (*Odocoileus virginianus*), in which there is a single main beam that sweeps upward and back from the pedicel, then curves forward. The points or *tines* arise from the dorsal surface of the main beam and in general are directed upward. The *brow tine,* however, which is a relatively short prong that arises a short distance above the enlarged base or *burr* of the antler, emerges from the median side of the beam. In the mule deer (*Odocoileus hemionus*) the antler is *dichotomously forked* so that the

**FIGURE 6.24.** Antler form in the principal members of the deer family (Cervidae) in North America: A, white-tailed deer (*Odocoileus virginianus*); B, mule deer (*Odocoileus hemionus*); C, moose (*Alces alces*); D, elk or wapiti (*Cervus canadensis*); E, caribou (*Rangifer tarandus*).

main beam, some distance above the brow tine, divides into two secondary beams. These secondary beams then each divide again. In elk (*Cervus*) we find that the antler consists of a main beam which sweeps up and backward with the tines emerging from the anterior surface of the beam. In addition to the brow tine near the base there is also a *bez tine* and both extend out above the muzzle. In adult bull elk the remaining tines usually number four.

In both deer and elk the number of tines or points tends to increase with maturity, although this cannot be used as a positive criterion to determine age. After the prime of life is past the number of points may be reduced, and in old age the antlers may be deformed.

The antlers of moose (*Alces alces*) differ from those of deer and elk in that in most adult bulls there is a main beam which is generally round, in cross section, only near the pedicel. The remainder of the beam becomes flat and *palmate* with points extending out from the anterior edge. There are certain individuals, however, which do not have the beam flattened, although the appearance of this so-called "cervina" type of antler is of more common occurrence in the Old World moose.

The antlers of caribou (*Rangifer*) are somewhat more complicated than those of other members of the deer family in North America. Each antler usually consists of a single, heavy main beam which sweeps upward and back for about half its length, then curves forward. The tip of the beam is generally laterally compressed and palmate with a number of points projecting out from the posterior edge. There is a pair of brow tines, one or occasionally both of which may be laterally compressed with points along the tip forming what is called the "shovel." Above the brow tine is the larger bez tine, which also is generally flattened near the tip and possesses points. The brow and bez tines are both directly forward and extend out above the muzzle.

True horns of the hollow type possessed by members of the Bovidae and Antilocapridae (Fig. 6.25), as opposed to the so-called horns of the rhinoceros, consist of an inner core of bone which is an outgrowth from the frontal bone. This core is a permanent structure and continues to grow throughout life. Encasing this structure, in the form of a sheath, is a keratinized, epidermal covering that is true horn. The size of the sheath increases with the size of the bony core. In the Bovidae the horn is unbranched and is usually a permanent structure throughout life, although it varies greatly in shape in different species as is readily seen on comparing the massive horns of the bighorn ram (*Ovis canadensis*) with the short prongs present on a mountain goat (*Oreamnos americanus*). Only a single pair of horns is possessed by most members of this family. The male of the so-called

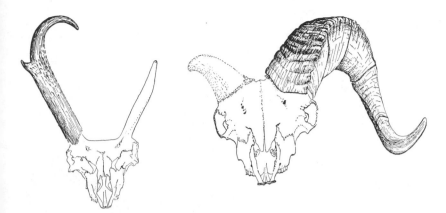

**FIGURE 6.25.** Diagram illustrating (*left*) the so-called "hollow" horns of the pronghorn (*Antilocapra americana*) and (*right*) the mountain sheep or bighorn (*Ovis canadensis*) with the outer, horny sheath removed from one side of each.

four-horned antelope (*Tetracerus quadricornis*) of India has two pairs of these structures. The anteriormost pair is the smaller and is situated above the orbital region. The female of this species lacks any horns. Of the 49 recognized genera of bovids there are only nine genera in which horns are restricted to the males.

The pronghorn antelope (*Antilocapra americana*), which belongs to a monotypic American family, possesses true horns which are unique in two respects. First, they are usually forked in adult males and, second, the outer sheath is shed annually. After the horn is shed it is replaced by another which grows from the skin that surrounds the core.

The giraffe and okapi, both members of a unique African family of artiodactyls, have hornlike structures that differ from those of any other kind of mammals. They develop from cartilaginous protrusions which are present at birth. These ossify and fuse to the top of the skull, where they appear as knobs permanently covered with living skin and hair. The giraffe possesses three of these knobs, one of which is median and anterior to the other two, and the okapi has but two.

## APPENDAGES

The limbs of mammals are very definitely adapted to the mode of life of the various species, whether they run, jump, or walk on the surface of the ground, burrow in the soil, climb trees, glide, fly through the air, or swim in the sea (Fig. 6.26).

Fast running mammals like the pronghorn antelope and deer are

**FIGURE 6.26.**   Limb adaptations in mammals: *A*, flying squirrel (volant); *B*, tree squirrel (arboreal); *C*, bat (aerial); *D*, kangaroo rat (saltatorial); *E*, mole (fossorial); *F*, polar bear (plantigrade); *G*, deer (cursorial); *H*, sea lion (aquatic); *I*, dolphin (marine). (Drawings by Jacqueline Schonewald.)

referred to as *cursorial*. Such species usually have proportionally long, slender limbs with the muscle masses concentrated close to the body. The femur and humerus are relatively short while the tibia and fibula as well as the radius and ulna are elongated. The distal skeletal elements, fused to some extent in the limb, are reduced in number and certain of them are elongated. The functional toes are often reduced to two or even one, only the nails or hooves of which touch the ground. The joints are grooved so as to restrict motion largely to forward and backward, thus reducing the danger of dislocation during periods of rapid movement.

The term *saltatorial* is applied to certain mammals that progress over the ground by means of jumping, such as the kangaroo rats (*Dipodomys*) and jumping mice (Zapodidae). In such forms the hind limbs are considerably larger than the front limbs and the hind feet are elongate as is the tail. The latter serves as an important organ of balance. The front feet play only a very minor part in locomotion in these mammals.

Many species walk about with their feet and toes on the ground. Good examples are the raccoon (*Procyon lotor*) and various kinds of bears (Ursidae). These mammals are *plantigrade* and have limbs of moderate size which are capable of motion in most directions. There is no marked fusion or reduction of the skeletal elements and the toes usually possess strong claws.

Other species, technically referred to as *fossorial*, are adapted to life beneath the surface of the ground, such as the moles (Talpidae). In these subterranean forms there is a shortening and strengthening of the appendages, especially the forelimbs. The latter tend to extend laterally from the body with the palms broadened and the claws extremely strong. In the pocket gophers (Geomyidae) the front limbs do not extend laterally from the body, but they show the same tendency to be short and broad with enlarged claws and highly developed shoulder musculature.

*Arboreal* mammals, such as many members of the squirrel family (Sciuridae) and many primates, spend much of their time in trees and are adapted to climbing and jumping from limb to limb. Such species do not have a reduction in the number of bones of the appendages, frequently have ball and socket joints to allow freedom of motion in all directions, and possess claws or nails on the ends of the digits. The front feet are generally capable of grasping objects such as branches. In arboreal species the tail often functions as an important accessory in locomotion. In the squirrels it serves partly as a rudder and as an organ of balance, while in the New World monkeys it is prehensile or grasping.

*Volant* mammals are gliders, as for example, the flying squirrels (*Glaucomys*) of North America. The front and hind limbs are squir-

FIGURE 6.27.  Ridges indicating the movements of a broad-footed mole (*Scapanus latimanus*) of western North America, which forages beneath the surface in search of worms and insects. The circle in the center is approximately 15 feet in diameter.

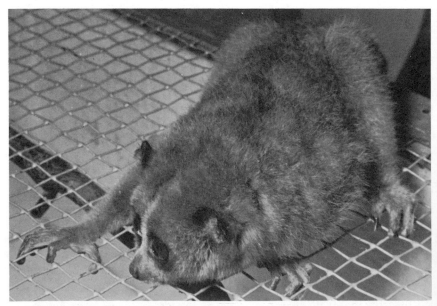

FIGURE 6.28.  The opposable thumb seen in the slow loris (*Nycticebus coucang*) of southeastern Asia is a decided aid to an arboreal animal, enabling it to grasp the branches of trees.

rel-like with well developed claws for climbing. A broad fold of furred skin, however, extends from each side of the body and connects the wrist with the ankle. A cartilaginous spur extends back from each wrist and serves as an additional support for the gliding membranes. The dense hairs of the tail extend laterally so that this organ appears very flattened, thus increasing the gliding surface and also providing a rudder and organ of balance.

The bats, representing *aerial* or flying mammals, show remarkable modification of the bones of the forelimb, especially the metacarpals and phalanges (Fig. 6.29). These bones, with the exception of those of the thumb, have become greatly elongated and are joined together by a double membrane of skin called the *chiropatagium* to form a wing. An extension of this membrane extends back as the *plageopatagium* to the hind leg. In most species of bats there is an additional membrane, the *uropatagium*, which extends from the hind legs to the tail and may encase the latter structure in whole or in part. Correlated with the modification of the forelimb for flight in this group is the development of powerful pectoral and shoulder muscles and the concentration of muscle masses within the body or at the base of the wing so as to lighten the distal part of this appen-

**FIGURE 6.29.** A pallid bat (*Antrozous pallidus*) photographed in flight to illustrate the wing membranes.

**FIGURE 6.30.** The bottle nosed dolphin (*Tursiops*) represents a specialized marine mammal with a streamlined body, front limbs which have become modified into flippers, a tail which has developed into a rudder (the flukes), and whose dorsal surface has acquired a fin that serves as a stabilizer. (Photograph courtesy of Bruce Markham.)

dage. The thumb retains its claw and in the fruit bats (Megachiroptera) of the Old World a remnant of the claw of the second digit of the forelimb is present.

The hind limbs are small and have rotated so that the bats may hang from vertical surfaces with the dorsal surface of the body outermost. In certain species, such as some of the fish-eating bats (*Pizonyx* and *Noctilio*), the hind claws have become remarkably large and may serve to impale small fish or crustaceans as the bats fly over the surface of water at night.

*Marine* and *aquatic* mammals show various appendicular modifications. In the cetaceans there is a complete loss of the pelvic limbs (except for tiny remnants of the girdle embedded in the body wall), a dorsoventral compression of the tail to form a flattened rudder called the fluke, and the development of the front limbs into flippers. No separate digits are present on the latter and nails or claws are lacking. In the pinnipeds the front and hind limbs are modified into *flippers,* although nails may still be present. The tail is greatly shortened, and in swimming the hind flippers extend back in such a way as to serve somewhat like the flukes of a whale. In the eared seals (Otariidae) the hind flippers can be brought forward under the body and assist travel over land, but in the hair seals (Phocidae) this is not possible.

**FIGURE 6.31.** The sea otter is a marine member of the family Mustelidae that makes use of a flat rock to break the shells of mollusks and crustaceans. The rock is placed on the chest for this purpose. (Photograph by Karl W. Kenyon.)

In aquatic species such as the otters and beavers, the limbs are not modified as flippers but the toes are webbed together as an aid to swimming. In other forms like water shrews we find that extensive rows of hairs on the sides of the toes serve much the same purpose.

## CLASSIFICATION OF MAMMALS

### CLASS MAMMALIA (MAMMALS)

Warm-blooded vertebrates possessing mammary glands; paired occipital condyles present; ramus of lower jaw composed of a single bone, the dentary; Triassic to Recent; three subclasses.

SUBCLASS PROTOTHERIA. Primitive egg-laying mammals of medium-small size (300 to 775 mm.)* possessing hair and mammary glands that lack nipples; pectoral girdle with separate interclavicle, coracoid, and precoracoid as well as scapula; epipubic or marsupial bones attached to pelvis; functional teeth absent in adults; oviducts separate and opening into a cloacal chamber; horny, medial spur that may be grooved for the passage of secretion from poison glands present on ankles of males; Pleistocene to Recent although undoubtedly represented in the Mesozoic; one order.

---

*Measurements given in this classification include only head and body length.

**ORDER MONOTREMATA (MONOTREMES).** Characters of the subclass; confined to the Australian region; two families.*

    *Family Tachyglossidae (Spiny Anteaters).* Mouth and rostrum extended into an elongate beak; tongue long and sticky; limbs and feet modified for digging; tail vestigial; body mostly covered with spines or quills; represented by two genera, *Tachyglossus* of Australia, Tasmania, and New Guinea and *Zaglossus* of New Guinea.

    *Family Ornithorhynchidae (Duckbilled Platypus).* Snout elongated into a flattened, ducklike bill which is covered with a hairless, moist, leathery skin; tail dorsoventrally flattened; fur short and coarse; feet webbed for swimming; lakes and streams of eastern Australia; one species, *Ornithorhynchus anatinus.*

**SUBCLASS ALLOTHERIA.** An ancient and aberrant group of mammals characterized by three upper incisors with the second much enlarged; canines rudimentary or missing; molars up to five, possessing cusps arranged in two or three longitudinal rows; one order.

    **ORDER MULTITUBERCULATA (MULTITUBERCULATES).** Characters same as those of the infraclass; Lower Jurassic to Eocene; North America, Europe, and Asia.

**MAMMALIA OF UNCERTAIN SUBCLASS:**

    **ORDER TRICONODONTA (TRICONODONTS).** Small, primitive mammals known principally by their lower jaws; molars with three cusps in a longitudinal row; Upper Triassic to Upper Jurassic; North America and Europe.

**SUBCLASS THERIA.** Mammals whose living representatives are viviparous and possess mammary glands with nipples; pectoral girdle consisting of a scapula, sometimes a coracoid and clavicle; no interclavicle or precoracoid; functional teeth usually present; oviducts usually partly fused; Jurassic to Recent; three infraclasses.

    *Infraclass Pantotheria.* Known only from the Middle and Upper Jurassic and thought possibly to have been ancestral to the higher orders of the subclass Theria; two orders.

    **ORDER PANTOTHERIA (PANTOTHERES).** Molar teeth characterized by a main inner and outer cusp with two or more accessory cusps; spatial relationship of upper and lower teeth basically like that in later mammals; premolars piercing and trenchant; canines present; incisors of a generalized type. Middle and Upper Jurassic; North America, Europe, and Africa.

    **ORDER SYMMETRODONTA (SYMMETRODONTS).** Crown of lower molars V-shaped with a high centro-external cusp and lesser anterior and posterior cusps; possibly an early offshoot from the pantotheres; Upper Jurassic; North America and Europe.

    *Infraclass Metatheria.* A primitive and old group of mammals whose young are born in a premature condition and then carried by the

---

*In the following classification only living families and genera are given unless otherwise indicated.

female, usually in a pouch or marsupium located on the abdomen; teats opening into marsupium; pectoral girdle consisting of a scapula, coracoid, and clavicle; epipubic or marsupial bones attached to pelvis; cheek teeth usually consisting of three premolars and four molars; Cretaceous to Recent; one order.

ORDER MARSUPIALA (MARSUPIALS). Characters of the infraclass; very small to fairly large mammals (50 to 1600 mm.); presently found in the Australian region (where different groups have evolved in a manner parallel to that of placental mammals in other parts of the world) and North and South America; nine living families.

Family Didelphidae (American Opossums). Size, small to medium (85 to 500 mm.); muzzle elongate, tail usually long and prehensile; foot pentadactyl with opposable great toe; marsupium present or absent; dental formula $\frac{5}{4}, \frac{1}{1}, \frac{3}{3}, \frac{4}{4}$; temperate North America to Argentina; 12 genera.

Family Dasyuridae (Dasyurids or Marsupial Mice, Rats, Cats, Wolves, and Devils). Very small to very large (95 to 1300 mm.) marsupials possessing a long, nonprehensile, haired tail; limbs of equal length, five toes on front foot, four or five toes on hind foot; marsupium poorly developed or absent; dental formula $\frac{4}{3}, \frac{1}{1}, \frac{2}{2-3}, \frac{4}{4}$; Australia, Tasmania, New Guinea, Aru, and Normanby islands; 19 genera.

Family Myrmecobiidae (Numbat). A medium-small (175 to 275 mm.) marsupial with a transversely striped body, a long, hairy, nonprehensile tail, and a long snout; four toes on front foot, five on hind foot; marsupium lacking; dental formula $\frac{4}{3}, \frac{1}{1}, \frac{3}{3}, \frac{5}{6}$; western and southern Australia; one species, *Myrmecobius fasciatus*. (Included by some in the family Dasyuridae.)

Family Notoryctidae (Marsupial "Moles"). Small (up to 180 mm.) marsupials with a short, stocky body, adapted to subterranean life; tail short and covered with a leathery skin; eyes vestigial; ears lacking an external pinna; claws greatly enlarged on digits three and four of front feet for digging; marsupium present and opening posteriorly; dental formula $\frac{3-4}{3}, \frac{1}{1}, \frac{2}{2}, \frac{4}{4}$; Australia; one genus, *Notoryctis*.

Family Peramelidae (Bandicoots). Small to medium-sized (200 to 550 mm.) marsupials with elongate muzzle, nonprehensile tail, and proportionately large hind legs; front and hind feet generally possessing five toes with second and third toes of hind feet syndactylous; marsupium present and opening posteriorly; dental formula $\frac{4-5}{3}, \frac{1}{1}, \frac{3}{3}, \frac{4}{4}$; Tasmania, Australia, New Guinea, and nearby islands; eight genera.

*Family Caenolestidae (Caenolestids).* Small (90 to 135 mm.) marsupials with an elongate muzzle and long tail, giving them a shrewlike appearance; five toes on each foot; marsupium absent; dental formula $\frac{4}{3-4}, \frac{1}{1}, \frac{3}{3}, \frac{4}{4}$; presently found in South America from Venezuela to Chile; three genera.

*Family Phalangeridae (Koala, Cuscuses, and Phalangers).* Small to moderately large (60 to 850 mm.) marsupials; tail frequently prehensile but essentially absent in the koala; feet pentadactyl with opposable big toe; gliding membranes connecting the front and hind limbs in the "flying phalangers"; marsupium present, usually opening anteriorly; dental formula $\frac{2-3}{1-3}, \frac{1}{0}, \frac{1 \text{ or } 3}{1 \text{ or } 3}, \frac{3-4}{3-4}$; the Solomon Islands west to Timor and the Celebes, also Australia, Tasmania, and New Guinea; 16 genera.

*Family Phascolomidae (Wombats).* Moderately large (800 to 1200 mm.), heavy-bodied, short-legged marsupials with short, coarse hair; tail rudimentary; feet possessing five digits; second and third toes of hind foot syndactylous; marsupium present and opening posteriorly; dental formula $\frac{1}{1}, \frac{0}{0}, \frac{1}{1}, \frac{4}{4}$; Australia, Tasmania, and Flanders Island; two living genera, *Phascolomis* and *Lasiorhinus.*

*Family Macropodidae (Kangaroos, Wallabies, and Wallaroos).* Medium to large (235 to 1600 mm.) marsupials with greatly enlarged hind limbs and long, strong tails which aid in locomotion; five digits on front foot; first digit usually lacking on hind foot with second and third digits syndactylous and fourth and fifth elongate; marsupium present and fourth and fifth elongate; marsupium present and opening anteriorly; dental formula $\frac{3}{1}, \frac{0-1}{0}, \frac{2}{2}, \frac{4}{4}$; Australia, Tasmania, New Guinea, and adjacent islands in the Australian region; 19 genera.

**Infraclass Eutheria.** Viviparous mammals possessing an embryonic allantoic placenta; both a marsupium and epipubic bones lacking; cloaca absent in all except a few primitive forms; molars, basically three on each side although reduced or absent in some species; Cretaceous to Recent; 17 living orders.

ORDER INSECTIVORA (INSECTIVORES). Primitive placental mammals of small or medium-small size possessing an elongated snout, small eyes, and small external ears; limbs pentadactyl; teeth primitive; worldwide except Australia, Greenland, Antarctica, and much of South America; eight families.

*Family Solenodontidae (Solenodons).* Relatively large (280 to 325 mm.), shrewlike insectivores with a very elongate snout and naked tail; dental formula $\dfrac{3}{3}, \dfrac{1}{1}, \dfrac{3}{3}, \dfrac{3}{3}$; confined to Cuba and Haiti; two genera, *Atopogale* and *Solenodon*.

*Family Tenrecidae (Tenrecs).* Small to relatively large (40 to 400 mm.) insectivores exhibiting a considerable range in characters; some genera possessing soft fur, others with hedgehog-like spines; toes separate or webbed; tail present or absent; dental formula $\dfrac{2}{3}, \dfrac{1}{1}, \dfrac{3}{3}, \dfrac{4}{3}$; islands of Madagascar and Comoro; nine genera.

*Family Potamogalidae (Otter Shrews).* Medium to relatively large (180 to 350 mm.) insectivores with short, dense fur and a long, strong tail adapted to aquatic life; dental formula $\dfrac{3}{3}, \dfrac{1}{1}, \dfrac{3}{3}, \dfrac{3}{3}$; tropical, equatorial Africa; two genera, *Potamogale* and *Micropotamogale*. (Otter shrews are placed in the Tenrecidae by some taxonomists.)

*Family Chrysochloridae (Golden Moles).* Medium-small (75 to 235 mm.) insectivores with a vestigial tail and long, leathery snout; eyes vestigial; external ears very small; pelage short and thick, often with a metallic golden luster; front and hind limbs short and stout, the former with the claw on the third digit greatly enlarged for digging; dental formula $\dfrac{3}{3}, \dfrac{1}{1}, \dfrac{3}{3}, \dfrac{2\text{-}3}{2\text{-}3}$; central and southern Africa; five genera.

*Family Erinacidae (Hedgehogs and Gymnures).* Medium to large (100 to 450 mm.) insectivores with a moderately long snout; eyes well developed; ears with an external pinna; tail rudimentary to moderately long; toes, generally five but reduced to four on the hind foot in some species; hairs developed into spines on the sides and upper parts of the body in hedgehogs; dental formula $\dfrac{2\text{-}3}{3}, \dfrac{1}{1}, \dfrac{3\text{-}4}{2\text{-}4}, \dfrac{3}{3}$; Europe, Asia, and Africa; 10 genera.

*Family Macroscelididae* (Elephant Shrews). Medium-small (100 to 315 mm.) insectivores with an elongate, movable snout; eyes large; external ears well developed; tail long and sparsely haired, often used in locomotion; hind legs larger than front legs; fur soft and short; dental formula $\dfrac{1\text{-}3}{3}, \dfrac{1}{1}, \dfrac{4}{4}, \dfrac{2}{2\text{-}3}$; Africa; five genera.

*Family Soricidae (Shrews).* Small (35 to 180 mm.) insectivores with a long, pointed muzzle; eyes small but functional; ears partly covered by short but dense body fur; tail short to moderately long, finely haired; limbs fairly short and generally unspecialized; dental formula $\dfrac{3}{1\text{-}2}, \dfrac{0\text{-}1}{0}, \dfrac{1\text{-}2}{1\text{-}2}, \dfrac{3}{3}$; found on all continents except Australia; 24 genera.

*Family Talpidae (Dasmans, Moles, and Shrew Moles)*. Medium-small (60 to 220 mm.) insectivores, mostly adapted to subterranean life; muzzle long and naked; eyes minute to vestigial; external ears small or with pinna absent; neck short; tail short or of medium length, often thickened; fur soft; front limbs in most species extending laterally from body with palms and claws greatly enlarged for digging; dental formula $\frac{2-3}{1-3}, \frac{1}{0-1}, \frac{3-4}{3-4}, \frac{3}{3}$; Europe, Asia, and North America; 15 genera.

**ORDER DERMOPTERA (GLIDING LEMURS).** Primitive gliding mammals of medium-small (375 to 420 mm.) size possessing furred membranes which not only extend outward from the sides of the body between the front and hind feet, but also forward to the neck and posteriorly to the tail, as in many kinds of bats; feet with five toes which possess sharply curved claws for hanging from trees; dental formula $\frac{2}{3}, \frac{1}{1}, \frac{2}{2}, \frac{3}{3}$; lower incisors ctenoid (comblike with a number of deep grooves); one family.

*Family Cynocephalidae*. Characters of the order; southeastern Asia, East Indies, and Philippine Islands; a single genus, *Cynocephalus*.

**ORDER CHIROPTERA (BATS).** True flying mammals of small to medium size (35 to 400 mm.); bones of the palms and fingers greatly elongated and covered by a double membrane which is connected to the body and extends posteriorly to the ankle; in many species an interfemoral membrane (uropatagium), extending between the hind legs, encases the tail partly or entirely; claws present on no more than the first two digits of the front limb; hind feet usually proportionately small, possessing five toes with strongly curved nails for hanging; teeth quite variable in number and shape; worldwide except certain remote island areas and the Antarctic; 17 families.

*Family Pteropidae (Fruit Bats)*. Usually considered as constituting a separate suborder, the Megachiroptera, with all the other families of bats placed in the suborder Microchiroptera; size medium-small to very large (50 to 400 mm.) for the order; external ears funnel-shaped with tragus absent; eyes large; a claw generally present on the second digit as well as the thumb; dentition simple; dental formula $\frac{1-2}{0 \text{ or } 2}, \frac{1}{1}, \frac{3}{3}, \frac{1-2}{2-3}$; Asia, Africa, Australia, and many of the islands of the central and western Pacific Ocean as well as the Indian Ocean; 38 genera.

*Family Rhinopomatidae (Mouse-tailed Bats)*. Small (60 to 80 mm.) bats lacking fur on the posterior part of the body; tail very long with only the extreme basal part encased in a narrow uropatagium; ear possessing a tragus; a leaflike

structure present on the nose; dental formula $\frac{1}{2}, \frac{1}{1}, \frac{1}{2}, \frac{3}{3}$;
northern Africa and southern Asia east to Sumatra; one genus, *Rhinopoma*.

**Family Emballonuridae (Sheath-tailed Bats).** Small to fairly large (37 to 100 mm.) bats; ears possessing a tragus; nose lacking a leaf; tip of tail free and situated on dorsal surface of uropatagium; glandular sacs often present on the dorsal surface of the wing near the shoulder; dental formula $\frac{1-2}{2-3}, \frac{1}{1}, \frac{2}{2}, \frac{3}{3}$; tropical Africa, Eurasia, Australia, Indonesia, the Pacific islands, and central and South America; 12 genera.

**Family Noctilionidae (Fish-eating Bats).** Large (100 to 130 mm.) bats lacking fur along the sides of the body; ears possessing a tragus; a nose leaf lacking; eyes small; tail about one-half as long as uropatagium, in which it is encased; wings long and narrow; dental formula $\frac{2}{1}, \frac{1}{1}, \frac{1}{2}, \frac{3}{3}$; northwestern Mexico south to Argentina, also on the Antilles and the Island of Trinidad; one genus, *Noctilio*.

**Family Nycteridae (Hollow-faced Bats).** Medium-small (45 to 75 mm.) bats with proportionately large ears united at their base; tragus present; nose leaf present; a deep pit on forehead; tail long and completely within the uropatagium; dental formula $\frac{2}{3}, \frac{1}{1}, \frac{1}{2}, \frac{3}{3}$; Africa, the Near East, Madagascar, southeastern Asia, and the East Indies; one genus, *Nycteris*.

**Family Megadermatidae (False Vampires).** Medium to large (65 to 140 mm.) bats lacking a tail; ears very large and united basally and possessing a tragus; a prominent nose leaf present; dental formula $\frac{0}{2}, \frac{1}{1}, \frac{1-2}{2}, \frac{3}{3}$; Africa, southern Asia, Malaysia, the Philippines, and Australia; three genera.

**Family Rhinolophidae (Horseshoe Bats).** Small to medium-sized (35 to 110 mm.) bats possessing a complicated nose leaf shaped somewhat like a horseshoe and composed of three parts; ears large and lacking a tragus; eyes small; tail short and uropatagium relatively small; wings broad and folded around body when at rest; dental formula $\frac{1}{2}, \frac{1}{1}, \frac{2}{3}, \frac{3}{3}$; Europe, Asia, Africa, and Australia; two genera, *Rhinolophis* and *Rhinomegalophis*.

**Family Hipposideridae (Old World Leaf-nosed Bats).** Small to medium-sized (30 to 110 mm.) bats with a complicated nose leaf structure somewhat like that of the horseshoe bats but possessing a transverse element which is often trident-shaped; ears small to medium-large, sometimes united at the base and lacking a tragus; tail present or absent, if present encased in the uropatagium; dental formula $\frac{1}{2}, \frac{1}{1}, \frac{1-2}{2}, \frac{3}{3}$; Africa, southern Asia, and Australia east to the Solomon Islands; nine genera. (Considered by some as a subfamily of the Rhinolophidae.)

*Family Phyllostomatidae (New World Leaf-nosed Bats)*. Small to large (40 to 135 mm.) bats usually possessing a simple leaf structure on the nose; ears generally narrow and pointed, joined together basally in some species; a tragus present; tail present or absent; teeth varying in number from 26 to 34; southwestern United States to Argentina; 51 genera.

*Family Desmodontidae (Vampire Bats)*. Small to medium-sized (65 to 90 mm.) bats lacking a tail and true nose leaf; ears small with tragus present; uropatagium reduced in size; thumb proportionately large; incisor and canine teeth specialized for cutting flesh; dental formula $\frac{1-2}{2}, \frac{1}{1}, \frac{1-2}{2-3}, \frac{0 \text{ or } 2}{0-2}$; northern Mexico to southeastern South America and the Island of Trinidad; three genera.

*Family Natalidae (Funnel-eared Bats)*. Small (35 to 55 mm.) bats with long, soft, often colorful fur and a proportionately long tail encased within the uropatagium; ears large and funnel-shaped; tragus present; a nose leaf absent; dental formula $\frac{2}{3}, \frac{1}{1}, \frac{3}{3}, \frac{3}{3}$; northern Mexico to Brazil and the Antilles; one genus, *Natalus*.

*Family Furipteridae (Smoky Bats.)* Small (35 to 60 mm.) bats with a moderately long tail encased in the uropatagium but terminating before reaching the posterior margin of this structure; ears relatively small and funnel-shaped; tragus present; thumb vestigial; dental formula $\frac{2}{3}, \frac{1}{1}, \frac{2}{3}, \frac{3}{3}$; the Island of Trinidad and Panama to Brazil and Chile; two genera, *Furipterus horrens* and *Amorphochilus schnoblii*.

*Family Thyropteridae (New World Sucker-footed Bats)*. Small (35 to 50 mm.) bats with a moderately long tail encased in the uropatagium, except the tip, which protrudes a short way beyond the posterior edge of this structure; ears funnel-shaped and separate; tragus present; stalked adhesive disks on wrists and ankles; thumb present; third and fourth toes syndactylous; dental formula $\frac{2}{3}, \frac{1}{1}, \frac{3}{3}, \frac{3}{3}$; Honduras to Brazil and Peru, and the Island of Trinidad; one genus, *Thyroptera*.

*Family Myzopodidae (Old World Sucker-footed Bats)*. Moderately small (50 to 60 mm.) bat with a long tail extending beyond the uropatagium; ears large; tragus present; adhesive disks lacking a stalk on wrists and ankles; thumb vestigial; dental formula $\frac{2}{3}, \frac{1}{1}, \frac{3}{3}, \frac{3}{3}$; confined to Madagascar; one species, *Myzopoda aurita.*

*Family Vespertilionidae (Vespertilionid Bats).* Small to medium-sized (30 to 105 mm.) bats with a moderately long tail encased in the uropatagial membrane; eyes small; ears variable in length and usually separated; tragus present; dental formula $\frac{1-2}{2-3}, \frac{1}{1}, \frac{1-3}{2-3}, \frac{3}{3}$; worldwide in distribution; 35 genera.

*Family Mystacinidae (New Zealand Short-tailed Bat).* Medium-sized (60 mm.) bat with a short tail which is free on dorsal surface of uropatagium; pelage very thick; ears separate; tragus long; thumbnail and claws of hind feet talonid in shape; dental formula $\frac{1}{1}, \frac{1}{1}, \frac{2}{2}, \frac{3}{3}$; New Zealand; one species, *Mystacina tuberculata.*

*Family Molossidae (Free-tailed Bats).* Small to large (40 to 130 mm.) bats with tail projecting considerably beyond the posterior border of the uropatagium; fur soft, short, and velvety; muzzle broad; ears fleshy and often united at base; tragus present; nose leaf absent; wings long and narrow; prominent bristles present on outer toes of foot; dental formula $\frac{1}{1-3}, \frac{1}{1}, \frac{1-2}{2}, \frac{3}{3}$; warmer parts of the world; 10 genera.

**ORDER PRIMATES (PRIMATES).** Small to large mammals (85 to over 2000 mm. in standing height) possessing highly developed cerebral hemispheres; orbits surrounded by bone and directed forward; limbs with ball and socket type of articulation, allowing for great mobility; basically pentadactyl with an opposable thumb and toe present in most species; nails usually flattened but sometimes clawlike; tail present or absent, prehensile in some forms; dental formula $\frac{1-2}{1-3}, \frac{0-1}{0-1}, \frac{1-3}{0-3}, \frac{3}{3}$; native to Asia, Africa, and North and South America, except for man, who is worldwide in distribution; 11 living families.

*Family Tupaiidae (Tree Shrews).* Very small (100 to 220 mm.) insectivore-like primates that superficially resemble small squirrels; upper incisors large; ears small; tail long and generally well haired; limbs pentadactyl; soles of feet naked and tuberculate; dental formula $\frac{2}{3}, \frac{1}{1}, \frac{3}{3}, \frac{3}{3}$; eastern Asia; five genera. (Frequently placed in the order Insectivora.)

*Family Lemuridae (Lemurs).* Small to medium-sized (120 to

440 mm.) primates with long, heavily furred tail; body fur soft and woolly; hind limbs larger than forelimbs; dental formula $\frac{0-2}{2}, \frac{1}{1}, \frac{3}{3}, \frac{3}{3}$; islands of Madagascar and Comoro in the Indian Ocean; five genera.

*Family Indridae (Woolly Lemurs).* Small to large (300 to 1050 mm.) lemurs with bare face and densely woolly pelage; tail short to long; hind limbs somewhat larger than forelimbs; thumb slightly opposable; salivary glands greatly enlarged; dental formula $\frac{2}{2}, \frac{1}{0}, \frac{2}{2}, \frac{3}{3}$; Madagascar; three genera.

*Family Daubentoniidae (Aye-Aye).* A small (360 to 440 mm.) primate with long, bushy tail; face short; ears large and membranous; hind toe opposable; third finger of front limb extremely slender; incisors rodent-like and permanently growing with enamel on the anterior surface; dental formula $\frac{1}{1}, \frac{0-1}{0}, \frac{1}{0}, \frac{3}{3}$; Madagascar; one species, *Daubentonia madagascariensis.*

*Family Lorisidae (Lorises and Galagos).* Small (170 to 390 mm.) primates with dense, woolly fur; tail short or absent in the lorises, long in the galagos; eyes very large; ears small to large and membranous; opposable thumb and toe; dental formula $\frac{2}{2}, \frac{1}{1}, \frac{3}{3}, \frac{3}{3}$; Africa, southern Asia, Indonesia, and Philippine Islands; six genera.

*Family Tarsiidae (Tarsiers).* Very small (85 to 160 mm.) primates with proportionately long, naked tail and very large eyes; ears moderately large and membranous; hind limbs very elongate; digits terminating in disklike pads; hind toe opposable; dental formula $\frac{1-2}{1}, \frac{1}{1}, \frac{3}{3}, \frac{3}{3}$; some of the Philippine Islands, Borneo, Celebes, and certain other Indonesian islands; one genus, *Tarsius.*

*Family Cebidae (Larger New World Monkeys).* Small to medium-sized (200 to 915 mm.) primates generally possessing a long tail, which in some species is prehensile; nostrils widely separated and opening on sides of nasal region; lacking ischial callosities and cheek pouches; thumb not opposable; big toe opposable; dental formula $\frac{2}{2}, \frac{1}{1}, \frac{3}{3}, \frac{3}{3}$; southern Mexico to Argentina; 11 genera.

*Family Callithricidae (Marmosets).* Small (130 to 370 mm.) primates with long, nonprehensile tail and soft, thick fur; ears, head, or shoulders often adorned with tufts of long hair; hind limbs longer than forelimbs; thumb not opposable; dental fomrula $\frac{2}{2}, \frac{1}{1}, \frac{3}{3}, \frac{2}{2}$; tropical South America; four genera.

*Family Cercopithecidae (Old World Monkeys).* Medium to large (325 to 1100 mm.) primates with very short to long, nonprehensile tail; face essentially naked; ears rounded; hind limbs generally larger than front limbs; thumb and big toe both opposable; dental formula $\frac{2}{2}, \frac{1}{1}, \frac{2}{2}, \frac{3}{3}$; temperate to tropical parts of Africa and Asia; east to Japan, Taiwan, the Philippines, and Indonesia; 11 genera.

*Family Pongidae (Anthropoid Apes).* Medium to very large (440 to 1700 mm.) primates lacking a tail; ears and face essentially hairless; arms longer than legs; both thumb and big toe opposable; dental formula $\frac{2}{2}, \frac{1}{1}, \frac{2}{2}, \frac{3}{3}$; central Africa, southeastern Asia, Java, Sumatra, and Borneo; four genera.

*Family Hominidae (Man).* A large primate with upright posture (height varying, with extremes in the Negro race; Wambute pigmies averaging 1450 mm. and members of the Watusi tribe averaging over 2100 mm.); tail absent; hind limbs longer than arms; thumb opposable but not big toe; dental formula $\frac{2}{2}, \frac{1}{1}, \frac{2}{2}, \frac{3}{3}$; presently worldwide; one species, *Homo sapiens.*

**ORDER EDENTATA (EDENTATES).** Small to moderately large mammals (150 to 1200 mm.) with at least some of the front claws greatly enlarged, either for digging or for hanging from the limbs of trees; incisors and canines always lacking; all teeth absent in some species; southern United States, south through most of South America; three living families.

*Family Myrmecophagidae (Anteaters).* Small to moderately large (150 to 1200 mm.) edentates with an elongated snout and long tongue covered with sticky mucus; teeth lacking; tail long and, in one species, prehensile; claws of front foot greatly enlarged for tearing open ant or termite nests; southern Mexico to Paraguay; three genera.

*Family Brachypodidae (Tree Sloths).* Medium-sized (500 to 640 mm.) edentates with rounded head, small ears, and no more than three syndactylous digits; six or nine cervical vertebrae; $\frac{5}{4-5}$ permanently growing cheek teeth present; Central America to Argentina; two genera, *Choloepus* and *Bradypus.*

*Family Dasypodidae (Armadillo-like Mammals).* Small to medium-sized (125 to 1000 mm.) mammals with horny shields on the dorsal surface of the body; snout moderately long; toes possess powerful claws for digging in ground; teeth peglike, permanently growing, and varying from $\frac{7-9}{7-9}$ to $\frac{25}{25}$; southern United States to Argentina; nine genera.

ORDER PHOLIDOTA (PANGOLINS). Medium-sized (300 to 900 mm.) mammals with imbricate scales covering dorsal surface of head, body, and tail as well as outer surface of legs; head very elongate; teeth absent; long claws on the five front and five hind toes; Africa and southeastern Asia; one family.

Family Manidae (Pangolins). Characters and range, those of the family; one genus, Manis.

ORDER LAGOMORPHA (PIKAS, HARES, AND RABBITS). Fairly small (125 to 750 mm.) mammals with very short or vestigial tail; external ear medium to large in size; five manual and four or five pedal digits; soles of feet haired; upper incisors distinguished by a large pair anteriorly with a very small pair directly behind; incisors completely surrounded by enamel; a diastema between incisors and premolars; dental formula $\frac{2}{1}, \frac{0}{0}, \frac{2-3}{2}, \frac{2-3}{2-3}$; worldwide except Australia (introduced), New Zealand (introduced), Madagascar, Antarctica, and most islands; two families.

Family Ochotonidae (Pikas). Small (125 to 300 mm.) diurnal lagomorphs with rounded ears and vestigial tail; hind limbs not markedly longer than front limbs; only four pedal digits; dental formula $\frac{2}{1}, \frac{0}{0}, \frac{3}{2}, \frac{2}{3}$; Eurasia and northern North America; one genus, Ochotona.

Family Leporidae (Hares and Rabbits). Medium to large-sized (250 to 760 mm.) lagomorphs that are principally crepuscular (active in the evening and morning hours) or nocturnal; short, usually well-furred, tail; ears elongated; hind legs markedly larger than forelegs; five pedal digits; dental formula $\frac{2}{1}, \frac{0}{0}, \frac{2-3}{2}, \frac{3}{3}$; distribution, that of the order.

ORDER RODENTIA (RODENTS). Very small to medium-sized mammals (40 to 1300 mm.) possessing only two incisor teeth above and two below; incisors permanently growing, with outer surface harder than inner surface; canines missing; premolars sometimes missing; molars usually $\frac{3}{3}$; worldwide; usually divided into three major suborders, the Sciuromorpha, the Myomorpha, and the Hystricomorpha, based upon a number of structural differences but principally the relationship of the masseter muscle to the infraorbital canal; 32 living families.

Suborder Sciuromorpha (squirrel-like rodents). Infraorbital canal very small and not transmitting any part of medial masseter muscle.

Family Aplodontidae (Mountain Beaver). A medium-sized (up to 450 mm.) rodent with a blunt head and coarse pelage; eyes and external ears relatively small; tail very short and hidden in fur; limbs of approximately equal length and each possessing five toes; cheek teeth permanently growing; dental formula $\frac{1}{1}, \frac{0}{0}, \frac{2}{1}, \frac{3}{3}$; extreme western North America; one species, *Aplodontia rufa*.

**Family *Sciuridae* (*Squirrel-like Animals*).** Small to medium-sized (65 to 600 mm.) rodents, usually possessing a moderately long tail and fur that is short and dense; eyes large and ears fairly small; front limbs smaller than hind limbs; five toes present on hind foot and four toes on front foot; postorbital process present on skull; first upper premolar very small, if present in the adult; dental formula $\frac{1}{1}, \frac{0}{0}, \frac{1-2}{1}, \frac{3}{3}$; essentially worldwide except Australia and Madagascar; about 50 genera.

**Family *Geomyidae* (*Pocket Gophers*).** Medium-small (90 to 350 mm.) rodents with a blunt muzzle and fur-lined cheek pouch on either side of head; eyes and external ears very small; tail rather short, blunt, and sparsely haired; limbs fairly short; five toes on each foot; claws of front feet enlarged for digging; lips so situated that they close behind the incisor teeth, which may be used for digging; dental formula $\frac{1}{1}, \frac{0}{0}, \frac{1}{1}, \frac{3}{3}$; western Canada south to northern Colombia; eight genera.

**Family *Heteromyidae* (*Kangaroo Rats and Mice and Pocket Mice*).** Small (55 to 180 mm.) rodents with external, fur-lined cheek pouches and a long tail covered with very short hair; hind legs larger, sometimes considerably so, than forelegs; ears small; nasal bones long and slender, extending beyond incisors; bullae of skull markedly enlarged; dental formula $\frac{1}{1}, \frac{0}{0}, \frac{1}{1}, \frac{3}{3}$; southwestern Canada to northern South America; five genera.

**Family *Castoridae* (*Beaver*).** Large (750 to 1300 mm.) aquatic rodents with a broad, paddle-like tail covered with scales; underfur soft and dense; five toes on each foot, those of the hind feet webbed; perianal scent glands, known as castor glands, present in both sexes; dental formula $\frac{1}{1}, \frac{0}{0}, \frac{1}{1}, \frac{3}{3}$; forested portions of Northern Hemisphere (original range much reduced by man); one genus, *Castor*.

**Family *Anomaluridae* (*Scaly-tailed Squirrels*).** Small to medium-sized (60 to 380 mm.) squirrel-like rodents (Sciuromorpha?) possessing a long tail with two rows of keeled scales on its underside near the base; a gliding membrane extending along either side of the body from the front to the hind limb in most species; ears and eyes squirrel-like; dental formula $\frac{1}{1}, \frac{0}{0}, \frac{1}{1}, \frac{3}{3}$; Africa; four genera.

*Family Pedetidae (Springhares).* Medium-sized (350 to 430 mm.) rodents (Sciuromorpha?) with hind legs greatly enlarged for jumping; tail long and bushy; eyes and ears proportionately large; dental formula $\frac{1}{1}, \frac{0}{0}, \frac{1}{1}, \frac{3}{3}$; central and southern Africa; one genus, *Pedetes*.

*Suborder Myomorpha (mouselike rodents).* Infraorbital canal somewhat enlarged for transmitting medial masseter muscle.

*Family Cricetidae (Cricetids).* Very small to fairly large (80 to 450 mm.) rodents adapted to many kinds of habitats but mostly terrestrial; tail present, ranging from long to quite short; premolars lacking; molars cuspidate, laminate, or prismatic; when cuspidate, cusps arranged in two rows; when laminate, laminae separated widely; dental formula $\frac{1}{1}, \frac{0}{0}, \frac{0}{0}, \frac{3}{3}$; essentially worldwide but absent from Australasia; about 100 genera.

*Family Spalacidae (Mole Rats).* Medium-sized (150 to 300 mm.) burrowing rodents with a blunt muzzle and lacking a tail; eyes vestigial and covered with skin; external ears reduced to ridges; fur short and velvety; incisors projecting from mouth and used in digging; dental formula $\frac{1}{1}, \frac{0}{0}, \frac{0}{0}, \frac{3}{3}$; southeastern Europe, Asia Minor, and parts of North Africa; one genus, *Spalax*.

*Family Rhizomyidae (Bamboo Rats).* Medium-sized (160 to 460 mm.) burrowing rodents resembling pocket gophers (Geomyidae) but lacking fur-lined cheek pouches; eyes and external ears small; limbs short and stout; tail moderately short and sparsely haired; southern Asia and tropical east Africa; three genera.

*Family Muridae (Old World Rats and Mice).* A very large assemblage (over 450 species) of small to medium-sized (110 to 800 mm.) rodents usually possessing a rather long, naked, scaly tail; soles of feet naked; lacking external, fur-lined cheek pouches; premolars absent; molars cuspidate or laminate; when cuspidate, cusps arranged in three rows although inner row may be very reduced; when laminate, laminae pressed together; dental formula $\frac{1}{1}, \frac{0}{0}, \frac{0}{0}, \frac{2-3}{2-3}$; native to Europe, Asia, Africa, Australasia, and Micronesia; about 100 genera; two genera, *Mus* and *Rattus*, widely distributed over the world by man.

*Family Gliridae (Dormice).* Small (60 to 190 mm.), squirrel-like mice, generally with a bushy tail; eyes large; ears rounded; four digits on front foot and five on hind foot; limbs adapted to climbing; dental formula $\frac{1}{1}, \frac{0}{0}, \frac{1}{1}, \frac{3}{3}$; Europe, Asia, and Africa; seven genera.

Family Platacanthomyidae *(Spiny Dormice)*. Small (70 to 270 mm.) rodents, somewhat resembling dormice but with proximal part of tail sparsely haired in contrast to terminal portion; five digits on front as well as hind foot; dental formula $\frac{1}{1},\frac{0}{0},\frac{0}{0},\frac{3}{3}$; southern Asia; two genera, *Platacanthomys* and *Typhlomys*.

Family Seleviniidae *(Desert Dormice)*. A short (up to 95 mm.), stout rodent with a long tail and dense, thick pelage; molting method similar to that of *Mirounga* (order Pinnipedia) in that the epidermis is shed in patches with the old hair; four digits on front foot and five on hind foot; anterior surface of upper incisors grooved; dental formula $\frac{1}{1},\frac{0}{0},\frac{2}{0},\frac{3}{3}$; central Asia; one species, *Selevina betpakdalaensis*.

Family Zapodidae *(Jumping Mice)*. Small (50 to 100 mm.) rodents with long, sparsely haired tail; internal cheek pouches present; auditory bullae relatively small; dental formula $\frac{1}{1},\frac{0}{0},\frac{0-1}{0},\frac{3}{3}$; holarctic in distribution; four genera.

Family Dipodidae *(Jerboas)*. Small (70 to 150 mm.) desert rodents with long tail that is well haired, at least terminally, and hind legs greatly enlarged for jumping; usually only three functional toes on hind foot; second, third, and fourth metatarsals fused; lacking fur-lined cheek pouches; molariform teeth cuspidate; dental formula $\frac{1}{1},\frac{0}{0},\frac{0-1}{0},\frac{3}{3}$; north Africa and deserts and steppes of Eurasia; 10 genera.

Suborder Hystricomorpha *(porcupine-like rodents and their relatives)*. Infraorbital canal greatly enlarged for transmitting medial masseter muscle.

Family Hystricidae *(Old World Porcupines)*. Medium to large (380 to 710 mm.) rodents with some hairs of upper surface of body modified as quills; body heavy-set; muzzle blunt; front and hind feet with five toes each; dental formula $\frac{1}{1},\frac{0}{0},\frac{1}{1},\frac{3}{3}$; southern Europe to Africa and eastern Asia and the Philippines; four genera.

Family Erethizontidae *(New World Porcupines)*. Medium to large (300 to 860 mm.), heavy-set rodents with a blunt muzzle and some hairs on upper surface of body, outer parts of legs, and usually top of tail modified as stiff, protective quills; limbs short; four functional digits on front feet and three on hind feet; dental formula $\frac{1}{1},\frac{0}{0},\frac{1}{1},\frac{3}{3}$; North and South America; four genera.

Family Caviidae *(Cavies)*. Moderately short (225 to 750 mm.), stout-bodied rodents either lacking a tail or with one that is very short; four toes on front and three on hind foot; pelage coarse but thick; dental formula $\frac{1}{1}, \frac{0}{0}, \frac{1}{1}, \frac{3}{3}$; South America; five genera.

Family Hydrochoeridae (Capybaras). The largest (1000 to 1300 mm.) of living rodents; tail very short; muzzle truncate; ears short; pelage sparse and coarse; four toes on front foot and three on hind foot; incisors grooved on anterior surface; dental formula $\frac{1}{1}, \frac{0}{0}, \frac{1}{1}, \frac{3}{3}$; eastern South America north to Panama; one genus, Hydrochoerus.

Family Dinomyidae (False Paca). A moderately large (up to 800 mm.) rodent with a thick-set body, a relatively large head, and a well-haired tail of medium length; front and hind feet each with four toes; longitudinal rows of white spots or stripes on body; ears small; dental formula $\frac{1}{1}, \frac{0}{0}, \frac{1}{1}, \frac{3}{3}$; northern and central South America along the lower parts of the Andes; one species, Dinomys branickii.

Family Dasyproctidae (Pacas and Agoutis). Medium to moderately large (320 to 690 mm.) rodents with elongated hind limbs and short tail; pelage coarse; four functional toes on front foot; three or five functional toes on hind foot; dental formula $\frac{1}{1}, \frac{0}{0}, \frac{1}{1}, \frac{3}{3}$; Mexico to Argentina; four genera.

Family Chinchillidae (Chinchillas and Viscachas). Moderately small to fairly large (225 to 660 mm.) rodents with long, fine fur and a well-haired tail; four toes on front foot; three to four toes on hind foot; dental formula $\frac{1}{1}, \frac{0}{0}, \frac{1}{1}, \frac{3}{3}$; central and southern Andes and the pampas of Argentina; three genera.

Family Capromyidae (Coypu and Hutias). Medium-small to fairly large (200 to 635 mm.) rodents; ears and eyes small; limbs relatively short; thumb greatly reduced in size; tail usually fairly long and often sparsely haired; dental formula $\frac{1}{1}, \frac{0}{0}, \frac{1}{1}, \frac{3}{3}$; South America and the West Indies; four living genera. (The coypu is sometimes placed in a separate family, the Myocastoridae.)

Family Octodontidae (Octodonts). Medium-small (125 to 200 mm.) rodents with silky fur and a tail of varying length; nose fairly pointed; thumb greatly reduced; dental formula $\frac{1}{1}, \frac{0}{0}, \frac{1}{1}, \frac{3}{3}$; western and southern South America; five genera.

Family Ctenomyidae (Tuco-Tucos). Medium-small (170 to 250 mm.) rodents adapted to fossorial life and superficially resembling North American pocket gophers (Geomyidae); external, fur-lined cheek pouches lacking; eyes and external ears small; tail of medium length; limbs short and claws adapted to digging; incisors much enlarged; dental formula $\frac{1}{1}, \frac{0}{0}, \frac{1}{1}, \frac{3}{3}$; Peru to Tierra del Fuego, South America; one genus, *Ctenomys*.

*Family Abrocomidae (Chinchilla Rats).* Medium-small (150 to 250 mm.) rodents with soft, chinchilla-like fur and a head resembling those of rats or mice (e.g., Muridae or Cricetidae); ears rounded; eyes large; tail moderately long; legs short with four toes on front foot and five on hind foot; dental formula $\frac{1}{1}, \frac{0}{0}, \frac{1}{0}, \frac{3}{3}$; the altiplano of Peru, Bolivia, Chile, and Argentina; one genus, *Abrocoma*.

*Family Echimyidae (Spiny Rats).* Small to medium-sized (180 to 480 mm.) rodents with a ratlike appearance and generally possessing bristle-like hairs on the body; tail short to long; thumb vestigial; dental formula $\frac{1}{1}, \frac{0}{0}, \frac{1}{1}, \frac{3}{3}$; Central America, northern South America, and Martinique in the Lesser Antilles; 14 living genera.

*Family Thryonomyidae (Cane Rats).* Medium-large (350 to 610 mm.) rodents with a heavy-set body; pelage coarse; ears small and rounded; tail of varying length but coarsely haired with short bristles; thumb and fifth toe of front foot vestigial; four toes on hind foot; dental formula $\frac{1}{1}, \frac{0}{0}, \frac{1}{1}, \frac{3}{3}$; Africa; one genus, *Thryonomys*.

*Family Petromyidae (Rock Mouse).* A medium-small (up to 200 mm.) rodent with a flattened body, a long, well-haired tail, large eyes, and small, rounded ears; legs short with four toes on front foot and five on hind foot; dental formula $\frac{1}{1}, \frac{0}{0}, \frac{1}{1}, \frac{3}{3}$; Africa; one species, *Petromys typicus*.

*Family Bathyergidae (African Mole-Rats).* Small to medium-sized (80 to 330 mm.) fossorial rodents with minute eyes and very short tail; external pinna of ear absent; skin hairless around opening of external auditory canal; limbs short and stout with five toes on each foot; dental formula $\frac{1}{1}, \frac{0}{0}, \frac{2\text{-}3}{2\text{-}3}, \frac{0\text{-}3}{0\text{-}3}$; Africa; five genera.

*Family Ctenodactylidae (Pectinators).* Medium-small (160 to 240 mm.) rodents with short or moderately short tail and stocky body; eyes large; ears rounded; fur soft; four toes on each foot; two inner toes of hind foot with comblike bristles; dental formula $\frac{1}{1}, \frac{0}{0}, \frac{1\text{-}2}{1\text{-}2}, \frac{3}{3}$; northern Africa; four genera.

**ORDER CETACEA (WHALES, PORPOISES, AND DOLPHINS).** Medium-sized to very large mammals (1.25 to over 30 m.) adapted to aquatic life; skin essentially lacking hair and overlying a thick layer of blubber; external ears lacking except for a minute auditory canal; nostrils modified as a blowhole or blowholes on top of head; front limbs developed into flippers; hind limbs absent; tail modified into flukes situated in a horizontal plane; teeth present in varying numbers (2 to about 260) in lower jaw, or both jaws in suborder Odontoceti; teeth absent in suborder Mysticeti, being replaced, from the functional standpoint, by plates of baleen which hang from the roof of the mouth and serve as strainers for securing food; 10 families.

*Suborder Mysticeti.*   Baleen whales; teeth absent; skull symmetrical; paired blowholes.

*Family Balaenidae (Right Whales).*   Medium-sized (6 to 18 m.) cetaceans with a stocky body and a very large head that may equal one third of the total length; grooves lacking on throat; baleen plates very long; flippers short and broad; dorsal fin present in one genus; oceans of the world but absent from tropical seas and south polar areas; two genera, *Balaena* and *Caperea.*

*Family Eschrichtidae (Gray Whale).*   A medium-large (up to 15 m.) whale lacking a dorsal fin; body slender; two to four shallow grooves on throat; general color, gray mottled with light spots; North Pacific Ocean along both American and Asiatic coasts; one species, *Eschrichtius gibbosus.*

*Family Balaenopteridae (Rorquals).*   Medium to very large (9 to 30 m.) whales possessing a dorsal fin and numerous parallel, pleated grooves on the throat and belly; oceans of the world; two genera, *Balaenoptera* and *Megaptera.*

*Suborder Odontoceti.*   Toothed cetaceans; skull asymmetrical; a single external blowhole.

*Family Ziphiidae (Beaked Whales).*   Medium to moderately large (4.5 to 13 m.) cetaceans with an elongate, beaklike snout; forehead enlarged to form a "melon" in front of the blowhole; a small, posteriorly situated dorsal fin present; a pair of grooves present on throat; one to two pairs of functional teeth present, at least in males, except in *Tasmacetus* where the dental formula is generally $\frac{19}{17}$; oceans of the world; five genera.

*Family Monodontidae (Narwhal and Beluga).*   Moderately small (up to 6 m.) whales with a blunt snout; gular grooves lacking; a definite dorsal fin absent; two teeth in the genus *Monodon*, with one (usually the left) extended into a long (up to 2.7 m.), straight, forwardly directed, helically grooved tusk in the male; dental formula in *Delphinapterus* about $\frac{9}{9}$; confined primarily to arctic seas; two genera.

*Family Physeteridae (Sperm Whales).*   Small to large (2.1 to over 18 m.) whales with functional teeth in the lower jaw only (8 to 25 in each ramus); snout projecting beyond lower jaw; a spermaceti organ or "case" located in the head; all oceans; two genera, *Physeter* and *Kogia*.

*Family Platanistidae (Long-snouted River Dolphins).* Freshwater, estuarine or coastal cetaceans of small size; beak long and slender; forehead enlarged to form a "melon"; eyes poorly developed; dorsal fin low and ridgelike; tooth rows close together and parallel; dental formula $\frac{26}{26}$ to $\frac{55}{55}$; eastern South America, India, and China; four genera.

*Family Stenidae (Rough-toothed, River and Coastal Dolphins).* Moderately small (1.6 to 2.5 m.) dolphins in which snout may or may not be sharply demarked from forehead; distinguished from ocean and long-snouted river dolphins in certain characters of air sinus system of head; dental formula varying from about $\frac{24}{24}$ to $\frac{32}{32}$; tropical and warm temperate seas; three genera.

*Family Phocoenidae (Porpoises).* Small (1.6 to 2.1 m.) cetaceans lacking a distinct beak; teeth spatulate or spadelike instead of conical; dental formula $\frac{15}{15}$ to $\frac{30}{30}$; all oceans and seas of the Northern Hemisphere as well as coastal waters of South America and rivers of southeastern Asia; three genera.

*Family Delphinidae (Ocean Dolphins).* Small to medium-sized (1.5 to 9.5 m.) cetaceans, most possessing a distinct beak; a dorsal fin present in all except one genus; males usually larger than females; teeth conical; dental formula $\frac{0}{2}$ to $\frac{65}{58}$; oceans of the world as well as certain rivers of southeastern Asia; 14 genera.

**ORDER CARNIVORA (CARNIVORES).** Very small to large mammals (135 to 3000 mm.) that are primarily terrestrial; at least four clawed toes on each foot; incisors small; canines large and pointed; dental formula $\frac{3}{2-3}, \frac{1}{1}, \frac{2-4}{2-4}, \frac{1-4}{1-5}$; teeth rooted and not permanently growing; essentially worldwide; seven families.

*Family Canidae (Doglike Carnivores).* Medium to fairly large-sized (340 to 1350 mm.) carnivores adapted to running; tail fairly long and generally bushy; ears large and pointed; muzzle elongate; first toe on fore and hind feet greatly reduced in size or vestigial; dental formula $\frac{3}{3}, \frac{1}{1}, \frac{4}{4}, \frac{1-4}{2-5}$; worldwide except New Zealand, Madagascar, and most Indonesian and Pacific islands; 15 genera.

*Family Ursidae (Bears and Giant Panda).* Large (1000 to 3000 mm.) plantigrade carnivores; ears proportionately small, rounded, and well furred; muzzle elongate; eyes moderately small; body stocky and heavily haired; feet with five toes each; soles of feet generally haired; tail very short; dental formula $\frac{2-3}{3}, \frac{1}{1}, \frac{4}{4}, \frac{2}{3}$; Asia, Europe, and North and South America; six genera.

*Family Procyonidae (Raccoon-like Animals).* Medium-small (300 to 670 mm.) carnivores with a long tail that is banded except in one genus which has a prehensile tail; ears either pointed or rounded; eyes fairly large; feet possessing five toes; dental formula $\frac{3}{3}, \frac{1}{1}, \frac{3-4}{3-4}, \frac{2}{2-3}$; North and South America and eastern Asia; seven or eight genera.

*Family Mustelidae (Mustelids).* Very small to medium-sized (135 to 1500 mm.) carnivores with well developed perianal scent glands; muzzle moderately long; ears short; body generally proportionately slender but very muscular; legs short; five toes on each foot; claws not retractile; dental formula $\frac{3}{2-3}, \frac{1}{1}, \frac{2-4}{2-4}, \frac{1}{1-2}$; worldwide except Australia and Madagascar; 25 genera.

*Family Viverridae (Civets).* Small to medium-sized (170 to 975 mm.) carnivores with long head and pointed muzzle; body proportionately long and slender; tail usually long and bushy; legs short with four or five toes on each foot; perianal scent glands often well developed; lower incisor 2 longer than 1 or 3; dental formula $\frac{3}{3}, \frac{1}{1}, \frac{3-4}{3-4}, \frac{1-2}{1-2}$; Africa, Madagascar, southern Europe, and Asia; 36 genera.

*Family Hyaenidae (Hyenas and Aardwolf).* Medium-large (550 to 1650 mm.) carnivores, somewhat doglike in appearance but front limbs larger and more developed than rear limbs; head proportionately large and jaws very powerful; ears of medium size and pointed; four or five toes present; claws dull and nonretractile; perianal scent glands present; dental formula $\frac{3}{3}, \frac{1}{1}, \frac{3-4}{1-3}, \frac{1}{1-2}$; Africa and southwestern Asia; three genera.

*Family Felidae (Cats).* Small to large (500 to 2700 mm.) carnivores possessing a proportionately round head and retractile or semi-retractile claws; ears of medium size and usually pointed; vibrissae well developed; limbs generally of equal length and well developed for springing at prey; five toes on front foot and four on hind foot; dental formula $\frac{3}{3}, \frac{1}{1}, \frac{2-3}{2}, \frac{1}{1}$; essentially worldwide except Australia and Madagascar; three genera.

**ORDER PINNIPEDIA.** Medium to large-sized mammals (1.5 to over 6 m.) with limbs modified as flippers for aquatic life; eyes large; pinna of ear either very small or entirely lacking; vibrissae well developed; tail very short or absent; molariform teeth mostly homodont; arctic and antarctic waters, continental coastlines and islands; few species occurring in tropical regions or bodies of fresh water; three families.

*Family Otariidae (Sea Lions and Fur Seals).* Medium-sized to large (1.5 to 3.5 m.) pinnipeds possessing an external pinna to the ear; eyes not unusually large; outer incisors some-what canine-like; front flippers large and naked; hind flip-pers reversible so that they can be brought under body for locomotion on land; tail short; males markedly larger than females; distribution similar to that of the order but absent from the North Atlantic and tropical Indian oceans; six genera.

*Family Odobenidae (Walrus).* A large (up to 3.7 m.) pinniped with a thick body, relatively hairless skin, and upper canine teeth that are long and tusklike; eyes small; external pinna to the ear essentially lacking; prominent bristles present on upper lip; tail absent; hind flippers reversible; circum-polar in the Arctic; one species, *Odobenus rosmarus.*

*Family Phocidae (Hair or Earless Seals).* Small to very large (1.25 to 6.5 m.) pinnipeds lacking an external pinna to the ear; eyes relatively large; front flippers haired and smaller than hind flippers; hind flippers not reversible; a short tail present; males not markedly larger than females, except in the genus *Mirounga;* distribution same as order; 13 genera.

**ORDER TUBULIDENTATA (AARDVARK).** A medium-sized (up to 1.6 m.) mammal with a stocky body, an elongate head, a piglike snout, and long tubular ears; body sparsely covered with coarse hair; legs short with four toes on front foot and five on hind foot; nails of front foot long, strong, and adapted for digging; tail tapered and muscular; teeth tubular and lacking enamel; perma-nent dentition $\frac{0}{0}, \frac{0}{0}, \frac{2}{2}, \frac{3}{3}$; Africa; one family.

*Family Orycteropodidae (Aardvark).* Characters and range same as those of the order; one species, *Orycteropus afer.*

**ORDER PROBOSCIDIA (ELEPHANTS).** Very large (5.5 to 7.5 m.) ter-restrial mammals with nose elongated into a proboscis or trunk; head massive; ears large and leathery; skin thick and nearly hairless; tail short; legs thick and columnar; nails forming three to five hooves; upper incisors developed into tusks; molariform teeth succeeding one another from behind; dental formula $\frac{1}{0}, \frac{0}{0}, \frac{3}{3}, \frac{3}{3}$; Asia and Africa; one family.

*Family Elephantidae (African and Indian Elephants).* Charac-ters and range same as those of the order; two genera, *Loxodonta* and *Elephas.*

**ORDER HYRACOIDEA (HYRAXES OR OLD WORLD CONIES).** Small mammals (300 to 600 mm.) with greatly reduced or vestigial tail and nails that are modified into small, hooflike structures on most of the toes; legs short; soles of feet possessing moist, naked suction pads for climbing; enamel present only on front of upper incisors, which are triangular in cross section; dental formula $\frac{1}{2}, \frac{0}{0}, \frac{4}{4}, \frac{3}{3}$; Africa and southwestern Asia; one family.

*Family Procaviidae (Hyraxes or Old World Conies.)* Characters and range same as those of the order; three genera.

ORDER SIRENIA (MANATEES, DUGONG, AND SEA COW). Large aquatic mammals (2.5 to 4 m. in living species, up to 7.5 m. in the now extinct Steller's sea cow) with hind limbs lacking and front limbs modified into paddle-like flippers; eyes very small; ears lacking a pinna; body essentially hairless except for thick bristles around mouth; tail modified into a horizontal flukelike structure; teeth reduced in number and highly modified, if present; Bering Sea (formerly), tropical seas, and coastal rivers; two families.

*Family Dugongidae (Dugong and Sea Cow).* Sirenians with tail fin notched in the center so as to divide it into two lobes or flukes; nostrils dorsal in position; nails lacking on flippers; two genera; island and coastal waters in tropical parts of the western Pacific and Indian Oceans (genus *Dugong*); formerly in Bering Sea (genus *Hydromalis*, now extinct).

*Family Trichechidae (Manatees).* Sirenians with a round, paddle-like tail fin lacking any median notch; nostrils at end of muzzle; vestigial nails on flippers; coastal areas of southeastern United States south to northeastern South America and along west coast of Africa; one genus, *Trichechus*.

ORDER PERISSODACTYLA (ODD-TOED HOOFED MAMMALS). Large to very large terrestrial mammals (1.8 to 4.2 m.) with main axis of the foot terminating on the third digit; head elongate; ears of moderate size and tubular; tail short; one or three functional toes; nails modified into hooves; dental formula $\frac{0-3}{0-3}, \frac{0-1}{0-1}, \frac{3-4}{3-4}, \frac{3}{3}$; Africa, Asia (including Java and Sumatra), and the Americas from southern Mexico to Argentina; three families.

*Family Equidae (Horselike Mammals).* Large (1.8 to 2.5 m. perissodactyls with a single functional toe on each foot; body covered with short hair; a mane present along back of neck; legs long and proportionally slender distally; hair on tail very long; incisors chisel-like; molariform teeth very high-crowned; dental formula $\frac{3}{3}, \frac{0-1}{0-1}, \frac{3-4}{3}, \frac{3}{3}$; Asia and Africa; one genus, *Equus*.

*Family Tapiridae (Tapirs).* Medium-sized (1.8 to 2.5 m. perissodactyls with nose elongated into a short proboscis eyes small; hair short; tail very short and essentially hairless; three functional toes on each foot with a small fourth toe present on the front feet; dental formula $\dfrac{3}{3}, \dfrac{1}{1}, \dfrac{4}{4}, \dfrac{3}{3}$; southeastern Asia including the Malayan area and the New World from southern Mexico to Argentina; one genus, *Tapirus.*

**Family Rhinocerotidae (Rhinoceroses).** Very large (2 to 4.2 m.) terrestrial mammals with three functional digits on each foot and one or two median horns on top of head; eyes small; skin very thick and essentially naked; legs proportionately short in contrast to massive body; stiff bristles on tail; dental formula $\dfrac{0-2}{0-1}, \dfrac{0}{0}, \dfrac{3-4}{3-4}, \dfrac{3}{3}$; Asia and Africa; four genera.

**ORDER ARTIODACTYLA (EVEN-TOED HOOFED MAMMALS).** Small to very large terrestrial mammals (less than 0.5 to 4.6 meters) with the main axis of the foot passing between toes 3 and 4; nails modified into hooves; tail short; upper incisors usually reduced or absent; dental formula $\dfrac{0-3}{1-3}, \dfrac{0-1}{0-1}, \dfrac{2-4}{2-4}, \dfrac{3}{3}$; essentially worldwide except the Australian region; nine families.

**Family Suidae (Swinelike Mammals).** Medium-sized (0.7 to 1.9 m.) artiodactyls with a large head and a nose that is elongated into a truncated snout with a disklike cartilage or bone in the tip; eyes small; hair bristle-like; tail slender and terminating in bristles; four toes on each foot; upper and lower canines developed into tusks which curve upward; dental formula $\dfrac{1-3}{3}, \dfrac{1}{1}, \dfrac{2-4}{2-4}, \dfrac{3}{3}$; Europe, Asia, Africa, and Madagascar; five genera.

**Family Tayassuidae (Peccaries).** Small (0.75 to 1 m.), piglike artiodactyls with four toes on the front foot and three on the hind foot; ears and eyes small; nose snoutlike with nostrils terminal; tail very short; body covered with coarse hair; musk gland present on back; legs proportionately slender in contrast to stocky body; upper canine tusks relatively small, sharp, and directed downward; dental formula $\dfrac{2}{3}, \dfrac{1}{1}, \dfrac{3}{3}, \dfrac{3}{3}$; southwestern United States to Patagonia; one genus, *Tayassu.*

**Family Hippopotamidae (Hippopotamuses).** Medium-sized to very large (1.5 to 4.5 m.), heavy-bodied artiodactyls with a large, rounded muzzle; eyes dorsal in position and bulging; ears small; body essentially hairless; four functional toes on each foot; incisors and canines tusklike; dental formula $\dfrac{2-3}{1-3}, \dfrac{1}{1}, \dfrac{4}{4}, \dfrac{3}{3}$; Africa; two genera, *Hippopotamus* and *Choeropsis.*

**Family Camelidae (Camels and Llamas).** Medium to large-sized (1.2 to 3 m.) artiodactyls with a long neck and only two toes on each foot; head small; upper lip cleft and proportionately large; hair on body soft and fine; dental formula $\frac{1}{3}, \frac{1}{1}, \frac{2-3}{1-2}, \frac{3}{3}$; Asia, Africa, and South America; two genera.

**Family Tragulidae (Chevrotains).** Very small (0.5 to 1 m.) artiodactyls with slender legs terminating in four functional digits; facial and foot glands as well as antlers lacking; upper canines elongated into small tusks; lower canines incisiform; dental formula $\frac{0}{3}, \frac{1}{1}, \frac{3}{3}, \frac{3}{3}$; central Africa and southeastern Asia; two genera, *Hyemoschus* and *Tragulus*.

**Family Cervidae (Deerlike Mammals).** Medium-small (0.75 to 3 m.) to large artiodactyls with long, slender legs bearing four toes, the middle pair being larger than the outer two; facial and metatarsal glands usually present; deciduous antlers worn by the male in all except two genera; females lacking antlers except in one genus; dental formula $\frac{0}{3}, \frac{0-1}{1}, \frac{3}{3}, \frac{3}{3}$; Europe, Asia, North Africa, and North and South America; about 17 genera.

**Family Giraffidae (Giraffe and Okapi).** Large to very large (2 to 4 m.) artiodactyls with long legs and two or three horn-like structures on the skull which are permanently covered with skin and hair; neck unusually long; feet possessing two hoofed toes; dental formula $\frac{0}{3}, \frac{0}{1}, \frac{3}{3}, \frac{3}{3}$; Africa; two genera, *Giraffa* and *Okapia*.

**Family Antilocapridae (Pronghorn).** A medium-sized (up to 1.5 m.), antelope-like artiodactyl possessing forked horns which have a deciduous outer sheath; head elongate; legs slender and possessing only two digits; tail short; dental formula $\frac{0}{3}, \frac{0}{1}, \frac{3}{3}, \frac{3}{3}$; plains and deserts of western North America; one species, *Antilocapra americana*.

**Family Bovidae (Cattle, Antelope, Sheep, and Goats).** Small to very large (0.5 to 3.5 m.) artiodactyls in which the males and usually the females possess a pair of unbranched horns (two pairs in the genus *Tetracercus*); toes 3 and 4 well developed; toes 2 and 5 small or entirely absent; dental formula $\frac{0}{3}, \frac{0}{1}, \frac{3}{2-3}, \frac{3}{3}$; North America, Europe, Asia, and Africa; about 44 genera.

## References Recommended

Allen, G. M. 1938, 1940. The Mammals of China and Mongolia. Natural History of Central Asia. New York, American Museum of Natural History, Vol. 10 (2 parts).

Allen, G. M. 1939. Bats. Cambridge, Mass., Harvard University Press.

Allen, G. M. 1942. Extinct and Vanishing Mammals of the Western Hemisphere. American Committee for International Wildlife Protection.

Armitage, K. B. 1961. Frequency of Melanism in the Golden-mantled Marmot. J. Mamm. *42*:100-101.

Bachrach, M. 1930. Fur. New York, Prentice-Hall, Inc.

Barbour, R. W. 1963. *Microtus:* A Simple Method of Recording Time Spent in the Nest. Science *141* (No. 3575):41.

Bartholomew, G. A., Leitner, P., and Nelson, J. E. 1964. Body Temperature, Oxygen Consumption and Heart Rate in Three Species of Australian Flying Foxes. Physiol. Zool. 37:179-198.

Beale, D. M. 1962. Growth of the Eye Lens in Relation to Age in Fox Squirrels. J. Wildlife Mgmt. *26*:208-211.

Bourliere, F. 1954. The Natural History of Mammals. New York, Alfred A. Knopf.

Brownlee, R. G., Silverstein, R. M., Müller-Schwarze, D., and Singer, A. G. 1969. Isolation, Identification and Function of the Chief Component of the Male Tarsal Scent in Black-tailed Deer. Nature *221*:284-285.

Buchner, C. H. 1964. Metabolism, Food Capacity, and Feeding Behavior in Four Species of Shrews. Canadian J. Zool. *42*:259-279.

Burt, W. H., and Grossenheider, R. P. 1964. A Field Guide to the Mammals. 2nd Ed. Boston, Houghton Mifflin Co.

Cahalane, V. H. 1947. Mammals of North America. New York, The Macmillan Co.

Chew, R. M. 1961. Water-metabolism of Desert-inhabiting Vertebrates. Biol. Rev. *36*:1-31.

Cowan, I. McT. 1936. Distribution and Variation in Deer (genus *Odocoileus*) of the Pacific Coastal Region of North America. Calif. Fish and Game *22*:155-246.

Dalquest, W. W., and Werner, H. J. 1954. Histological Aspects of the Faces of North American Bats. J. Mamm. 35:147-160.

Eadie, W. R. 1938. The Dermal Glands of Shrews. J. Mamm. *19*:171-174.

Ellerman, J. R. 1940-1941. The Families and Genera of Living Rodents. London, British Museum of Natural History, 2 Vols.

Erickson, A. W., Nellor, J., and Petrides, G. A. 1964. The Black Bear in Michigan. Mich. State Univ. Agric. Exp. Sta., Res. Bull. No. 4:1-102.

Fisler, G. F. 1963. Effects of Salt Water on Food and Water Consumption and Weight of Harvest Mice. Ecology *44*:604-608.

Flower, W. H., and Lydekker, R. 1891. An Introduction to the Study of Mammals Living and Extinct. London, Adams and Black.

Gregory, W. K. 1951. Evolution Emerging; A Survey of Changing Patterns from Primeval Life to Man. New York, The Macmillan Co., 2 Vols.

Griffin, D. R. 1958. Listening in the Dark. New Haven, Yale University Press.

Griffin, D. R., Webster, F. A., and Michael, C. R. 1960. The Echolocation of Flying Insects by Bats. Animal Behavior 8:141-154.

Hall, E. R. 1946. Mammals of Nevada. Berkeley, University of California Press.

Hall, E. R., and Kelson, K. R. 1959. The Mammals of North America. New York, The Ronald Press, 2 Vols.

Hamilton, W. J., Jr. 1939. American Mammals. New York, McGraw-Hill Book Co., Inc.

Hatt, R. T. 1932. The Vertebral Columns of Ricochetal Rodents. Bull. Am. Mus. Nat. Hist. 63:599-738.

Hill, W. C. O. 1953-1962. Primates; Comparative Anatomy and Taxonomy. New York, Interscience Publishers, Inc., 5 Vols.

Howell, A. B. 1930. Aquatic Mammals. Springfield, Ill., Charles C Thomas.

Howell, A. B. 1944. Speed in Animals. Chicago, University of Chicago Press.

Hudson, J. W. 1964. Water Metabolism in Desert Rodents. Proc. 1st Internat. Symp. on Thirst in the Regulation of Body Water, pp. 211-235. Oxford, London, New York, and Paris, Pergamon Press.

Ingles, L. G. 1965. Mammals of the Pacific States. Stanford, Calif., Stanford University Press.

Kavanau, J. L. 1961. Identification of Small Animals by Proximity Sensing. Science *134*:1694-1696.

Kellogg, R. 1928. The History of Whales—Their Adaptation to Life in the Water. Quart. Rev. Biol. 3:29-76, 174-208.

Kellogg, W. N. 1961. Porpoises and Sonar. Chicago, University of Chicago Press.

King, J. E. 1964. Seals of the World. London, Trustees of the British Museum (Natural History).

Lyman, C. P. 1943. Control of Coat Color in the Varying Hare *Lepus americanus* Erxleben. Bull. Mus. Comp. Zool. Harvard 93:393–461.

MacMillen, R. E., and Lee, A. K. 1967. Australian Desert Mice: Independence of Exogenous Water. Science 158:383-385.

Manning, T. H. 1964. Age Determination in the Polar Bear. Canadian Wildlife Service, Occasional Paper No. 5.

Mayer, W. V. 1952. The Hair of California Mammals with Keys to the Dorsal Guard Hairs of California Mammals. Am. Mid. Nat. 48:480-512.

McCue, J. J. G. 1964. A Portable Receiver for Ultrasonic Waves in Air. IEEE Transactions Sonics and Ultrasonics Group, Vol. SU-11, No. 1, pp. 41-49.

Miller, G. S., Jr., and Kellogg, R. 1955. List of North American Recent Mammals. U.S. Nat. Mus. Bull. 205.

Murie, O. J. 1935. Alaska-Yukon Caribou. North American Fauna, No. 54.

Murie, O. J. 1951. The Elk of North America. Harrisburg, Pa., The Stackpole Co., and Washington, D.C., Wildlife Management Institute.

Norman, J. R., and Fraser, F. C. 1938. Giant Fishes, Whales and Dolphins. London, G. P. Putnam's Sons, Inc.

Norris, K. S., Ed. 1966. Whales, Dolphins and Porpoises. Berkeley, University of California Press.

Novick, A. 1958. Orientation in Paleotropical Bats. I. Microchiroptera. J. Exper. Zool. 138:81-154.

Novick, A. 1962. Orientation in Neotropical Bats. I. Natalidae and Emballonuridae. J. Mamm. 43:449-455.

Orr, R. T. 1963. Porpoises and Dolphins. Pacific Discovery 16 (No. 4):22-26.

Osterberg, D. M. 1962. Activity of Small Mammals as Recorded by a Photographic Device. J. Mamm. 43:219-229.

Palmer, R. S. 1954. The Mammal Guide. New York, Doubleday & Co., Inc.

Peterson, R. L. 1955. North American Moose. Toronto, University of Toronto Press.

Pocock, R. I. 1939-1941. The Fauna of British India, Including Ceylon and Burma. London, Taylor and Francis, Ltd., 2 Vols.

Poulter, T. C. 1963. Sonar Signals of the Sea Lion. Science 139:753-755.

Quay, W. B. 1965. Integumentary Modifications of North American Desert Rodents, p. 59-74. *In*: Proc. Symposium on Biology of the Skin and Hair Growth. Sydney, Angus and Robertson.

Rice, D. W. 1963. Pacific Coast Whaling and Whale Research. Trans. 28th N. Am. Wildlife and Nat. Resources Conf., pp. 327-335.

Ridgway, S. H., Scronce, B. L., and Kanwisher, J. 1969. Respiration and Deep Diving in the Bottlenose Porpoise. Science 166:1651-1654.

Romer, A. S. 1970. The Vertebrate Body. 4th Ed. Philadelphia, W. B. Saunders Co.

Rust, C. C., and Meyer, R. K. 1969. Hair Color, Molt, and Testis Size in Male, Short-tailed Weasels Treated with Melatonin. Science 165:921-922.

Scheffer, V. B. 1958. Seals, Sea Lions, and Walruses. Stanford, Calif., Stanford University Press.

Schevill, W. E. 1961. Cetacea. *In* Gray, P., Ed. The Encyclopedia of the Biological Sciences. New York, Reinhold Publishing Corp., pp. 205-209.

Schevill, W. E., Watkins, W. A., and Ray, C. 1963. Underwater Sounds of Pinnipeds Science 141(No. 3575): 50-53.

Scott, W. B. 1924. A History of the Land Mammals of the Western Hemisphere. New York, The Macmillan Co.

Seton, E. T. 1929. Lives of Game Animals. Garden City, N.Y., Doubleday, Doran and Co., 4 Vols.

Simpson, G. G. 1945. The Principles of Classification and a Classification of Mammals. Bull. Am. Mus. Nat. Hist., Vol. 85.

Van Heel, W. H. D. 1962. Sound and Cetacea. Netherlands J. Sea Research 1 (No 4):407-507.

Walker, E. P. 1964. Mammals of the World. Baltimore, Johns Hopkins Press.

Weichert, C. K. 1965. Anatomy of the Chordates. 2nd ed. New York, McGraw-Hill Book Co., Inc.

Young, J. Z. 1957. The Life of Mammals. New York, Oxford University Press.

*Chapter Seven*

# SYSTEMATICS

The term systematics is currently applied to the study of classification of plants and animals and is in a sense synonymous with taxonomy. A systematic zoologist, therefore, is interested in the differences as well as the similarities between various kinds of animals.

In order to refer satisfactorily to particular kinds of organisms, names have become a necessity. For many centuries man has applied common or vernacular names to the creatures that he has come to know and these are used today by both the layman and the scientist. Vernacular names, however, present many problems. They vary in different countries just as does the language. Sometimes in different parts of the same country one name may be applied to several unrelated species. The facts that there is little uniformity in the use of common names and, furthermore, that many kinds of obscure and little known organisms have no common names necessitate the use of scientific names.

## THE BINOMIAL SYSTEM

The binomial system employed today, consisting of Latin generic and specific names, was first introduced by Linnaeus. In pre-Linnaean times names were given to many different species, but they were often long and cumbersome, frequently describing the animal or plant. Quite a few of our North American species were first described in this manner by Mark Catesby in his *Natural History of Carolina, Florida, and the Bahama Islands,* published between 1731 and 1743. The Linnaean System reduced the number of words used to designate a species to two—one for the genus and one for the species—and advocated the use of Latin, a universal language, for this purpose. This system was soon accepted and the tenth edition of Linnaeus' *Systema Natura,* which appeared in 1758, is considered as the starting point for taxonomic nomenclature. Only scientific names published in or since this work are accepted.

For many years there has been an International Commission on Zoological Nomenclature, which has established certain rules to which systematic zoologists the world over adhere. A few of these rules are given in the following paragraph.

A specific name must consist of more than one letter and must be either Latin or Latinized. In any one genus there cannot be two specific or subspecific names the same and, in the animal kingdom, no two generic names can be the same, although botanical names may duplicate zoological names. According to the so-called Law of Priority the first name given must be accepted for all times. A synonym is a name given to a species after the accepted name was given, and that name must never be used again for another species in the same genus. The same holds true for genera and higher categories. Actual publication of a name is necessary and, while there is no definite ruling on the number of copies that must be distributed, it is generally accepted that it must be 10 or more.

Most of the work of the early systematists, that is, those engaged in the scientific classification of plants and animals, was in the nature of cataloguing. Their interest was largely concerned with applying names to species and, in this sense, they were merely nomenclators. Today zoological nomenclature is looked upon as a means, not an end. Although nomenclature and classification interact in many ways they are really two distinct subjects. The former provides a ready means of reference to a particular kind of organism and permits the museum specialist to file his specimens away in a neat and systematic manner. Classification, however, indicates the relationships and differences between various organisms and may be based on anatomy, physiology, paleontology, or even behavior.

The specific name indicates a certain homogeneity and is used for a group of animals or plants that have characters in common. Thus it indicates likeness for a population and also serves to show that this population differs from those of other species. The generic name, likewise, indicates a degree of relationship. The several species that may be included within a genus are more closely related to one another than they are to species in another closely related genus. So also are various degrees of similarity and difference indicated by families, orders, and higher taxonomic categories.

# GEOGRAPHIC VARIATION

## SUBSPECIES

For some time now, vertebrate zoologists have employed a trinomial system which necessitates the use of a subspecific name in addition to the generic and specific name for animals that exhibit what is called geographical variation. Actually the use of the trinomial system was begun over 100 years ago, although it was not given great emphasis until the beginning of the present century.

The subspecies or geographic race is, in a sense, a subjective category, but it does serve to express a concept. There are many species that exhibit great uniformity, so much so that statistical analyses of sample populations from various localities throughout their ranges fail to show the occurrence of significant differences. On the other hand there are a large number of species that show marked variation in different regions. These differences are not to be confused with color phases, which occur in a number of species. Black bear (*Ursus americanus*) cubs belonging to the same litter may be brown and black. In central United States screech owls (*Otus asio*) may be red or gray in the same locality. These are color phases, and young of each type may be found in the same nest. In western United States screech owls are gray. The shade of gray, however, varies with the locality. It is darker along the coast and paler inland. Deer mice (*Peromyscus maniculatus*) in the humid coastal forests of the Pacific Northwest are dark in color, whereas members of the same species inhabiting the Sonoran desert of the Southwest are pale. These geographic differences are not directly the result of environment in a Lamarckian sense. Experiments by Sumner (1932) in breeding generations of pale, desert-taken mice in a coastal region where the same species is much darker have shown that they remain pale, thus indicating the hereditary nature of these trends.

The fact that many different species show an inclination to vary in a somewhat similar manner under similar environmental conditions has resulted in several so-called "rules" of variation. Not all species, however, adhere to these rules. Many endotherms tend to be larger in colder climates than do their relatives in warmer environments, according to Bergmann's Rule. This means that in cold regions there is a tendency toward less body surface in proportion to weight, which would appear to be a means of conserving body heat. This trend toward larger body size in the north and smaller size farther south is shown in many species of birds and mammals. One of the best examples is seen in the song sparrow (*Melospiza melodia*) which ranges widely over much of North America (Fig. 7.1). Song sparrows from the Aleutian Islands are almost the size of towhees. As

**FIGURE 7.1.** Song sparrows (*Melospiza melodia*) in western North America follow Bergmann's Rule of decreasing in size from north to south, as well as Gloger's Rule with respect to color. The four specimens shown above, from left to right, represent breeding birds from the Aleutian Islands, southeastern Alaska, central California, and the Colorado Desert.

one proceeds inland and then south along the Pacific coast there is a gradual but definite reduction in body size until in southwestern United States we find song sparrows so much smaller that they hardly seem related to the large Aleutian Islands form. In fact, without being aware of the various intermediate populations, one would unquestionably consider these extremes to be specifically distinct. Studies made on populations of woodrats of the genus *Neotoma* in western North America have shown that body size is inversely correlated with environmental temperature and that this is decidedly adaptive. Large woodrats have a selective advantage over smaller individuals in a cold climate because of the smaller surface-to-mass ratio which provides greater insulation and thereby aids in conserving body heat. In desert regions where temperatures are high, smaller size facilitates heat dissipation.

Some ectotherms such as lizards and snakes show quite an opposite trend as far as size is concerned. For a number of species there is a gradual reduction in body size as one proceeds from warmer southern regions to cooler northern regions. Since these animals depend upon their environment for heat it is evident that a small body that could build up temperature rapidly during brief periods of sunshine would be advantageous.

According to Allen's Rule, the extremities of many endotherms show a tendency to vary inversely to body size. In other words, structures such as the ears, feet, and tail in many species are smaller in the north and larger in the south. This again may be another means of conserving body heat in cold climates and also of radiating heat in regions where it is warm. It is possible, however, that there are other plausible explanations for such geographic trends. Along the Pacific coast of North America certain species of rabbits, especially the brush rabbit (*Sylvilagus bachmani*), adhere to Allen's Rule with respect to size of the external ear, and the size of the pinna is directly correlated with the size of the auditory bulla of the skull (Fig. 7.2). As the mean temperature of the environment increases, the size of the auditory apparatus increases. If we assume that the latter indicates greater ability to hear we might expect that brush rabbits living farther south hear better than do their relatives to the north. It has been demonstrated, however, that heat interferes with

**FIGURE 7.2.** The brush rabbit (*Sylvilagus bachmani*) illustrates both Allen's Rule and Gloger's Rule. The dark, small-eared specimen on top comes from the cool, humid coast of northern California, and the pale, large-eared specimen is from the hot, arid interior of Baja California.

sound transmission. Cold air is rather transparent to sound, but as the temperature increases the air becomes more and more opaque to sound. Humidity, likewise, affects sound transmission. Sound travels best in an atmosphere of high humidity. As the humidity decreases the atmosphere becomes more and more opaque to sound. It seems possible, therefore, that the greater development of the auditory apparatus in the southern part of the range of the brush rabbit, where the temperature is high and the humidity low, is an adaptation to an environment in which it is more difficult to hear and, therefore, to detect enemies. This tendency toward larger ears in warm regions is exhibited by many other kinds of mammals.

Another very obvious trend shown by many kinds of birds and mammals is expressed in Gloger's Rule, which infers that races living in arid regions are lighter in color than those living in humid regions. The rich red color of the fox sparrow (*Passerella iliaca*) of eastern United States or the dark reddish brown color of these birds in coastal Alaska contrasts markedly with the pale, grayish color of fox sparrows found breeding on rather arid mountains in western United States. Even greater contrast is seen in certain species, such as the black-tailed jack rabbit (*Lepus californicus*) that ranges in western North America from the humid Pacific coast to the arid Sonoran deserts. In many such instances the correlation of the color of the animal with that of the environment seems quite obvious and of adaptive significance both for predator and prey. Humidity here may be considered as playing an indirect role in the sense that plant growth is generally more dense and luxuriant in humid regions and the soil darker owing to the presence of humus. In arid desert regions plant growth is sparse and the soil is largely lacking in humus. Desert plants often have reduced leaf structures and tend to be pale or glaucous in color. That humidity is not a direct factor in producing paler pigmentation in animals is demonstrated in certain desert areas where black lava beds of considerable extent exist, notably in the Tularosa Basin of southern New Mexico. Here, under desert conditions, certain small mammals tend to be very dark, thus paralleling the color of the environmental background (Benson, 1933).

Color, in addition to affording a protective value by enabling certain vertebrates to blend with their background and thereby be less conspicuous to either predator or prey, may also function to some extent in thermoregulation. Light colors in hot, desert regions reflect more heat from the sun, whereas dark colors have the opposite effect and may serve to maintain high body temperatures in endotherms or to raise body temperatures in ectotherms. There are obvious exceptions to this, however, since a number of arctic species are white, at least in winter, when it is coldest, and there are many terrestrial vertebrates in the tropics that are very dark in coloration. Furthermore, many of the small mammals that conform to Gloger's Rule are nocturnal. In such instances coloration would appear to be largely protective in nature.

Although it has been shown that lizards and snakes reverse Bergmann's Rule for warm-blooded vertebrates (that is, they are smaller in cold regions), this does not hold true for fishes in general. According to Jordan's Rule, fishes in colder waters tend to have more vertebrae than do those in warmer waters. It has been suggested that the lower temperatures retard sexual maturity and tend to produce larger forms. The number of fin rays in certain species tends to be reduced in warmer waters.

As has already been intimated, there are many exceptions to these general rules regarding geographic variation, and there are many species that fail to show any marked tendency to break up into local races. However, the significance of subspeciation is considered important from the evolutionary standpoint. Scientists are convinced that slight genetic mutations are constantly appearing within populations of various kinds of animals and that environmental selection is at work, weeding out those changes that are detrimental to the species and favoring those that are beneficial. The cumulative effect of these slight mutations would seem to account for the differences shown in various parts of the range of a geographically varying species. If the differences became sufficiently great and the intermediate populations were to be eliminated through climatic or geological changes, we could conceivably have separate species develop which would be isolated from one another or *allopatric*. If, after a period of time, their ranges later were again to become continuous or *sympatric* it is probable that they would not interbreed with one another. Dice (1931) has shown that in Washington County, Michigan, two races or subspecies of the deer mouse (*Peromyscus maniculatus bairdii* and *P. m. gracilis*) both occur in the same locality but behave as separate species. They represent, in a sense, opposite ends of a chain and are linked together by a series of intermediate races. If the intermediates were eliminated, *P. m. bairdii* and *P. m. gracilis* would probably be regarded as separate species.

Indications of a similar circular or graduated chain-link overlap are found among certain birds. For example the herring gull (*Larus argentatus*) of eastern Asia, northern North America, and western Europe overlaps the range of the lesser black-backed gull (*L. fuscus*) in the latter area. This European species is believed to have evolved from *L. argentatus* during the Pleistocene and subsequently to have become isolated from it.

Another good example of this is seen in the lungless salamander, *Ensatina eschscholtzi*, which ranges from southern British Columbia south to southern California (Stebbins, 1957). Seven geographic races of this species are currently recognized. Throughout a considerable part of California the coastal populations of *E. eschscholtzi* tend to be brown or reddish above, whereas those occurring farther east along the western slopes of the Sierra Nevada are marked by blotches ranging from cream to orange. The Great Valley between the ranges of these two color types is too arid to support this species of salamander and so constitutes a barrier at present. In the mountainous parts of northern California there is gradual intergradation between the coastal and inland color patterns. In southern California, however, the two color types come together and become sympatric, but interbreeding has not been found to occur here even though individuals of the two color types have been found within 150 yards of each other. This suggests that southern California represents an area in which the extreme ends of the species come together and behave as though they were separate species.

For many years the presence or absence of intermediate populations or intergrades has served as a criterion to determine whether closely related forms are geographic races of the same species or represent distinct species. When there is continuity of the breeding ranges of such forms with demonstrable intergradation, over either a broad or a limited area, the answer is obvious, although distribution maps sometimes seem to oversimplify the situation by merely showing a line separating one subspecies from another. One might even gain the impression that specimens from one side of the line are readily distinguishable from those on the other side. Actually, in nature, intergradation is usually gradual and occurs over a relatively broad area. Since animals care little for man-made rules, the systematist is frequently confronted with major problems that do not conform to what might be termed the "normal" pattern. These problems often arise in species whose ranges are discontinuous, such as those restricted to the higher parts of separate mountain ranges or to the upper parts of various river systems, or those that occur on both continental land masses and adjacent islands. When the range of variation between closely related allopatric populations is found to overlap, one concludes that these populations are conspecific. It is when overlapping does not occur with respect to certain characters that complications present themselves. Here it becomes a matter of degree and the taxonomist must make his own decision, which often becomes a subject fit for debate by his fellow workers.

There are a number of examples among vertebrates of related forms that, though their ranges overlap, behave as separate species, except in certain localities where hybridization occurs. Here again it is the experience and judgment of the taxonomist, as well as his familiarity with the morphology and ecology of the forms in question, that play a major part in deciding whether such populations are to be treated as specifically distinct.

In recent years there have been a number of objections to the subspecies concept. It has been asserted by some that it is too subjective. In other words there is no set criterion as to what constitutes a geographic race. One systematist, known as a "splitter," may consider a minute but constant variation within part of a population sufficient to justify nomenclatural recognition, while another systematist, known as a "lumper," may insist that far greater differences must be apparent before nomenclatural recognition is justified. While these arguments may be true for subspecies they are also true for genera, families, orders, classes, and even phyla. Actually the species is the only objective category in nature, and this is true only in a static sense.

The purpose of the subspecific name is to indicate incipient changes that may ultimately result in the production of new species and to correlate these slight differences in structure, behavior, or physiology in various parts of the range of a species with differences in the environment. Although the taxonomic method may be subject to some valid criticism, the purpose or aim is above reproach.

## CLINES

Another approach to the problem of geographic variation was suggested in 1939 by Huxley, who proposed the term *cline*. Clines are long geographic character gradients. Thus by means of graphs one may express geographic trends in weight or in linear measurements within the range or a part of the range of a species. For example, a species adhering to Bergmann's Rule might show a north to south cline of decreasing body length (Fig. 7.3). Clines, however, are not to be confused with subspecies, since a cline refers to a single character. A population from one place can belong to a single subspecies but may belong to a number of clines. A subspecies, therefore, represents a population occupying a part of the range of a species that differs in one or more characters from other populations occurring in other parts of the range of the same species. A cline, on the other hand, since it refers to a single character may show the geographic trend of this character in a number of subspecies. Both systems have merit and serve to express evolutionary trends. To date no attempt has been made to name clines, since they can be referred to descriptively.

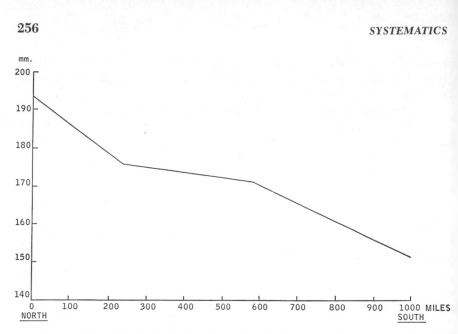

**FIGURE 7.3.**   A theoretical cline illustrating the decreasing length in a vertebrate structure from north to south.

## SEXUAL DIMORPHISM

In Chapter 12 sexual dimorphism is discussed in relation to sex recognition and courtship behavior. This appears to be the primary reason for most secondary sexual differences in vertebrates. However, it is likely that sexual dimorphism in size and in the feeding apparatus may in some instances have a definite adaptive value in reducing competition between sexes for food. In many birds, for example, the males are considerably larger than the females, with the result that there may be little or no intersexual competition in feeding. Each may occupy a separate subniche, thus permitting greater population density than would otherwise be possible. This has been demonstrated in insular populations of woodpeckers in the West Indies. Divergence of this sort is essentially a form of adaptive radiation within a single population.

Even sexual differences in behavior would seem to have definite adaptive advantages in certain vertebrates. Among some pinnipeds, such as sea lions, elephant seals, and the northern fur seal, there are not only marked size differences between the sexes but in many of these species the males do not feed while on the breeding rookeries. Females periodically do leave for food, since they must produce a rich milk for the young. Food competition with males at this time would be to their disadvantage.

## ISOLATION

The existence of barriers—geographical, ecological, physiological, and behavioral—obviously has played a major role in the evolutionary development of subspecies, species, and higher taxonomic categories. In the case of the two subspecies of *Peromyscus*, previously mentioned, that occur in the same locality but behave as separate species, the isolating mechanism is ecological (see p. 253). One form lives in the forest or woodland and the other inhabits more open grassland. *Larus argentatus* and *L. fuscus* are separated by ecological and behavioral differences in their area of overlap in western Europe so that hybridization is quite rare (see p. 253).

Mountain ranges, rivers, or unfavorable habitat within the range of a species may reduce interbreeding and gene flow within a population, thereby breaking it up into local groups. There are instances in which there may be complete isolation of some populations within a species. This is seen in the pikas (*Ochotona princeps*) found on certain mountain ranges in the Great Basin region of western North America. Desert valleys, presently surrounding some of the ranges inhabited by these lagomorphs, effectively prevent any possibility of interbreeding between populations on adjacent ranges. In some instances this isolation has been sufficiently long to bring about subspecific differentiation. Isolation of a similar nature is exhibited by

**FIGURE 7.4.** The pika (*Ochotona princeps*) of western North America. (Courtesy of U.S. National Park Service.)

certain cyprinodontid fish in the desert areas of western North America. Complete isolation of this sort, however, is relatively uncommon.

So far we have considered only geographic barriers on continental land masses. Islands, because of their physical isolation, often develop a distinct endemic biota. It is there too that one has a better opportunity to analyze the processes of evolution. Islands are usually classified either as oceanic or continental. The former may be of volcanic origin or basically coral reefs. They are separated from continental land masses by deep oceanic troughs as well as distance. Continental islands usually have a history of connection with continental areas to which they are more or less adjacent.

The Galápagos and Hawaiian Islands of the Pacific Ocean and the Tristan da Cunha group of the South Atlantic represent some well known examples of volcanic oceanic islands. On the Hawaiian Islands a separate family of birds, the Drepanididae, has evolved. Nine distinct genera and 43 species and subspecies of these endemic honeycreepers are recognized. Somewhat similar avian differenti-

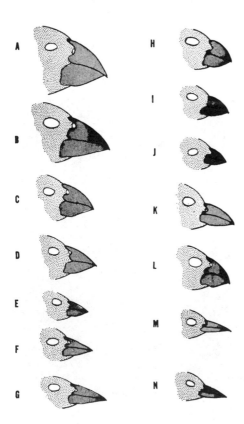

FIGURE 7.5. Variations in the shapes of the bills of different species of Darwin's finch indicate adaptations for different feeding habits. A to G, *Geospiza*, the seed-eaters; H to J, *Camarhynchus*, insect-eaters; K, *Cactospiza*, the tool-using finch; L, *Platyspiza*, a vegetarian; M, *Pinaroloxias*, the Cocos Island finch; N, *Certhidea*, the warbler-finch. (Bowman, R. I.: Pacific Discovery, Vol. 18, No. 5, 1965.)

ation has taken place on the Galapagos Islands, where a separate subfamily of finches, the Geospizinae, has developed and undergone marked divergent evolution, resulting in four or five genera and 13 species. In addition the Cocos Island finch (*Pinaroloxias inornata*) is regarded as a member of this subfamily. On the Tristan da Cunha group there is an endemic genus of finch (*Nesospiza*) containing two species that occur on the same islands. There are a number of distinct species of birds on oceanic islands in the South Pacific. However, unlike the previously mentioned islands, usually no more than a single species of a genus occurs on any one island.

There are many examples of continental islands. These range from relatively large land masses such as Madagascar, the islands of Japan, Tasmania, and Taiwan to small archipelagos such as the Channel Islands of southern California. Such continental islands, if of any significant size and distance from the mainland, usually possess a number of endemic species or subspecies of terrestrial vertebrates. The island populations that differ from related continental populations sometimes appear to be a result of divergent evolution, which has occurred subsequent to either invasion or separation. In some instances, however, they may be relict forms which once also occurred on the mainland but have since become extinct or modified in the latter area. In general, the number of species found on continental islands tends to be directly proportional to the size of the land mass. Furthermore, in an archipelago such as the Galapagos the number of endemic species found on an island seems to be correlated with the degree of that island's isolation from others in the group.

There are still other isolating mechanisms which must be considered. Among insects differences in the genitalia of closely related species have long been considered an important means of preventing hybridization. Such differences also occur in vertebrates and undoubtedly play an important role. In many places closely related species of rodents occur in the same general habitat. They may exhibit different ecologic preferences but still come into close contact with one another. This is true of members of such rodent genera as *Eutamias, Oryzomys, Peromyscus,* and *Microtus.* Recent studies of the genitalia of males in such closely related forms, which often resemble each other superficially, have shown marked structural differences in the shape and size of the baculum, as well as other parts of the penis, which could possibly prevent interbreeding.

Another very important isolating mechanism is the time of the reproductive season. A good example of this is shown by pinnipeds, which are mammals with a peripheral or linear distribution, being restricted to coastlines and islands. Because of the limited available habitat, several species frequently are found together during much or

FIGURE 7.6.   Three species of pinnipeds at Año Nuevo Island: *right*, the dark animals are bull California sea lions (*Zalophus californianus*); *lower middle*, a sleeping harbor seal (*Phoca vitulina*); *upper left*, a female Steller sea lion (*Eumetopias jubata*) in the water.

even all of the year. Even so, such species rarely breed at the same time, so that there is little chance for any hybridization to occur. On Año Nuevo Island, along the central California coast, Steller sea lions (*Eumetopias jubata*), California sea lions (*Zalophus californianus*), harbor seals (*Phoca vitulina*), and elephant seals (*Mirounga angustirostris*) live in close contact with one another. The Steller sea lions reproduce in June and July, the harbor seals from March to May, and the elephant seals from late December to the end of February. Only the California sea lions and the Steller sea lions have concurrent breeding seasons. However, in June and July, when the Steller sea lions are reproducing on Año Nuevo Island, the California sea lions leave and move to southern California and northwestern Mexico, where their breeding rookeries are located. In eastern North America the blue-winged warbler (*Vermivora pinus*) and the golden-winged warbler (*V. chrysoptera*), closely related species, are extensively sympatric and fertile hybrids are common. This is attributed to similarities in habitat selection and courtship behavior as well as unbalanced sex ratios. However, hybridization may be reduced in this complex, according to Ficken and Ficken (1968), because conspecific pairs of one species are formed before the other species arrives. Furthermore, there may be species differences in the releasers involved in pair formation as well as reduced mating success of the hybrids.

Three species of salamanders of the genus *Taricha* occur in western United States, *T. torosa*, *T. rivularis*, and *T. granulosa*. In certain localities *T. granulosa* is sympatric with the other two species. Courtship is very similar in all three species and there is considerable overlap in the time of breeding. It has been suggested, therefore, that the isolating mechanism may be sex attractants released by the females which only stimulate males of their own species. This is probably true also of certain mammals in which skin glands seem to play an important role in sex recognition within species.

Both territorial and courtship display may be very important as isolating mechanisms in closely related species. There are many groups of lizards in which the display differs considerably. In the lava lizard, *Tropidurus*, an iguanid genus of the Galapagos Islands, there are 12 species, each restricted to a separate island and each having a different type of display pattern. This has been attributed to genetic drift. It seems likely that if, through chance, a species from one island were to become established on one of the other islands where there is already an established species, this behavioral difference would inhibit hybridization. In the anole lizards (genus *Anolis*) of the Lesser Antilles, differences between sympatric species in territorial display by the males play a very important part in sexual selection by the females. The latter depend upon these differences to select a conspecific mate. This has also been observed among certain closely related species of birds. For example, in areas where the boat-tailed grackle (*Cassidix major*) and the great-tailed grackle (*C. mexicanus*) occur sympatrically, the males of each species solicit females of either species and mutually exclude males of the other species from their territories. The females, therefore, must select mates on the basis of differences in display, which here serves as an isolating mechanism.

## THE RATE OF EVOLUTION

Since vertebrates have a relatively slow rate of reproduction in contrast to many lower organisms, both plant and animal, we are inclined to assume that long periods of time have been involved in the production of geographic variants. Some evidence tends to support this concept. There have been studies made of certain late Pleistocene deposits which fail to show any significant differences between the populations of some species then and their modern representatives. On the other hand we find that subspecific differentiation has clearly occurred since the conclusion of the last Ice Age in a number of species. Some of the northern parts of the Holarctic region are now inhabited by animals which could not have survived there in the glacial periods. Furthermore, certain of these species presently exhibit clinal trends in these regions or are subspecifically differentiated from more southerly populations of the same species. Even specific differentiation has occurred in some species in this same period of time. For example, there are two species of collared lemmings recognized in North America today. One of these is *Dicrostonyx hudsonius,* which is now confined to the Ungava peninsula of eastern Canada. Occurring widely over many other parts of arctic North America as well as northern Eurasia is *D. torquatus.* It has been suggested that the formation of Hudson Bay during the period of ice recession in late Pleistocene may have served to split an east-west cline in what was a single species, thus isolating the eastern North American population so that it became specifically distinct. Cave remains from the late Pleistocene in Wyoming and Pennsylvania show that both the *torquatus* and *hudsonius* types existed then south of the Wisconsin Ice Sheet.

Island species sometimes provide information on the rate at which evolution may occur on isolated land masses. The nearly two dozen diversified species of Hawaiian honeycreepers constituting the avian family Drepanididae are believed to have come from ancestors that arrived in the Hawaiian Islands in the middle of the Pliocene, when the first forests developed there. This was only about five million years ago. Similarly the many endemic Galapagos finches are thought to have arisen from common ancestors that arrived in that archipelago in late Pliocene, possibly no more than one and one-half million years ago.

A very interesting study of the house sparrow (*Passer domesticus*) was recently made by Johnston and Selander (1964) which dramatically illustrates that evolutionary changes may occur far more rapidly than was heretofore believed possible. The first known introduction of this Old World species into North America occurred in 1852. Since then house sparrows have spread very widely over the continent. Presently these birds are exhibiting an adaptive differentiation throughout their new range that parallels very closely that of many native species. In accordance with Gloger's Rule, populations of these sparrows in the Pacific Northwest, where it is moist and humid, have darkly pigmented plumage, whereas those from the southwestern desert area are pale. Likewise, they follow Bergmann's Rule, since larger size is apparent in the north and smaller size to the south. Although it is but little more than 100 years since the first introductions were made in North America, this species has expanded into its present range only in this century. These genetic

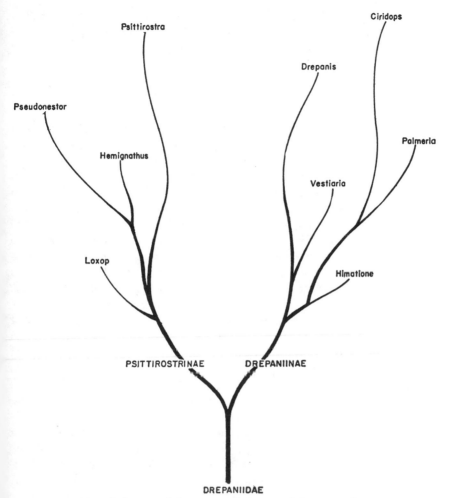

**FIGURE 7.7.** Phylogeny of the genera of Drepanididae. (Amadon, D.: Bull. Am. Mus. Nat. Hist., No. 95, 1950.)

changes, therefore, have probably occurred largely within the last 50 years or so.

Although, on the basis of morphological evidence, the rate at which evolutionary changes occur varies greatly between different organisms as well as between different characters in the same organism, this is not borne out by studies on protein structures. Sarich and Wilson (1967) have shown that protein molecules, which are undergoing evolutionary changes like the organism of which they are a part, often change to a similar degree in different groups of organisms over a given period of time. In studying the evolution of primates Sarich and Wilson concluded that man, the apes, and the

Old and New World monkeys shared a common ancestor 45,000,000 years ago and that the various lines leading to modern species have experienced a similar amount of albumin evolution.

## MODERN SYSTEMATICS

The vertebrate systematist of today is concerned both with refining our knowledge of the relationships between various taxonomic groups and with analyzing variations that occur within populations. During the past 200 years nearly all of our living North American vertebrates have been named, and the many collections assembled in museums have provided fairly accurate information on the distribution of most species. Within the past 50 years or so there has also been accumulated a host of subspecific names which indicate the geographic variability that occurs within many specific groups. The groundwork, therefore, has been laid for the student who wishes to delve deeper into relationships and variation. This does not imply, however, that we are in the age of "armchair science." Field studies are necessary so that one may become familiar with the habits of various species and especially with the particular environments to which they have adapted themselves. To monograph a group of animals that one has never studied in the field is comparable to writing a so-called authoritative account about a region that one has never visited. That a particular character varies within parts of the range of a bird is of interest, but the significance of this variation is the important point, and this can frequently be determined only by a detailed knowledge of the habits and the environment of the species concerned. Nevertheless, laboratory studies are essential. Anatomical research is constantly changing our concepts of relationships between groups.

Systematic vertebrate zoologists make use of many different characters in working out the differences and relationships of various groups. In general it might be said that every structure in the body, as well as the function of each organ and the behavior of the organism itself, should be given careful study by the systematist. This, although ideal, is rarely possible.

Taxonomists tend to place great emphasis on the vertebrate skeleton. The reasons for this are manifold. Students of vertebrate classification, particularly those working with the larger taxonomic units, must necessarily make use of paleontological material which, of course, consists largely of fossilized skeletal parts. We know little of the behavior, physiology, or even the structure of many of the organ systems of animals that lived in the past, but their skeletons are often perfectly preserved. The problem of housing the study material also often necessitates the preservation of skeletal elements rather than the entire specimen. Museums can scarcely afford the space and funds that would be required to keep the entire bodies of certain recent vertebrates such as large sharks, alligators, elephants, bears, whales, and many others. The skeletons of these animals, however, when properly prepared, can readily be housed and made available for comparative research. The smaller vertebrates do not present such a problem and are frequently preserved in their entirety. Ichthyologists and herpetologists customarily preserve their smaller specimens in alcohol or formalin. Most ornithological and mammalogical collections consist of study skins supplemented by skeletal material and alcoholic specimens. The skull of a mammal is also customarily preserved with the skin and serves as a useful tool for the taxonomist.

Although the skeletal parts of vertebrates are of prime importance in classification, every attempt is made by those working with these groups to gather information on the structure and function of the various other organ systems so as to give a more complete picture of relationships.

The recent trend among most modern vertebrate systematists has been to study variation within limited groups such as species. This entails adequate collecting to determine geographic ranges and sufficiently large samples of populations so that biometrical methods may be employed. Specialists within each class of vertebrates make use of various characters which best serve to show variation.

Cellular and molecular biology have recently added what might be called new tools in the systematic interpretation of relationships of vertebrates at all levels from subspecies to orders. For decades we have known of chromosomal differences between species, differences not only in the genes but in the number, size, and shape of the chromosomes themselves. Little use was made of this knowledge by vertebrate zoologists, although botanists have long used it in taxonomic studies. Perhaps the detailed laboratory techniques involved served as a deterrent. Currently, however, chromosomal studies are proving very important at the specific, subgeneric, and generic level. Questionable island species such as *Peromyscus guardia* found on several islands in the north end of the Gulf of California and closely related to *Peromyscus eremicus* on the nearby mainland of the peninsula of Baja California have been confirmed as specifically distinct on the basis of chromosomal differences. Even the specific status of other very closely related species belonging to this complicated genus is being clarified as a result of such studies. Among reptiles, closely related groups of whiptail lizards (genus *Cnemidophorus*) of the family Teiidae have been specifically sorted out on the basis of chromosomal structure in addition to the very few gross morphological differences that are apparent. Furthermore, some lizards of this genus are parthenogenetic. Two of these whose chromosomal structures have been studied are believed to have arisen through hybridization.

Investigations on the diploid number and karyotypes of North American ground squirrels of the subgenus *Spermophilus*, in combination with other morphological characters, have proved most useful in reappraising the relationships of the eight Nearctic species recognized. This group is believed to be of quite recent origin and its members show a nearly continuous sequence of gradation, suggesting that differentiation at the specific level has not long been underway. The chromosomal studies corroborate this.

In chromosomal studies the cells usually used are either from bone marrow or from the gonads. To insure mitotic division where the karyotypes can be readily seen, hormonal stimulation is sometimes given prior to removal of the material to be studied. The karyotypes are generally divided on the basis of size into macrochromosomes and microchromosomes. Where there is terminal fusion of paired chromosomes the term *acrocentric* is used, whereas if there is central fusion they are referred to as *metacentric*. Each of these categories is further subdivided by cytologists. For example, metacentric chromosomes that cross more nearly terminally than at the midpoint are called *submetacentric*.

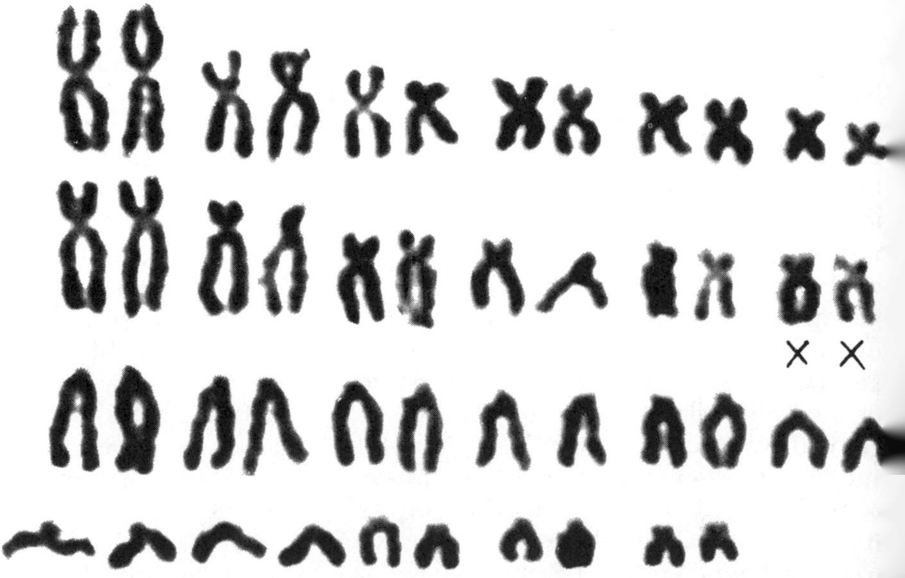

**FIGURE 7.8.** Karyotype of a female Townsend ground squirrel (*Spermophilus townsendi vigilis*) 2600×. (Nadler, C. F., Jour. Mamm. *47*, No. 4, 1966.)

Electrophoretic studies of proteins are shedding considerable light on the relationships of some groups of vertebrates whose systematic position was uncertain and are confirming or disproving previous conclusions reached with regard to the relationships of others. The principle involves the movement of charged particles in an electrical field when they are suspended in a liquid medium. Since proteins in blood serum or egg albumen are charged particles, they will undergo such movement. Those with a positive charge move toward the cathode while those with a negative charge move toward the anode. The speed of the movement depends upon the electrical charge in the particular protein molecule, as well as to some extent upon the size and shape of the molecule. As a result, when such animal proteins as blood serum or egg albumen are placed on a strip of filter paper or some other medium (such as a gel of starch saturated with a buffer solution), and an electric current is passed through the medium, the various kinds of proteins involved become separated along the axis of the current. When the result is placed in a dye which attaches only to proteins, the relative amounts and positions of the latter can easily be seen.

Although differences may result from concentration or denaturation, when these are accounted for the significant taxonomic characters relate to differences in the number of proteins and in their mobility. The latter presumably indicates genetic difference in one or more amino acids involved. Some of the most important work in this field has been carried on by Dr. Charles G. Sibley, Director of the Peabody Museum of Natural History at Yale University. He and his associates, working with egg albumen of several thousand species of birds, have shed much light on the familial classification of members of the class Aves and have discovered various evolutionary trends. Similar studies on the plasma proteins of fishes, amphibians, reptiles, and mammals are providing a new tool in taxonomy. Even electrophoretic studies on the venom of various kinds of rattlesnakes have added to our knowledge of the relationship of the many different species in North America.

One of the many problems that have confronted taxonomists for generations has been the subjective methods that have been employed. Even the definitions of the various taxa presented many problems and were, therefore, subject to different interpretations by different individuals. The New Systematics, which was proposed by Huxley and others around 1940, attempted much greater objectivity. Furthermore, from then on more effort was made to understand evolutionary mechanisms. Even so, the demand for greater objectivity continued. Today we have a new school of numerical taxonomists who are taking advantage of modern electronic computers by programing vast amounts of data into these machines. This develop-

Denaturation

Sample concentration

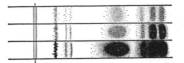

Polymorphism

a
b
ab

Differences in number

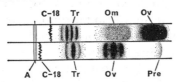

Differences in mobility

FIGURE 7.9. Some sources of variation in electrophoretic patterns and their effects. Denaturation: pattern of undenatured egg white above, denatured below. Sample concentration: dilute sample above, optimal concentration middle, concentrated sample below. Polymorphism: samples from three individuals of the same population showing three ovoconalbumin phenotypes, a, b, and the heterozygote ab. Differences in the number of proteins in two species. Differences in the mobilities of homologous proteins in two species. (Sibley, C. G., Bulletin 32, Peabody Museum of Natural History, 1970.)

ment means the achievement of results presently which would have taken vastly longer to attain in the past.

According to Sokal and Sneath (1963), "Numerical taxonomy is the numerical evaluation of the affinity or similarity between taxonomic units and the ordering of these units into taxa on the basis of their affinities." The system is based on statistical treatment of many phenotypic (usually morphological) characters. None of these characters is assumed *a priori* to be of greater or lesser value. In general, numerical taxonomy rejects phylogenetic considerations. This rejection is based on the idea that phylogeny is often conjectural, or at least imperfectly known, and also that when it is known it is really based on phenetic resemblances. The methods of numerical taxonomists are still not accepted by many and there remains a wide gap between the phenetic and phyletic schools of thought.

While behavioral, physiological, histological, and molecular studies of vertebrates all add greatly to our understanding of speciation and evolutionary development, a concept of the morphology of the organism as a whole is still a basic requirement. In the following pages the principal gross morphological characters used by specialists in each group of vertebrates are given. Measurements, whenever possible, are made with vernier calipers to insure accuracy and are usually recorded in millimeters or centimeters.

## MAMMALS

Mammalogists record the total length (from tip of nose to end of last tail vertebra), the tail length (from base of tail to tip of last vertebra), the length of the hind foot (from heel to tip of longest claw), and the length of the ear (from notch or crown to tip of pinna) on their specimen labels. Such data are secured from animals in the flesh and are extremely useful in determining average external body dimensions. The most useful cranial measurements (Fig. 7.10) used by mammalogists to show infraspecific as well as interspecific variation include the following:

*Condylobasal Length.* Distance from the anteriormost projections of the premaxillary bones to the posteriormost part of the exoccipital condyles.

*Basilar Length.* Distance from the posterior edge of the alveoli of the first upper incisors to the anteriormost inferior border of the foramen magnum.

*Zygomatic Breadth.* Greatest distance between the outermost borders of the zygomatic arches.

*Mastoid Breadth.* Greatest breadth of skull between the external margins of the mastoid bones.

*Palatilar Length.* Distance from posteriormost margins of the alveoli of the first upper incisors to the anteriormost point on the posterior border of the palate.

*Interorbital Constriction.* Least distance between the orbits across the top of the skull.

*Length of Tooth Row.* Distance, parallel to longitudinal axis of skull, from front of anteriormost tooth to back of posteriormost tooth.

*Length of Nasals.* Greatest length of nasals along median line of skull.

*Length of Incisive Foramina.* Greatest length from anterior border to posterior border of the incisive (anterior palatine) foramina.

Many other cranial measurements are made use of by specialists in various groups. For example, among rodents, the length of the diastema or space between the outermost incisors and the first molarform teeth is often a valuable measurement. Among many mammals

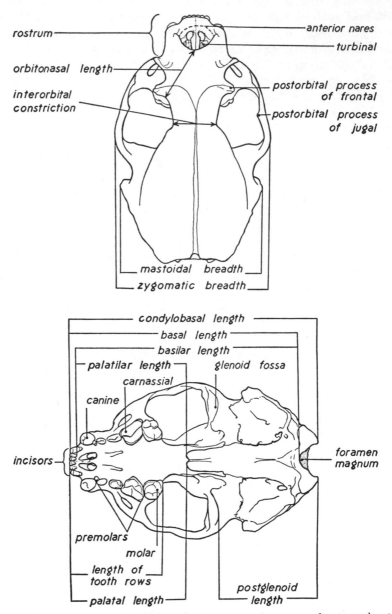

**FIGURE 7.10.**  Parts of the skull of the river otter, *Lutra canadensis*, and points, or parallels, as the measurement may require, between which cranial measurements were taken (×¹/₂). *Top,* dorsal view; *bottom,* ventral view. (From Hall, E. R.: Mammals of Nevada. University of California Press, 1946.)

with hypsodont or high-crowned teeth, the shape and conformation of the enamel folds is diagnostic for specific determination. For that matter variation in tooth structure and size in most mammals is of considerable taxonomic importance. Cetacean skulls are so highly modified that many measurements not applicable to other types are useful to show differences between species as well as larger categories.

Another character which, when present, has been shown to be of taxonomic importance is the baculum, or os penis, which frequently differs markedly between species that are otherwise but slightly distinguishable from one another.

As already indicated earlier in this chapter, hair color is subject to considerable geographic variation in many mammalian species and, as such, proves useful in systematic studies. In comparing coat color of populations within species, it is important that all specimens be of the same age category, such as juvenile, subadult, or adult, since the color of the hair of most mammals changes with age. Likewise among adults one must compare specimens in comparable pelages. There is little value to be derived from comparing a series of skins in fresh pelage with another series in worn, faded pelage. Relatively few mammals exhibit sexual dimorphism in coat color, so that in most instances color comparisons may be made regardless of sex. There are, however, a few exceptions. Sexual dimorphism in external and cranial measurements, on the other hand, is of frequent occurrence and must be given consideration.

## BIRDS

It is not customary to take external measurements of birds in the flesh, although it is important that the color of the soft parts such as the feet, legs, bill, eyes, wattles, combs, and other such structures be recorded from freshly secured specimens, since they are subject to rapid postmortem changes.

There are a number of standard measurements that avian taxonomists obtain from preserved skins, some of which are as follows:

*Length of Culmen.* This may be taken in several different ways. The most common method employed is to measure the length of the total culmen; that is, from the tip of the upper mandible to the point where the horny covering of the bill meets the integument of the forehead. This, therefore, is the chord of the entire culmen. Frequently, only the exposed culmen is measured. This is the distance, in a straight line, from the tip of the upper mandible to the point where the tips of the feathers of the forehead cease to hide the culmen. This measurement is less accurate than the first, since abra-

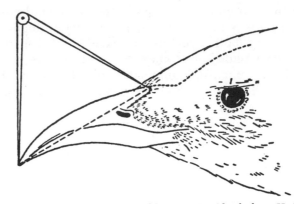

**FIGURE 7.11.**   Length of culmen. (Baldwin, S. P., Oberholser, H. C., and Worley, L. G.: Measurements of Birds. Scientific Publications of the Cleveland Museum of Natural History, 1931.)

sion or loss of feathers on the forehead may account for error. In various birds that possess a cere at the base of the bill, the culmen may be accurately measured from the tip to the anterior edge of the cere.

*Length of Bill from Nostril.*   The distance in a straight line from the tip of the upper mandible to the anterior margin of the nostril.

*Depth of Bill.*   This may be measured in several different ways. One method is to measure the distance from the base of the culmen to the lower margin of the lower mandible immediately below. Another method is to measure the vertical height of the bill at a point bisecting the anterior edge of the nostrils.

*Length of Wing.*   The distance from the tip of the longest primary to the anteriormost point on the bend of the wing (actually the wrist). Since the primaries are usually curved, this measurement represents the chord of the wing. Occasionally certain workers prefer to flatten the feathers to obtain the wing length. When this is done it should be clearly indicated that the measurement was secured in this manner.

*Length of Tail.*   This represents the distance from the base of the tail at a point between the emergence of the two innermost retrices from the skin to the tip of the longest feather when the tail is closed. Sometimes the proximal point of measurement is designated as the posterior base of the oil gland.

*Length of Tarsus.*   The distance from the middle of the joint between the tibia and the metatarsus on the posterior side to the junction of the metatarsus with the middle toe on the front side (usually the lower edge of the lowest undivided scute).

*Length of Middle Toe.*   The distance from the dorsal base of the middle toe at its junction with the metatarsal joint (or the lower edge of the lowest entire dorsal scute) to the distal end of the toe where the integument ends on the dorsal base of the nail. Most birds have three toes in front and one behind so that there is little difficulty in determining the middle one. In some which have two toes in front and two behind, such as most woodpeckers, the outermost front toe is measured.

There are many additional measurements that may be used, but those given here are the ones most commonly employed by ornithologists. A pair of fine pointed dividers proves extremely useful in securing accurate measurements on bird skins and, of course, the exact method employed in obtaining such data should be carefully recorded. Furthermore, it is important to indicate whether a specimen from which measurements have been obtained is in fresh or worn plumage. Old and frayed wing or tail feathers obviously are not going to measure as long as new feathers. Care must also be exerted to be sure that the feathers of birds that have recently molted are completely grown.

Ornithologists, like mammalogists, rely to a large extent on external coloration for identification and find considerable geographic variation in this regard occurring within many species. Segregation of specimens therefore by sex, age, and season of the year is important when color comparisons are made.

## REPTILES

Herpetologists working primarily with infraspecific, specific, or even generic groups rely to a large extent on integumentary structures for taxonomic characters, although overall length, snout-vent length, tail length, and certain limb measurements are frequently used.

Since herpetological material (except for very large species) is usually preserved, entire, in alcohol or formalin, it is not generally customary to record measurements on freshly collected material. Color notes taken at such times, however, are highly desirable. The fact that reptiles are covered with horny scales that vary greatly in size, shape, and number provides taxonomists with readily accessible characters for study. They may or may not be arranged in rows.

Names have been applied to the scales or groups of scales that cover the heads of most reptiles (Fig. 7.12). These include rostral, postrostrals, nasals, internasals, loreals, prefrontals, frontals, supraoculars, circumorbitals, superciliaries, parietals, interparietals, postoculars, temporals, auriculars, nuchals, and upper labials. On the lower jaw there are the mental, lower labials, postmentals, chin shields, and gulars. These names, of course, are based upon the

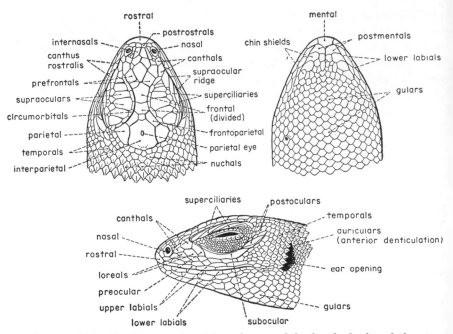

**FIGURE 7.12.** Dorsal, ventral and lateral view of the head of a lizard showing scale patterns. (After Stebbins, R. C.: Amphibians and Reptiles of Western North America. McGraw-Hill Book Co., Inc., 1954.)

particular region on the head on which the scales are located. The number of scale rows on the body is also of taxonomic importance and may vary considerably among closely related forms. In turtles the groups of large scales or shields on the carapace and plastron are given names and are subject to considerable variation. The number of large scales on the ventral surface of the body of most snakes are counted. The scales anterior to the anal plate are called the *gastrosteges*. Those posterior to the anal plate are the *urosteges*. The latter are sometimes divided. The shape and position of the hemipenis in male reptiles is another character used. The number and position of femoral and preanal glandular pores is important in the classification of many lizards. Since sexual dimorphism is not uncommon among reptiles, all specimens studied should be carefully sexed.

## AMPHIBIANS

Amphibians lack such definite ectodermal structures as hair, feathers, and scales; therefore taxonomists must rely on other charac-

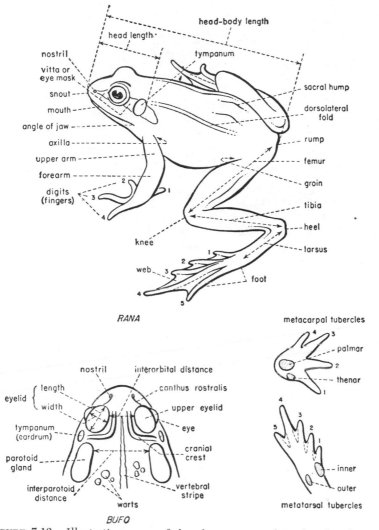

FIGURE 7.13. Illustrating some of the characters used in the classification of anurans. (After Stebbins, R. C.: Amphibians and Reptiles of Western North America. McGraw-Hill Book Co., Inc., 1954.)

ers. In addition to size, shape, and coloration, systematists make use of variation in the number of toes and of phalanges; the presence or absence of tubercles on the foot; the presence or absence of glands, especially the parotoid gland; the condition of the skin, whether smooth or rough; the absence or presence and location of teeth (Fig. 7.13). For many salamanders the presence and number of costal grooves on the side of the body are useful characters (Fig. 7.14).

**FIGURE 7.14.** The number and degree of development of costal and caudal grooves, such as those seen on the side of this arboreal salamander (*Aneides lugubris*), are important in the classification of certain caudate amphibians.

## FISHES

Ichthyologists frequently make use of a number of external measurements which may be obtained from preserved material in the case of smaller fish. Since usually only parts of very large fishes are saved, measurements must be made at the time the specimen is secured.

The measurements used most frequently are:

*Standard Length.* Distance from tip of snout to end of last vertebra.

*Total Length.* Distance from anteriormost part of head to posteriormost part of body (usually tip of caudal fin).

*Length of Snout.* Distance from tip of snout to anterior edge of orbit.

*Length of Upper Jaw.* Distance from anteriormost part of premaxillary to posteriormost part of maxillary.

*Length of Head.* Distance from tip of snout to posteriormost edge of opercular membrane.

*Length of Dorsal Fin or Anal Fin.* Distance between origin and insertion.

*Length of Pectoral, Pelvic, or Caudal Fin.* Distance from base to tip of longest ray.

*Height of Dorsal or Anal Fin.* Length of longest spine or ray.

Since the size of many kinds of fishes is dependent on age, more emphasis is usually placed on ratios of one length to another than on absolute measurements themselves.

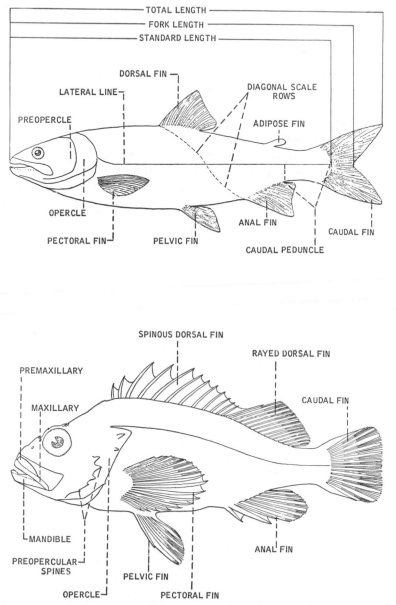

**FIGURE 7.15.** Some external characters and methods of measuring fishes. *Top,* ake trout (*Salvelinus*); *bottom,* rockfish (*Sebastodes*).

Certain other data are of considerable systematic value to ichthyologists, especially the number of fin rays and certain scale counts (Fig. 7.15). Customarily Roman numerals are used to denote the number of spines in a fin and Arabic numerals for the rays. The former, when present, are given before the latter. Thus dorsal XII, 12 to 16 means one dorsal fin with 12 spines followed by from 12 to 16 rays. The caudal fins of some fishes have spines above and below. Since counts are made from the dorsal to the ventral margin of this fin, Roman numerals would both precede and follow the Arabic numerals, which represent the rays in the middle.

Scale counts are almost as important to ichthyologists as they are to herpetologists. The conventional scale formula consists of three sets of figures, such as 8 + 90 + 10. The first figure represents the number of scales in a diagonal row from the origin of the dorsal fin down and back to the lateral line. The second figure represents the number of scales from above the opercular opening to the base of the caudal fin, either on the lateral line or, if this is not possible, along the first longitudinal row above the lateral line. The third figure represents the number of scales from the origin of the anal fin and forward in a diagonal row to the lateral line.

The number of gill rakers, the position and number of teeth, and many other characters too numerous to mention here are used by various specialists in the field. As is true in other groups of vertebrates, the color of fishes is subject to variation, but accurate data must be secured from fresh material as changes occur in many species immediately after death.

### References Recommended

Allee, W. C., Emerson, A. E., Park, O., Park, T., and Schmidt, K. P. 1949. Principles of Animal Ecology. Philadelphia, W. B. Saunders Co.

Amadon, D. 1950. The Hawaiian Honeycreepers (Aves, Drepaniidae). Bull. Am. Mus. Nat. Hist. 45(4):151-262.

Baldwin, S. P., Oberholser, H. C., and Worley, L. G. 1931. Measurements of Birds. Scientific Publ. Cleveland Mus. Nat. Hist. 2:1-165.

Benson, S. B. 1933. Concealing Coloration Among Some Desert Rodents of the Southwestern United States. Univ. Calif. Publ. Zool. 40:1-70.

Blackwelder, R. E. 1967. Taxonomy; A Text and Reference Book. New York, John Wiley and Sons, Inc.

Blair, W. F., and Littlejohn, M. J. 1960. Stage of Speciation of Two Allopatric Populations of Chorus Frogs (Pseudacris). Evolution 14:82-87.

Brown, J. H. 1968. Adaptation to Environmental Temperature in Two Species of Woodrats, Neotoma cinerea and N. albigula. Misc. Publ. 135, Mus. Zool. Univ. Michigan, 48 pp.

Brown, J. H., and Lee, A. K. 1969. Bergmann's Rule and Climatic Adaptation in Woodrats (Neotoma). Evolution 23:329-338.

Carlquist, S. 1965. Island Life. Garden City, N.Y., The Natural History Press.

Clemens, W. A., and Wilby, G. V. 1961. Fishes of the Pacific Coast of Canada. 2nd Ed. Bull. 68, Fisheries Res. Board of Canada.

Davis, W. C., and Twitty, V. C. 1964. Courtship Behavior and Reproductive Isolation in the Species of Taricha (Amphibia, Caudata). Copeia 1964:601-610.

Dice, L. R. 1931. The Occurrence of Two Subspecies of the Same Species in the Same Area. J. Mamm. 12:210-213.

Durrant, S. D. 1952. Mammals of Utah; Taxonomy and Distribution. Univ. Kansas Publ., Mus. Nat. Hist. 6:1-549.

Ficken, M. S., and Ficken, R. W. 1968. Reproductive Isolating Mechanisms in the Blue-winged Warbler–Golden-winged Warbler Complex. Evolution 22:166-179.

Fitch, H. S. 1940. A Biogeographical Study of the *ordenoides* Artenkreis of Garter Snakes (Genus *Thamnophis*). Univ. Calif. Publ. Zool. 44:1-150.

Gorman, G. C. 1968. The Relationships of *Anolis* of the *roquet* Species Group (Sauria: Iguanidae)—III. Comparative Study of Display Behavior. Breviora No. 284, 31 pp.

Gorman, G. C. 1970. Chromosomes and Systematics of the Family Teiidae (Sauria, Reptilia). Copeia 1970:230-245.

Gorman, G. C., and Atkins, L. 1968. New Karyotypic Data for 16 Species of *Anolis* (Sauria: Iguanidae) from Cuba, Jamaica, and the Cayman Islands. Herpetologica 24:13-21.

Guilday, J. E. 1968. Pleistocene Zoogeography of the Lemming *Dicrostonyx* (Cricetidae: Rodentia), a Reevaluation. Univ. Colorado Studies, Ser. Earth Sci. No. 6:61-71.

Hall, E. R. 1946. Mammals of Nevada. Berkeley, University of California Press.

Hall, E. R. 1951. American Weasels. Univ. Kansas Publ., Mus. Nat. Hist. 4:1-466.

Hooper, E. T. 1952. A Systematic Review of the Harvest Mice (Genus *Reithrodontomys*) of Latin America. Misc. Publ. Zool., Univ. Mich., No. 77.

Hooper, E. T., and Musser, G. T. 1964. The Glans Penis in Neotropical Cricetines (Family Muridae) with Comments on Classification of Muroid Rodents. Misc. Publ., Mus. Zool. Univ. Mich., No. 123:1-57.

Hubbs, C. L. 1922. Variations in the Number of Vertebrae and Other Meristic Characters of Fishes Correlated with the Temperature of Water During Development. Am. Nat. 56:360-372.

Hubbs, C. L. 1940. Fishes of the Desert. Biologist 22:61-69.

Hubbs, C. L. 1943. Criteria for Subspecies, Species and Genera, as Determined by Researches on Fishes. Ann. N.Y. Acad. Sci. 44:109-121.

Hubbs, C. L. 1953. Hybridization in Nature Between the Fish Genera *Catostomus* and *Xyrauchen*. Papers Mich. Acad. Sci. Arts and Letters 38:207-233.

Huxley, J. S., Ed. 1940. The New Systematics. New York, Oxford University Press.

Johnston, R. F., and Sclander, R. K. 1964. House Sparrows: Rapid Evolution of Races in North America. Science 144:548-550.

Klein, H. G. 1960. Ecological Relationships of *Peromyscus leucopus noveboracensis* and *P. maniculatus gracilis* in Central New York. Ecol. Monog. 30:387-407.

Lack, D. 1969. Subspecies and Sympatry in Darwin's Finches. Evolution 23:252-263.

Levan, A., Fredga, K., and Sandberg, A. A. 1964. Nomenclature for Centromeric Position on Chromosomes. Hereditas 52:201-220.

Lowe, C. H., Wright, J. W., Cole, C. J., and Bezy, R. L. 1970. Chromosomes and Evolution of the Species Groups of *Cnemidophorus* (Reptilia: Teiidae). Syst. Zool. 19:128-141.

Mayr, E. 1942. Systematics and the Origin of Species. New York, Columbia University Press.

Mayr, E., Ed. 1957. The Species Problem. Washington, D.C., Am. Assn. Adv. Sci., Publ. 50.

Mayr, E. 1963. Animal Species and Evolution. Cambridge, Mass., Harvard University Press.

Mayr, E., 1965. Numerical Phenetics and Taxonomic Theory. Syst. Zool. 14:73-97.

Mayr, E. 1969. Principles of Systematic Zoology. New York, McGraw-Hill Book Co.

Mayr, E., Linsley, E. G., and Usinger, R. L. 1953. Methods and Principles of Systematic Zoology. New York, McGraw-Hill Book Co., Inc.

Mengel, R. M. 1964. The Probable History of Species Formation in Some Northern Wood Warblers (Parulidae). The Living Bird, Third Annual of Cornell Laboratory of Ornithology, 1964:9-43.

Michener, C. D. 1964. The Possible Use of Uninominal Nomenclature to Increase the Stability of Names in Biology. Syst. Zool. 13:182-190.

Miller, A. H. 1941. Speciation in the Avian Genus *Junco*. Univ. Calif. Publ. Zool. 44:173-434.

Muul, I. 1968. Behavioral and Physiological Influences on the Distribution of the

Flying Squirrel, *Glaucomys volans*. Misc. Publ. 134, Mus. Zool. Univ. Michigan, 66 pp.

Nadler, C. F. 1966. Chromosomes and Systematics of American Ground Squirrels of the Subgenus *Spermophilus*. Jour. Mamm. *47*:579-596.

Orr, R. T. 1940. The Rabbits of California. Occasional Papers Calif. Acad. Sci., No. 19.

Orr, R. T. 1964. Interspecific Behavior Among Pinnipeds. Zeitschrift für Saugetierkunde *30*:163-171.

Petersen, M. K. 1968. Electrophoretic Blood-serum Patterns in Selected Species of *Peromyscus*. Amer. Midl. Nat. *79*:130-148.

Rausch, R. L., and Rausch, V. R. 1965. Cytogenetic Evidence for the Specific Distinction of an Alaskan Marmot, *Marmota broweri* Hall and Gilmore (Mammalia: Sciuridae). Chromosoma (Berl.) *16*:618-623.

Rivas, L. R. 1964. A Reinterpretation of the Concepts "Sympatric" and "Allopatric" with Proposal of the Additional Terms "Syntopic" and "Allotopic." Syst. Zool. *13*:42-43.

Sarich, V. M., and Wilson, A. C. 1967. Rates of Albumin Evolution in Primates. Proc. Nat. Acad. Sci. *58*:142-148.

Selander, R. K. 1966. Sexual Dimorphism and Differential Niche Utilization in Birds. Condor *68*:113-151.

Selander, R. K., and Giller, D. R. 1961. Analysis of Sympatry of Great-tailed and Boat-tailed Grackles. Condor *63*:29-86.

Sheppe, W., Jr. 1961. Systematic and Ecological Relations of *Peromyscus oreas* and *P. maniculatus*. Proc. Am. Phil. Soc. *105*:421-446.

Short, L. L., Jr. 1969. "Isolating Mechanisms" in the Blue-winged Warbler-Golden winged Warbler Complex. Evolution *23*:355-356.

Sibley, C. G. 1954. Hybridization in the Red-eyed Towhees of Mexico. Evolution *8*:252-290.

Sibley, C. G. 1960. The Electrophoretic Patterns of Avian Egg-white Proteins as Taxonomic Characters. Ibis *102*:215-284.

Sibley, C. G. 1962. The Comparative Morphology of Protein Molecules as Data for Classification. Syst. Zool. *11*:108-118.

Sibley, C. G. 1964. The Characteristics of Specific Peptides from Single Proteins as Data for Classification. *In* Leone, C. A., Ed. Taxonomic Biochemistry and Serology. New York, The Ronald Press.

Sibley, C. G. 1967. Proteins: History Books of Evolution. Discovery *3*(1):5-20.

Sibley, C. G. 1970. A Comparative Study of the Egg-white Proteins of Passerine Birds. Peabody Mus. Nat. Hist., Yale Univ., Bull. 32, 131 pp.

Sibley, C. G., and Brush, A. H. 1967. An Electrophoretic Study of Avian Eye-lens Proteins. Auk *84*:203-219.

Sibley, C. G., and Hendrickson, H. T. 1970. A Comparative Electrophoretic Study of Avian Plasma Proteins. Condor *72*:43-49.

Sibley, C. G., and Johnsgard, P. A. 1959. Variability in the Electrophoretic Patterns of Avian Serum Proteins. Condor *61*:85-95.

Simpson, G. G. 1945. The Principles of Classification and a Classification of Mammals. Bull. Am. Mus. Nat. Hist., Vol. 85.

Simpson, G. G. 1965. Current Issues in Taxonomic Theory. Science *148*:1078.

Sokal, R. R., and Sneath, P. H. A. 1963. Principles of Numerical Taxonomy. San Francisco, W. H. Freeman and Co.

Stebbins, R. C. 1949. Speciation in Salamanders of the Plethodontid Genus *Ensatina*. Univ. Calif. Publ. Zool. *48*:377-526.

Stebbins, R. C. 1954. Amphibians and Reptiles of Western North America. New York, McGraw-Hill Book Co., Inc.

Stebbins, R. C. 1957. Intraspecific Sympatry in the Lungless Salamander *Ensatina eschscholtzi*. Evolution *11*:265-270.

Sumner, F. B. 1932. Genetic, Distributional, and Evolutionary Studies of the Subspecies of Deer Mice (*Peromyscus*). The Hague, Bibliographia Genetica, 9.

Wilson, A. C., and Kaplan, N. O. 1964. Enzyme Structure and Its Relation to Taxonomy. *In* Leone, C. A., Ed. Taxonomic Biochemistry and Serology. New York, The Ronald Press, pp. 321-346.

Zweig, G., and Crenshaw, J. W. 1957. Differentiation of Species by Paper Electrophoresis of Serum Proteins of *Pseudemys* Turtles. Science *126*:1065-1067.

# Chapter Eight

# DISTRIBUTION

## GENERAL FEATURES

The word distribution may be used in several different ways. The paleontologist is apt to think of the range of species through time, whereas the herpetologist or ornithologist is primarily interested in the spatial distribution of presently existing species. It is necessary to bear in mind, however, that although we may think of the present range of any form as static this undoubtedly is not the case.

For vast periods of time the earth has been undergoing climatic and geologic changes, and there is no indication that these changes are not continuing and will not continue in the future. Recent evidence indicates that no more than ten or twelve thousand years have elapsed since the end of the last glacial period in North America. Since then the distribution of many kinds of vertebrates has undergone considerable change. Arctic species which formerly occurred as far south as the United States have had their ranges compressed to the far North, while species of the temperate zones have expanded northward into areas in which they could not have survived a few thousand years ago.

Some of our knowledge of these changes has been learned by a study of the bones of vertebrates as well as pollen grains found in caves and sinkholes, where they had been deposited in late Pleistocene times. Recently, carbon dating of vertebrate material from a sinkhole in southwestern Pennsylvania has shown that most of the species that occurred there about 12,000 years ago (11,300 ± 1000) were inhabitants of the Hudsonian or Canadian zones or taiga. The remains of collared lemmings (*Dicrostonyx hudsonius*), which are presently found on the Ungava Peninsula of Quebec and in Labrador, indicate that the arctic tundra came much farther south not too many thousands of years ago. Remains of *Dicrostonyx torquatus* in Box Elder Cave in Wyoming, dating back to the late Wisconsin Ice Age, also prove the subsequent northern movement of this tundra species since the late Pleistocene.

Climatic changes not only have a direct effect upon the distribution of vertebrate species, most of which have very definite ranges of tolerance, but usually have a greater effect upon them indirectly. Temperature, rainfall, and humidity to a large extent control the environment. A change in any one of these factors is likely to change the environment and make it unsuitable for members of some species. The latter must either move into an environment more suited to their needs, adapt themselves to the changed environment, or be eliminated. There is little doubt that all three alternatives have been taken by different species in the past.

There is considerable geological as well as zoogeographic evidence to show that Siberia and Alaska were connected intermittently by a land bridge in the Pleistocene. No more than 35,000 years ago when there was a glacial maximum, the Bering-Chukchi platform between these two continents was exposed over a great distance north and south as a result of a lowering of the sea level by more than 300 feet. This is believed to account for the arrival in North America of such big herbivores as moose (*Alces*), elk (*Cervus*), caribou (*Rangifer*), sheep (*Ovis*), and possibly man himself. There are no records of any of these species in deposits in North America that are older than Pleistocene, and the moose that is presently found in Alaska (*Alces alces gigas*) is so closely related to the Siberian form (*A. a. pfizenmayeri*) that it is possible it may have crossed the Bering Strait in fairly recent times. A number of smaller mammals such as the brown lemming (*Lemmus lemmus*), the collared lemming (*Dicrostonyx torquatus*), several voles (*Microtus oeconomus* and *M. gregalis*), the red-backed mouse (*Clethrionomys rutilus*), and the arctic ground squirrel (*Citellus undulatus*) are Holarctic species whose arrival from Siberia probably occurred via the Bering land bridge not very many thousands of years ago.

There is no record of the grizzly bear, which is now regarded as the same species as the Eurasian brown bear (*Ursus arctos*), occurring in North America prior to the late Pleistocene. After it reached this continent from Asia, semi-arctic conditions favored its rapid expansion. Late Wisconsin Ice Age fossils of grizzlies occur at least 800 miles east of their known range within historic times, which is the 100th meridian. Apparently the territory of this species has constricted since the late glacial period.

The Pleistocene not only brought an influx of large Palearctic mammals to North America, but it eliminated a great number of native species. Not many thousands of years ago the Great Plains of this continent were rich in big herbivores and carnivores, including mastodonts, mammoths, sloths, camels, horses, bison much larger than our present species, dire wolves, lions, sabre-toothed cats, and many others. North America in a sense was like present-day Africa, with a great reservoir of large species. In Recent times, however, only a few species of herbivores remain, such as the bison, pronghorn, deerlike animals, sheep, mountain goat, and muskox; and among the fairly large carnivores, the wolf, coyote, and mountain lion.

Ice bridges, which are the result of climate, have, like land bridges, aided in the movement of some vertebrates. Even today we find certain terrestrial animals that are isolated on islands during the summer months but have free access over the ice in winter to adjacent land masses. Such movement may lead to an expansion of the range of a species and probably accounts for the presence of many terrestrial forms on coastal islands as well as islands on inland bodies of water. The evidence of repeated immigrations of certain mammals, even possibly in recent times, across the winter ice bridge in the Bering Strait is believed to account for the Holarctic distribution of some species. It is probable also that many species are transported to new areas on ice floes. If suitable habitats are available for the survivors of these trips such means of dispersal may occasionally be of significance.

Many plants and animals are rafted down rivers to new areas as a result of heavy rains (Fig. 8.1). Under certain circumstances they may be able to survive and become established in the new environment. Ocean currents, too, have played an important part in aiding the distribution of plants and animals. Many of the terrestrial species inhabiting the Galápagos Islands, about 600 miles off the coast of Ecuador, are believed to have arrived there or been derived from ancestors that came there by drift. The Peruvian Current flows northward along the west coast of South America, turns west when it reaches Ecuador, and passes by these volcanic islands. Between the mainland and the islands, logs and natural vegetation rafts may drift with this current at speeds of 50 miles per day, or faster if the prevailing southeasterly winds are blowing. Numerous organisms, including small mammals, could survive a 10 to 14 day trip of this sort.

Geological changes, like climatic changes, have greatly influenced the distribution and evolutionary development of vertebrates. The elevation and submergence of the land, the rise and fall of mountain masses, the development of river systems, and the deposition of alluvium are but some of the more obvious changes on the

**FIGURE 8.1.** Numerous islands of vegetation seen floating down the Guayas River toward the Gulf of Guayaquil in Ecuador after heavy rains in the Andes. These islands may harbor many animal organisms. The native dugout canoe gives some idea of the size of these floating masses.

surface of the earth that have played such important parts in this regard. Although most geological changes require long periods of time, others are more rapid. There are areas where depression of the land can be measured by the year and old surveys are no longer accurate. Some old harbors in Chesapeake Bay are now many hundreds of yards from water. In San Francisco Bay, remains of Indian shellmounds which once were obviously back from the shoreline are now submerged beneath 4 feet of water.

Certain geological phenomena, such as volcanic eruptions, may be classified as catastrophic and almost instantly affect the distribution of all living things in the immediate vicinity. The sudden eruption of a volcano on San Benedicto Island in the Revillagigedo group, off the west coast of Mexico, in August 1952 is a good example of a natural catastrophe that brought about the elimination of nearly all the resident birds. The only avian species represented by an endemic population restricted to this island, however, was the rock wren (*Salpinctus obsoletus exsul*). Subsequent investigations seemed to indicate that this subspecies had been completely eliminated. Other species, formerly represented, subsequently reinvaded San Benedicto Island from adjacent islands and the mainland.

Within historic times there have been observed changes in ver

**FIGURE 8.2.** *Top,* San Benedicto Island as it appeared in 1925 with considerable vegetation present at lower elevations. (Photograph by G. Dallas Hanna.) *Bottom,* San Benedicto Island showing the crater shortly after the volcanic eruption in 1952. (Official photograph of U.S. Navy.)

tebrate distribution. Some have been natural; others have been affected directly or indirectly by man. Certain species have benefited while others have become extinct.

Although the enormous impact of man and his activities are making it more and more difficult to determine whether or not any of the more obvious recent changes in distribution are natural, some appear to be so. For example, the black-headed gull (*Larus ridibundus*) of Europe was first reported along the Atlantic coast of the United States in 1930. In recent years it has occurred regularly along the coast of New England. Since the New England coast has been a favorite area for competent bird watchers for generations, it is obvious that the presence of this gull was not overlooked until recently, but rather that the species is gradually invading a new region.

It is probably safe to say that most vertebrate species in North America have been affected in one way or another, either directly or indirectly, as a result of human activity. Foreign species, such as the ring-necked pheasant (*Phasianus colchicus*), have been purposefully and successfully introduced. Others, for example the Norway rat (*Rattus norvegicus*), have been introduced accidentally and resulted in millions of dollars of economic loss annually. Some species from the east coast of North America, such as the striped bass (*Roccus saxatilis*), have been successfully introduced to west coast waters. Many species, such as the sea otter (*Enhydra lutris*) and the trumpeter swan (*Olor buccinator*), were barely saved from extinction before it was too late, while others, like the passenger pigeon (*Ectopistes migratorius*) and the Labrador duck (*Camptorhynchus labradorium*), did not fare as well. These are but a very few of the examples of vertebrates that have been affected directly by man. The great majority of changes in the distribution of vertebrates have resulted from man's ability to bring about environmental modification. Cultivation of the land, the drainage of marshes, the construction of dams, and deforestation are but some of the well known examples so often stressed by conservationists.

In recent years pollution has progressively affected the distribution of vertebrate animals as well as other forms of life. Some of this is caused by industrial wastes fouling lakes and rivers to such an extent that fishes suffer serious periodic die-offs or in some areas are eliminated completely. Sludge from mining operations and excessive surface soil runoff as a result of lumbering operations likewise eliminate native populations of fishes and amphibians in some river systems. Leaks from offshore oil wells may have a very serious effect particularly on sea birds, whose feathers become soaked with oil. Wrecked oil tankers have discharged millions of gallons of crude oil along coastlines, resulting in the nearly complete elimination locally of such birds as alcids.

Of even greater significance today is the effect of insecticides used in agriculture and in controlling mosquitoes and other insect pests. The most serious of these are the chlorinated hydrocarbons whose chemical structure does not break down for years. Such compounds are widely distributed in the ocean and have entered the food chain of many organisms, since they accumulate in fatty tissue. There are now catastrophic declines evident in populations of some vertebrate species at the end of such food chains. The most obvious effect has been on certain fish-eating birds such as brown pelicans and gulls, and on raptors. In affected species increases in body content of chlorinated hydrocarbons are correlated with decreases in calcium content of the egg shell. In some species, such as the brown pelican along the coasts of Louisiana and California, reproduction has almost ceased because the birds cannot incubate eggs without breaking them. The reproduction of peregrine falcons has essentially ceased in eastern North America for the same reason. Studies on the thickness of egg shells in museum collections have shown that the decreases began around 1947.

Directly or indirectly, island populations have fared worse than continental species as a result of man. Some species were completely exterminated; for example, the dodo (*Raphus cucullatus*), which lived on the island of Mauritius in the Indian Ocean until the seventeenth century and was systematically hunted by European colonists.

FIGURE 8.3. Nesting brown pelicans (*Pelecanus occidentalis*) on Isla Granite, Sea of Cortez, Mexico.

This large, flightless, pigeon-like bird was unable to defend itself against such a formidable aggressor. At the time of Darwin's visit to the Galapagos Islands in 1835 there presumably were six endemic species of rice rats (*Oryzomys*) living in various parts of this archipelago. Although he collected only one of these, the presence of the others was discovered by later explorers. The visitation of these islands by ships, however, resulted in the introduction of European rats and mice. Four of the species of native rodents are now believed to be extinct because of their inability to compete with the more dominant feral species.

Sometimes several factors are involved in the elimination of island species. Guadalupe Island, off the west coast of Baja California, Mexico, presents a good example of this. This remote island was once inhabited by 10 endemic species of birds and mammals. In addition to this it provided one of the most important rookery areas for the northern elephant seal (*Mirounga angustirostris*) and the Guadalupe fur seal (*Arctocephalus townsendi*). The dominant plants consisted of native pine, cypress, and a palm. Some of the early sailing vessels used by Russian sealers were responsible for introducing the house mice and domestic cats. These introductions were probably accidental, but sailors purposely introduced goats, as was the custom of that time, to insure a supply of meat on subsequent visits. The goats multiplied and eliminated much of the native vegetation

**FIGURE 8.4.**   The Guadalupe fur seal (*Arctocephalus townsendi*). (Photograph b George E. Lindsay.)

as well as seedling conifers. The Guadalupe towhee (*Pipilo consobrinus*) and the Guadalupe wren (*Thryomanes brevicauda*) finally became extinct when the underbrush, which was their necessary habitat, was consumed by the goats. The endemic species of petrel (*Oceanodroma macrodactyla*), which nested under rocks, and the flicker (*Colaptes rufipileus*), a ground-feeding woodpecker, were exterminated by feral house cats. The sealers, who came to the island to secure hides and blubber, eliminated the only native raptorial bird, the Guadalupe caracara (*Caracara lutosus*), because it preyed upon the young goats.

Meanwhile they almost brought about the extinction of the native fur seal and the elephant seal. Both, however, barely managed to survive, as did four species of birds: a house finch (*Carpodacus amplus*), a junco (*Junco insularis*), a rock wren (*Salpinates obsoletus guadeloupensis*), and a kinglet (*Regulus calendula obscurus*). The goats and cats still remain on this now very barren island and the status of most of the surviving land vertebrates is precarious. Some old pines and cypress survive but there are no young trees. The palms have fared better because they are probably unpalatable even to most goats.

In New Zealand, where the only native land animals are bats and a rat (*Rattus exulans*), it has been clearly demonstrated that the introduction of non-native herbivores may have a very serious effect. Some years ago the red deer (*Cervus elaphus*) and the Australian opossum (*Trichosurus vulpecula*) were brought to this island. They have been responsible for the elimination in certain areas of many kinds of plants that were not resistant to browsing. At the same time unpalatable species of plants have replaced them, and in places severe erosion of the soil has occurred.

One of the most outstanding examples of serious economic loss and ecological disturbance of the land is to be found in Australia, where the European rabbit (*Oryctolagus cuniculus*) became firmly established in the middle of the nineteenth century. Because of lack of competition from other small herbivores and the absence of several important natural enemies, these lagomorphs increased rapidly and by 1928 were estimated to have spread over two-thirds of that continent. In 1953 there was said to be between 500 million and 1000 million feral rabbits in Australia in an average year. Their competition for food with sheep and cattle resulted in an economic loss of almost $900,000 annually.

We must not conclude, however, that all vertebrates have suffered from human activity. Some have expanded their ranges and are more abundant and widespread now than they were before the white man came to this land. The cutting or burning of forests may be detrimental to forest-dwelling animals, but it often proves to be decidedly advantageous to non-forest dwellers. Careful study in certain parts of the country has shown that at least in some areas deer are more numerous today than they were before the settlers came. This is partly attributable to the reduction of their natural enemies as well as to proper game management and legislation. However, since deer are not primarily forest inhabitants, fire and lumbering have produced more brushland and second growth, which are favored by these animals, and thereby increased the carrying capacity of many deer ranges.

Artificial lakes and reservoirs have proved favorable for certain kinds of fishes, amphibians, reptiles, birds, and mammals. Some species have even benefited by the planting of shade trees, lawns, orchards, and crops.


## FAUNAL REGIONS

The problems relating to the distribution of animals have been of keen interest to zoogeographers for a long time. As a result a number of systems involving various categories have been proposed. Some of these approach the subject from a worldwide basis, whereas others are far more limited in nature.

The division of the world into major zoogeographic regions was first proposed by P. L. Sclater in 1858 and later adopted by Alfred Russel Wallace in 1876. According to this system six major regions were suggested because of certain broad similarities in plant and animal life within each.

The *Nearctic Region* includes the whole of North America south to and including the tableland of Mexico. The *Palaearctic Region* includes Europe, extreme northern Africa, and northern Asia. The *Neotropical Region* extends from Mexico south through Central and South America. The *Ethiopian Region* includes Arabia and all of Africa except the extreme northern part. The *Oriental Region* comprises most of southern Asia from the Persian Gulf east to southern China, the Philippines, Borneo, and Java. The *Australian Region* extends from Wallace's Line, which passes through the Strait of Makassar and the deep sound between the islands of Bali and Lombok, eastward to include Australia, New Guinea, New Zealand, and many of the Pacific islands.

The use of the term region today is, for the most part, only of broad general interest to the student of distribution. It is convenient, nevertheless, in quickly conveying a general idea of range. For example, species that are restricted to North America north of the Isthmus of Tehuantepec are commonly referred to as Nearctic species. Since a number of kinds of plants and animals occur both in the Nearctic and the Palaearctic, the term Holarctic is frequently used. This implies circumpolar distribution.

The very broad aspect of this regional concept necessitated the subdivision of each region into subregions. Three such subregions were recognized in the Nearctic: the Canadian or Cold Subregion, the Western or Arid Subregion, and the Eastern or Humid Subregion. One, however, gains little idea of the climatic requirements of a particular species even by the use of these terms, since within the Western Arid Subregion, for example, we find humid coastal forests, deserts, and alpine peaks. Such difficulties no doubt were instrumental in the development of the life zone concept.

## LIFE ZONES

Near the close of the last century the late Dr. C. Hart Merriam, then chief and also founder of the United States Bureau of Biological Survey, which was the forerunner of our present Fish and Wildlife Service, proposed the life zone concept. This was based upon the idea that North America is divisible into a number of transcontinental belts or life zones succeeding each other from north to south (Fig. 8.5). The differences between these belts, as expressed by the plants and animals living in each one, were considered to be governed primarily by temperature, although modified to some degree by certain other factors as we shall see. According to Dr. Merriam, the northward distribution of terrestrial animals and plants is governed by the sum of the positive temperatures for the entire season of growth and reproduction, and the southward distribution is governed by the mean temperature for a brief period during the hottest part of the year.

In general, therefore, if the North American continent were perfectly level, there would be a constantly changing set of life zones as we proceed from north to south or let us say from the arctic to the tropics. However, altitude effects some drastic modifications in this theoretical picture, since each rise of about 250 feet above sea level means a reduction of 1° F. in temperature. High mountains in the United States, therefore, result in the southward extension of the more northern zones. For example, essentially arctic conditions prevail in the middle of summer on some of the higher peaks of the Rocky Mountains in Colorado. Many other factors may modify life zones locally. To mention a few, there is slope exposure (whether facing the north or the south), prevailing air currents, proximity to the ocean or even to large inland bodies of water which may have a moderating effect, isolation (isolated mountains are less apt to exhibit marked zonation than extensive mountain masses), humidity,

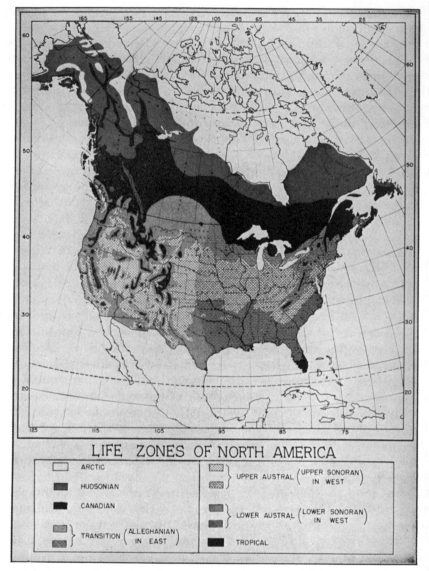

**FIGURE 8.5.** Life zones of North America. (Courtesy of C. F. W. Muesebeck and Arthur D. Cushman, U.S. Bureau of Entomology and Plant Quarantine.)

proximity to deserts, abundance or scarcity of water, and the presence of rocky outcrops which may reflect heat and light.

According to the life zone concept, the North American continent may be divided into three primary transcontinental regions, the Boreal, the Austral, and the Tropical. The Boreal region extends from the north polar area south to southern Canada with southern extensions coming down the Appalachian Mountains, the Rocky Mountains, the Cascade–Sierra Nevada axis, and, to a lesser extent, other smaller continental ranges. The Austral region includes most of the United States and a large portion of Mexico. The Tropical region is limited to the extreme southern parts of the United States, some of the lowlands of Mexico, and most of Central America.

Each of these regions is subdivided into life zones: the Boreal into Arctic-Alpine, Hudsonian, and Canadian; the Austral into Transition, Upper Austral (Carolinian in eastern North America, Upper Sonoran in western North America), and Lower Austral (Austroriparian in eastern North America, Lower Sonoran in western North America); the Tropical into Arid Tropical and Humid Tropical.

Merriam considered 43° F. as marking the threshold of physiological activity for plants and animals. By computing the mean normal temperatures for the six hottest consecutive weeks of the year for various localities over North America, it was possible to draw *isotherms* or lines of climatic belt which limited each of the life zones. In general each zone was considered to have a range of about 7° F. — Arctic-Alpine ranges from 43° F. to 50° F., Hudsonian from 50° F. to 57.2° F., Canadian from 57.2° F. to 64.4° F., Transition from 64.4° F. to 71.6° F., Upper Austral from 71.6° F. to 78.8° F., and Lower Austral from 78.8° F. to a point short of Tropical.

Since plants in general are more stable than animals they were selected as the primary life zone indicators. Furthermore, for this purpose, trees and shrubs were given preference over herbaceous plants because of their greater permanence. Mammals and birds, although not as important as plants, were considered next as zonal indicators.

The Arctic-Alpine Zone lies north or above the limit of tree growth and includes the polar region and arctic tundra, as well as the parts of mountains farther south that are above timberline. It is characterized by plants such as dwarf willow, arctic poppy, various saxifrages and gentians (Fig. 8.6). Some mammal and bird indicators are the arctic fox, arctic hare, lemming, snow bunting, snowy owl and rosy finch.

The Hudsonian Zone consists of the northern coniferous forests composed principally of various species of spruce and fir as well as the higher Boreal forests on many mountain ranges farther south Fig. 8.7). The wolverine, woodland caribou, great gray owl, and pine grosbeak are some vertebrate indicators.

The Canadian Zone represents the southern part of the great transcontinental Boreal forests and comparable coniferous forests on large mountain ranges extending much farther south (Fig. 8.8). Vertebrate inhabitants include the varying hare, marten, lynx, spruce grouse, Canada jay, and white-throated sparrow.

The Transition Zone extends across northern United States and

**FIGURE 8.6.** Arctic Alpine Life Zone in the Sierra Nevada of California at approximately 11,000 feet. Some of the vegetation in this swampy meadow consists of alpine willow (*Salix anglorum*), which rarely attains a height of more than 6 inches.

**FIGURE 8.7.** Hudsonian Life Zone in the Sierra Nevada of California. The trees near lake level are lodgepole pine (*Pinus contorta* var. *Murrayana*) and mountain hemlock (*Tsuga mertensiana*). Prostrate forms of the white-bark pine (*Pinus albicaulis*) are seen on the peak in the distance.

**FIGURE 8.8.**   A Canadian Life Zone forest consisting of red fir (*Abies magnifica*) in the Sierra Nevada of California.

south on the major mountain ranges (Fig. 8.9). As the name implies, it represents a zone in which there is a broad intermingling of elements representing both the Boreal forest and the Austral. The plant indicators include chestnut, walnut, hemlock, beech, birch, yellow pine, many oaks, and sage-brush, depending to a great extent on locality. Typical vertebrate inhabitants include the Columbian ground squirrel, the sage grouse, and sharp-tailed grouse.

The Upper Austral Zone extends broadly, although interruptedly, across the central part of United States from the Atlantic to the Pacific and south into Mexico. The Carolinian area, from the Great Plains eastward, is humid in summer and contains many plant indicators, including hazelnut, hickory, sweet gum, and sycamore. Vertebrate inhabitants include the fox squirrel, prairie vole, eastern wood rat, worm-eating warbler, lark sparrow, cardinal, and Carolina wren. The Upper Sonoran of western North America is arid in summer and is characterized by numerous plant indicators including piñon pine, digger pine, Mexican tea, buckeye, antelope brush, and chamise (Fig. 8.11). Vertebrate inhabitants include the gray fox, ring-tailed cat, northern grasshopper mouse, valley quail, scrub jay, Bewick wren, plain titmouse, and bushtit.

FIGURE 8.9. Over much of western North America ponderosa pine (*Pinus ponderosa*) is the dominant conifer in the Transition Life Zone.

The Lower Austral Zone extends across southern United States from the Carolinas, Florida, and the Gulf States to central California. The humid eastern or Austroriparian area is marked by such plant indicators as the magnolia, long-leaf pine, and bald cypress and is inhabited by such vertebrates as the cotton rat, rice rat, boat-tailed grackle, chuck-will's widow, and Swainson's warbler (Fig. 8.12). The extremely arid Lower Sonoran of the southwest is indicated by such plants as the ocotillo, creosote bush, mesquite, and many types of cacti (Fig. 8.13). Vertebrate indicators include the California leaf-nosed bat, kit fox, many species of kangaroo rats and pocket mice, desert quail, vermilion flycatcher, cactus wren, and verdin.

The Tropical Zone of the Americas includes southern Florida, much of the coastal and southern parts of Mexico, Central America, and northern South America. It is sometimes divided into Arid Tropical and Humid Tropical. This represents an oversimplification of a number of complicated and extensive plant formations. A much better classification for these vegetation zones has been given by Leopold (1950) for Mexico. Seven major tropical zones are recognized. These are as follows: arid tropical scrub, thorn forest, tropical deciduous forest, savanna, tropical evergreen forest, rain forest, and cloud forest.

**FIGURE 8.10.** Carolinian Life Zone in Noble County, Ohio. The dominant trees are white and red oaks with an understory of hickory. (Photograph courtesy of the U S. Forest Service.)

**FIGURE 8.11.** Upper Sonoran Life Zone in western Nevada. Sagebrush (*Artemisia tridentata*) and Utah juniper (*Juniperus ostrosperma*), common Great Basin plants, are both shown in this picture. (Photography courtesy of the U.S. Forest Service.)

**FIGURE 8.12.** Austroriparian Life Zone in the Francis Marion National Forest, South Carolina, showing a stand of cypress along Wadboo Creek. (Photograph courtesy of the U.S. Forest Service.)

**FIGURE 8.13.** Lower Sonoran Life Zone on the Amargosa Desert of Southern Nevada. The dominant perennial plant here is creosote bush (*Larrea divaricata*).

*Arid tropical scrub* is found in regions of low rainfall and consists of xeric vegetation of a desert type intermingled with some plants of the thorn forest and deciduous forest. Both figs (*Ficus*) and bald cypress (*Taxodium*) may be found along streams.

The *thorn forest* is, likewise, found in regions of low rainfall where precipitation occurs principally in the late summer. The vege-

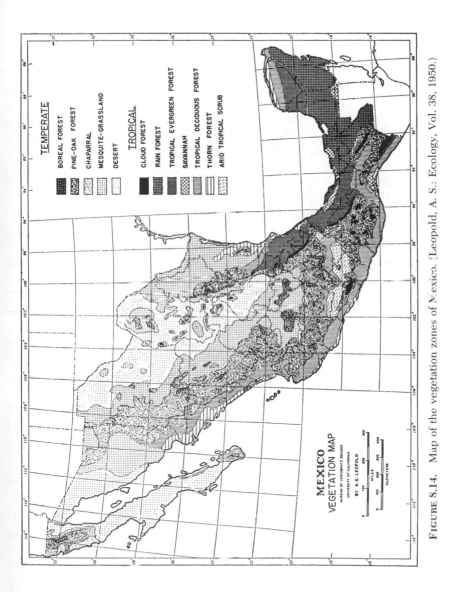

FIGURE 8.14. Map of the vegetation zones of Mexico. (Leopold, A. S.: Ecology, Vol. 38, 1950.)

tation is dominated by thorny leguminous trees along with such species as the kapok tree (*Ceiba pentandra*), the morning glory tree (*Ipomoea arborescens*), and cardon (*Pachycereus pecten-aboriginum*).

The **tropical deciduous forest** is more humid than either the thorn forest or arid tropical scrub and is usually found more inland, frequently just below the montane pine-oak woodland. Several species of copal (*Bursera*), yellow trumpet (*Tecoma stans*), morning glory tree, cardon, mala mujer (*Jatropha urens*), various members of the genus *Acacia* and *Ficus,* as well as the silk cotton tree (*Bombax palmeri*) are a few of the indicators.

The **tropical savanna** is rather limited in occurrence and is the result of local conditions inhibiting the production of a climax type of vegetation. Various tropical grasses with scattered trees consisting of palms and species of the tropical deciduous and thorn forest are characteristic.

The **tropical evergreen forest** is one of relatively high rainfall. It contains many broadleaf evergreen trees such as pumpwood (*Cecropia peltata*), kapok tree, cigarbox tree (*Cedrela mexicana*), and oil palms. *Wigandia caracasana*, various figs, tears of blood (*Bocconia frutescens*), *Anthurium xanthosomifolium*, poor man's raincoat (*Gunnera insignis*), ceriman (*Monstera deliciosa*), *Philodendron radiatum*, as well as orchids, bromeliads, and other kinds of epiphytes are abundant.

**FIGURE 8.15.** A short-tree, arid tropical forest in the Cape Region of Baja California, Mexico. *Bursera, Pachycereus, Jatropha,* and *Tecoma* are some of the common plant genera of this tropical deciduous forest.

FIGURE 8.16. A tropical evergreen forest dominated by such genera of trees as *Cecropia* and *Ficus,* with numerous epiphytic species. (Photographed in the Sierra Madre del Sur, Oaxaca, Mexico.)

The *rain forest* is somewhat like the tropical evergreen, but the annual precipitation is much higher and the arboreal vegetation taller. Very large trees such as mahogany (*Swietenia macrophylla*), chicle (*Achras zapota*), canshan (*Terminalia obovata*), and ramon breadnut tree (*Brosium alicastrum*) are a few of the several hundred kinds of arborescent plants found in this forest. Some species exceed 250 feet in height. Palms, bamboos, many epiphytes, and lianas are abundant in the underforest.

The *cloud forest* is a product of altitude, precipitation, and fog or low clouds. It occurs at elevations of 5000 to 7000 feet and consists of pines and oaks intermingled with an understory of plants such as tree ferns (*Cyathea arborea*), sweet gum (*Liquidamber styraciflua*), and giant horsetails (*Equisetum*). Bromeliads and orchids are abundant. The climate is cool but does not reach freezing in winter.

The life zone concept has many good as well as many bad features. It was widely accepted and used by students of vertebrate distribution for a long time. Even now it proves useful in certain respects. The term Arctic-Alpine immediately brings to mind the

**Figure 8.17.** A cloud forest in southern Mexico, where pines and oaks mingle with liquidambar and tree ferns.

treeless tundra of the North or the high mountain tops above timber-line, and, of course, those vertebrates associated with this zone such as the snowy owl, the snow bunting, or the rosy finch. The Lower Sonoran presents a picture of desert with mesquite and creosote bush, kangaroo rats, cactus wrens, and gridiron-tailed lizards. These zones present convenient means of expressing a great deal concerning the climate and plant and animal life of various parts of the country in a very few words. The oversimplification of this system, however, is one of its major drawbacks.

The uninitiated might easily gain the impression that, because a particular area comes within certain temperature limits, the plants and animals there should be the same as in areas of similar temperature in other parts of the country. This, however, is far from true. A New England bird student will think of the Transition Zone as primarily one of hardwoods or mixed hardwoods and conifers where in summer one may hear the songs of the oven bird, the American redstart, and the veery. A similar student in the Rocky Mountains region thinks of the Transition Zone as the great belt of yellow pine where the solitary vireo, western tanager, and chipping sparrow are so familiar. On the other hand a bird student in the northwestern coastal part of California is apt to associate the Transition Zone with the cool dark forests of coast redwood where it is possible to listen for minutes and hear nothing but the song of the winter wren or the occasional call of the varied thrush.

Merriam's Life Zone concept has been criticized by many, especially by Kendeigh (1932) and Shelford (1932). It has been pointed out that temperature at times of the year other than of growth and reproduction may be important in limiting the northward distribution of animals and plants. This is obviously true for many kinds of organisms, but our knowledge in this matter, as far as vertebrates are concerned, is too limited to permit any general conclusions. Studies by Salt (1952) on the relation of metabolism to climate and distribution in three species of finches of the genus *Carpodacus* occurring in western North America agree to some extent with Merriam's first law of temperature. The Cassin finch (*C. cassini*) is primarily a species of the arid Canadian Zone in the mountains of the Great Basin in summer. The purple finch (*C. purpureus*) along the Pacific coast occurs largely in the relatively humid Transition Zone forests at this season, while the house finch (*C. mexicanus*) frequents the dry warm Sonoran Zones. Both temperature and air moisture appear to be critically limiting factors controlling the summer distributions of these birds. In winter, however, temperature and humidity tend to become more uniform over extensive areas that are dissimilar with regard to vegetation and geography, and this is correlated with the

FIGURE 8.18. Although the Galapagos Islands are virtually on the equator, the shoreline vegetation is of a desert type because of the cooling effect of the Peruvian Current and the relatively low rainfall. (Photograph by George E. Lindsay.)

lessening of the segregation of the ranges of these three finches. There are times when all three species may be found together in central California. In the spring, when the temperatures begin to rise, the birds once again segregate to their respective breeding ranges.

It has been pointed out also that daily maximum temperatures are more important than daily mean temperatures in controlling the southern distribution of certain animals. Fluctuations in the mean temperature of the air from day to day have been found to have relatively little effect on the physiology of certain birds that have been studied, but high maximum temperatures greatly upset their metabolism, interfere with nesting behavior, and cannot be endured for any extensive period of time.

Considerable criticism has also been made of Merriam's statistical data on daily temperatures, which were the basis for his isotherms. It would appear that, due to misinterpretation of information provided by the U.S. Weather Bureau, there were errors in his computations. Even the selection of 43° F. as the temperature marking the threshold of physiological activity for plants and animals is considered too arbitrary, since various animals and plants differ considerably in this regard.

## BIOMES

The difficulties encountered in the life zone concept led to the advancement of a number of new ecologic divisions, one of which was proposed by Clements and Shelford in 1939 and promulgated later by other ecologists. It is obvious that there are broad natural biotic units, dominated by a climax type of vegetation, that extend over large areas on the North American continent. Most extensive are the tundra of the north (Fig. 8.20), the coniferous forests or taiga (Fig. 8.22), the deciduous forests, and grasslands (Fig. 8.23). The term *biome* has been applied to these natural units, each of which represents a plant and animal formation of fully developed and developing communities. These communities attain a climax which is controlled by climate, with vegetation the dominant element (Fig. 8.19).

Since there are extensive plant and animal communities in western North America that cannot suitably be included in any of the major continental biomes, it has been suggested that they be given equal rank. Such communities would include the creosote bush desert, the sagebrush desert, the piñon-juniper and interior chaparral community, the coastal chaparral, and the humid coniferous forest.

In both the New and Old World tropics and subtropics there are

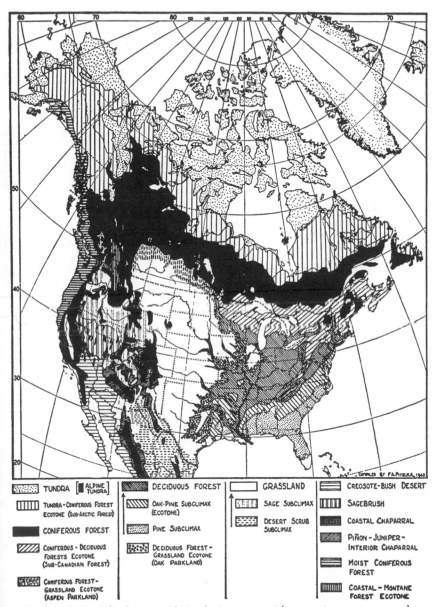

| | | | | | |
|---|---|---|---|---|---|
| ⬚ TUNDRA | ▓ ALPINE TUNDRA | ⫴ DECIDUOUS FOREST | ☐ GRASSLAND | ☰ CREOSOTE-BUSH DESERT |
| ⫴ TUNDRA-CONIFEROUS FOREST ECOTONE (SUB-ARCTIC FOREST) | | OAK-PINE SUBCLIMAX (ECOTONE) | SAGE SUBCLIMAX | SAGEBRUSH |
| ■ CONIFEROUS FOREST | | PINE SUBCLIMAX | DESERT SCRUB SUBCLIMAX | COASTAL CHAPARRAL |
| CONIFEROUS - DECIDUOUS FORESTS ECOTONE (SUB-CANADIAN FOREST) | | DECIDUOUS FOREST - GRASSLAND ECOTONE (OAK PARKLAND) | | PIÑON - JUNIPER - INTERIOR CHAPARRAL |
| CONIFEROUS FOREST- GRASSLAND ECOTONE (ASPEN PARKLAND) | | | | MOIST CONIFEROUS FOREST |
| | | | | COASTAL - MONTANE FOREST ECOTONE |

FIGURE 8.19. The biomes of North America, with extensive ecotones and certain subregions indicated. (After Pitelka, in Odum, F. P.: Fundamentals of Ecology. 2nd Ed., W. B. Saunders Co., 1959.)

**FIGURE 8.20.** Aerial view of the arctic tundra near Pt. Barrow, Alaska, showing the polygonal configuration of the land and the numerous small lakes so characteristic of this region. (Photograph by G. Dallas and Margaret M. Hanna, California Academy of Sciences.)

other biomes which are very extensive and of great significance to vertebrates. These would include arid tropical thorn scrub, tropical deciduous forest, and tropical evergreen forest. Traveling down the west coast of Mexico from the United States border, one gradually leaves the creosote desert of Sonora and northern Sinaloa and comes into arid tropical thorn scrub. From Nayarit on south, both tropical deciduous and tropical evergreen forests are found in the lower coastal regions. Similar biomes occur south to western Ecuador and Peru along the west coast of South America and are much more widely distributed in eastern South America.

It must be borne in mind that the major biomes in general do not terminate abruptly but rather blend with one another over broad transitional belts which are termed *ecotones*. For example, in the subarctic of northern Canada there is an extensive area where the tundra and the coniferous forests gradually blend with one another. This represents the tundra-coniferous forest ecotone. Similar ecotones occur between other major biomes (Fig. 8.25).

It is also important to remember that the biome is concerned with the final or climax vegetation. However, in the course of the development of the climax there is a series of successional states that must be passed through. Thus, if deforestation occurs through fire or

**FIGURE 8.21.** The tundra itself is rich in plant life, which provides food for lemmings, voles, arctic hares, and other northern species. Labrador tea (*Ledum*), lichens, alpine willow, and other low-growing species are seen in the above picture, taken along the west side of Hudson Bay.

**FIGURE 8.22.** The coniferous biome extends across Canada as well as southward on the larger mountain ranges.

**FIGURE 8.23.** The grassland biome occupies much of central North America. The above photograph shows pronghorns (*Antilocapra americana*) on the short grass prairie of southern Manitoba.

**FIGURE 8.24.** The tundra-taiga ecotone in northern Manitoba.

FIGURE 8.25. An example of the coniferous-deciduous forest ecotone in the Adirondack Mountains of New York (Photograph by Tracy I. Storer.)

human activity in a coniferous forest, a new forest of the same type does not immediately develop. Usually a series of communities will arise, each succeeding the other. There may be invasion by grasses and herbs, followed by brush or chaparral, then perhaps by deciduous trees that again will ultimately be replaced by conifers, which represent the climax and are essentially self-perpetuating.

A good many species of land vertebrates show definite restriction to single biomes. For example, barren ground caribou and lemmings are largely restricted to the tundra, pine marten to the coniferous forests, the wood thrush to the deciduous forests, and the prairie chicken and prairie dog to the grassland. Others such as the long-tailed weasel or the coyote show little restriction of this sort to particular biomes.

## BIOTIC PROVINCES, AREAS, COMMUNITIES, AND ECOSYSTEMS

Another method of classifying plant and animal communities and their distribution is by biotic provinces. According to Dice (1943), a *biotic province* "covers a considerable and continuous geographic area and is characterized by the occurrence of one or more important ecologic associations that differ, at least in proportional area covered, from the associations of adjacent provinces." Such a distributional system, therefore, involves a relatively large and continuous geographic area. Although usually climaxed by a single biome, more than one such climax association may be present.

Biotic provinces may be subdivided into *biotic districts* and *life belts*. The biotic district represents a part of a biotic province that is distinguished from other districts by ecologic differences that are of lesser importance than those distinguishing the provinces. Life belts are based primarily on altitude. On a single mountain within one biotic province one may have an alpine belt, a montane belt, and a grassland belt.

Ecologists realize that no one organism or even community of organisms exists independently or is self-sufficient. A woodland or a grassland community is dependent upon the interaction of a great many factors that comprise the whole environment. The amount of sunlight available throughout the year, the rainfall, the temperature, barometric pressure, the moisture content in the air, and the composition of the soil (including minerals as well as microscopic plant and animal organisms) are but some of the factors that go to produce this community. The term *ecosystem* is used to express the combination and interaction of these living and nonliving parts.

Although each of these systems has merit and may be of significance in defining the ranges of certain species and in pointing out various centers of differentiation, none is completely satisfactory. Since no standards can be set as to degree of distinctness that must prevail, the number of species whose ranges must conform to the arbitrary lines of such areas, or even the climatic characteristics, the entire concept of these regions must be largely subjective. Most vertebrate species occupy more than one biotic province or area, since their distribution is most frequently governed by the presence or absence of suitable habitat and such habitats are rarely restricted to a single region. From the standpoint of geographically varying species of wide distribution, however, there is often a definite coincidence of the ranges of biotic areas and subspecies and in this respect such systems sometimes are of considerable value. Perhaps the term ecosystem is the most useful of any of these, but there are still many kinds of vertebrates that are not restricted to special ecosystems.

## HABITAT AND NICHE

Within a biome, life zone, biotic province, or ecosystem there are even more precisely defined ecologic formations which are determined not only by major factors such as climate, latitude, or altitude but by the type of soil, drainage, erosion, water, wind, and many other environmental features that might be considered of a rather local nature. Such formations are referred to as *habitats* (Fig. 8.26).

The number of habitats available to vertebrates in North America are too numerous and varied to be considered here. However, a few selected examples may illustrate the importance of such ecologic units from the standpoint of distribution. Along the ocean shore there may be coastal cliff, sand dune, rocky beach or sandy beach, intertidal reefs, and, farther out, the pelagic or open ocean. Each of these may be considered as a potential habitat for certain vertebrates. Duck hawks and ravens may nest in the coastal cliffs. Limbless lizards may burrow in the sand dunes over which snowy plover forage. Some shorebirds show preference for sandy beaches while others like rocky beaches. Intertidal reefs, enclosing pools of seawater, are inhabited by many kinds of fishes and also foraged over by certain kinds of sea birds. The open ocean is in itself divisible into numerous habitats ranging from abyssal depths, inhabited by strange fishes with luminescent organs, to the surface, on which albatrosses and petrels spend much of their lives.

From the standpoint of distribution of vertebrate species as well

**FIGURE 8.26.** Talus slopes adjacent to patches of low-growing vegetation provide the specific habitat occupied by the pika (*Ochotona princeps*) throughout most of its range in North America.

as individuals, the habitat is of prime importance and the most nearly definable ecologic formation. For example, to say that the desert kangaroo rat (*Dipodomys deserti*) is an inhabitant of the Nearctic Region or even the Western Subregion merely informs us that it occurs on the North American continent or, more specifically, in the western part of the continent. To further state that it occurs in the Creosote Bush Desert Biome or in the Artemisian, Mohavian, and Sonoran Provinces only intimates that it ranges over some of the desert areas of the West. This would also be presumed if it were described as a resident of the Lower Sonoran Life Zone. A knowledge of its habitat, however, which consists of fairly deep, wind-drifted desert sand, immediately brings to mind some of the specific environmental requisites that must be present wherever this species is found

There are ubiquitous species whose distribution seems governed less by habitat than by available food supply. A well known example is the deer mouse (*Peromyscus maniculatus*) that ranges over much of North America, irrespective for the most part of life zones, altitude, or habitat. A great many species of vertebrates, however, not only are restricted to certain habitats but occupy a definite *niche* or place within that habitat. In a woodland formation some species are restricted to the upper story, others to the main trunk or larger limbs of trees, some to the undergrowth, some to the litter beneath, and others even below the surface of the ground. For all practical purposes, therefore, in considering vertebrate distribution the particular niche that a species occupies may be looked upon as the smallest definable ecologic unit. It is necessary, however, that it be referred to in connection with a habitat; otherwise it tells us little

## References Recommended

Allee, W. C., Emerson, A. E., Park, O., Park, T., and Schmidt, K. P. 1949. Principles of Animal Ecology. Philadelphia, W. B. Saunders Co.

Axelrod, D. I. 1958. Evolution of the Madro-Tertiary Geoflora. Botan. Rev. 24:433–509

Banfield, A. W. F. 1954. The Role of Ice in the Distribution of Mammals. J. Mamm 35:104–107.

Brattstrom, B. H., and Howell, T. H. 1956. The Birds of the Revilla Gigedo Island Mexico. Condor 58:107–120.

Carlquist, S. 1965. Island Life; A Natural History of the Islands of the World. Garden City, N. Y., The Natural History Press.

Clements, F. E., and Shelford, V. E. 1939. Bio-ecology. New York, John Wiley & Sons Inc.

Darlington, P. J., Jr. 1957. Zoogeography: The Geographical Distribution of Animals New York, John Wiley & Sons, Inc.

Davis, W. B. 1939. The Recent Mammals of Idaho. Caldwell, Idaho, Caxton Printer Ltd.

Dice, L. R. 1943. The Biotic Provinces of North America. Ann Arbor, University Michigan Press.

Dice, L. R. 1952. Natural Communities. Ann Arbor, University of Michigan Press.

Gentry, H. S. 1942. Rio Mayo Plants; A Study of the Flora and Vegetation of the Valley of the Rio Mayo, Sonora. Carnegie Instit. Wash. Publ. 527.

Goldman, E. A., and Moore, R. T. 1946. The Biotic Provinces of Mexico. J. Mamm. 26:347-360.

Greenway, J. C., Jr. 1958. Extinct and Vanishing Birds of the World. Am. Comm. Internat. Wildlife Protection, Special Bull. No. 13.

Grinnell, J. 1914. Barriers to Distribution as Regards Birds and Mammals. Am. Nat. 48:248-254.

Guilday, J. E. 1967. Differential Extinction During Late-Pleistocene and Recent Times, pp. 121-140. In Martin, P. S., and Wright, H. E. [Eds.], Pleistocene Extinctions: The Search for a Cause. Proc. VII Congr. Internat. Assoc. Quaternary Res., vol. 6.

Guilday, J. E. 1968. Grizzly Bears from Eastern North America. Amer. Midl. Nat. 79:247-250.

Guilday, J. E., Martin, P. S., and McGrady, A. D. 1964. New Paris No. 4: A Late Pleistocene Cave Deposit in Bedford County, Pennsylvania. Bull. Nat. Speleological Soc. 26:121-194.

Hagmeier, E. M., and Stults, C. D. 1964. A Numerical Analysis of the Distributional Patterns of North American Mammals. Syst. Zool. 13:125-155.

Hershkovitz, P. 1958. A Geographical Classification of Neotropical Mammals. Fieldiana: Zool. 36:581-620.

Hickey, J. J., and Anderson, D. W. 1968. Chlorinated Hydrocarbons and Eggshell Changes in Raptorial and Fish-eating Birds. Science 162:271-273.

Howard, W. E. 1964. Introduced Browsing Mammals and Habitat Stability. J. Wildlife Mgmt. 28:421-429.

Howard, W. E. 1965. Control of Introduced Mammals in New Zealand. New Zealand Dept. Sci. Indust. Res., Info. Ser. 45.

Ingersoll, J. M. 1964. The Australian Rabbit. Am. Scientist 52:265-273.

Kendeigh, S. C. 1932. A Study of Merriam's Temperature Laws. Wilson Bull. 44:129-143.

Leopold, A. S. 1950. Vegetative Zones of Mexico. Ecology 31:507-518.

Merriam, C. H. 1892. The Geographic Distribution of Life in North America. Proc. Biol. Soc. Wash. 7:1-64.

Merriam, C. H. 1899. Life Zones and Crop Zones of the United States. Bull. U.S. Biol. Surv. 10:1-79.

Miller, A. H. 1951. An Analysis of the Distribution of the Birds of California. Univ. Calif. Publ. Zool. 50:531-644.

Miller, R. S. 1964. Ecology and Distribution of Pocket Gophers (Geomyidae) in Colorado. Ecology 45:256-272.

Myers, G. S. 1967. Zoogeographical Evidence of the Age of the South Atlantic Ocean. Studies in Tropical Oceanography, Miami, No. 5:614-621.

Odum, E. P. 1959. Fundamentals of Ecology. 2nd Ed. Philadelphia, W. B. Saunders Co.

Pitelka, F. A. 1941. Distribution of Birds in Relation to Major Biotic Communities. Am. Mid. Nat. 25:113-137.

Rausch, R. L. 1963. A Review of the Distribution of Holarctic Recent Mammals, pp. 29-43. In Pacific Basin Biogeography. Proc. 10th Pac. Sci. Congr. Honolulu, Bishop Museum Press.

Salt, G. W. 1952. The Relation of Metabolism to Climate and Distribution in Three Finches of the Genus Carpodacus. Ecol. Monog. 22:121-152.

Sclater, W. L., and Sclater, P. L. 1899. The Geography of Mammals. London, Kegan Paul, Trench, Trübner and Co., Ltd.

Shelford, V. E. 1932. Life Zones, Modern Ecology, and the Failure of the Temperature Summing. Wilson Bull. 44:144-157.

Shreve, F., and Wiggins, I. L. 1964. Vegetation and Flora of the Sonoran Desert. Stanford, Calif., Stanford University Press.

Wallace, A. R. 1876. The Geographical Distribution of Animals. New York, Harper & Bros., 2 Vols.

# Chapter Nine

# TERRITORY AND HOME RANGE

## THE TERRITORIAL CONCEPT

In the previous chapters the general and special morphological characters of vertebrates have been discussed. Consideration has also been given to speciation, subspecific differentiation, and clines as well as to factors relating to the distribution of species, both worldwide and local. Most of this information has accumulated as a result of the study of preserved, nonliving material. One can also examine a preserved vertebrate specimen and fairly accurately determine whether it belongs to a species that flies, walks, or swims, whether it is a herbivore or a carnivore, and many other features that may characterize it in life. One cannot tell, however, whether it is migratory or resident, whether it is monogamous, polygamous, or polyandrous, what sort of sounds it makes, or how it reacts to numerous external stimuli. Students of behavior, therefore, must work with living things and accurately record and interpret their observations.

The present chapter is concerned with the spatial distribution of the individual and the relationship of the latter to other members of the same species. The distribution of a species may be determined by collecting and collections, but the distribution of the individual is determined only by observation.

For centuries man has been aware that within certain species of animals individuals may have special areas or territories in which they live and which they will defend against intrusion by other individuals of the same species or of other species. It was not until 1920, however, that H. Eliot Howard presented a clear and concise

314

picture of this territorial concept based upon studies of various English birds. Since that time there have been many studies made on this subject as well as criticisms and modifications of the concept. It has been said by some that Howard set ornithology back a number of years because so many students subsequently tried to fit the pattern of behavior of species that they studied to that which he described. This certainly was not Howard's fault. In *An Introduction to the Study of Bird Behavior* (1929, p. xii) he makes the following statement: "The subject is the behavior of birds, not of all birds but of some birds that I happen to have watched."

As we shall see there are many different kinds of territory and not all birds have territory. Furthermore, the concept of territory is by no means restricted to the class Aves.

Although Howard did not specifically define territory in birdlife, he implied that it was an area in which the male confined his movements, made himself conspicuous and also became intolerant of other males of the same species. The purpose of this is to prevent overcrowding and to insure the availability of an adequate supply of food near the nest for the young. The matter of rapid access to nearby food is stressed because it is necessary for the parent or parents to devote a great deal of time to brooding, especially when the young are small and when the weather is inclement.

In 1935 Mayr proposed the following definition: "Territory is an area occupied by one male of a species which it defends against intrusions of other males of the same species and in which it makes itself conspicuous." This is very similar to Howard's concept. A short time later Tinbergen (1936) defined sexual territory as an area that is defended by a fighting bird against individuals of the same species and sex shortly before and during the formation of a sexual bond. In Tinbergen's opinion this fighting serves to secure objects or situations necessary for reproduction. Noble (1939) gives a much broader definition of territory when he states that it "is any defended area."

It has been suggested by Pettingill (1946) that territory performs four major functions: (1) it guarantees essential cover, nesting materials, and food for the developing young; (2) it protects the nest, the sex partner, and young against other males; (3) it limits the population of a bird community so that its carrying capacity will not be exceeded; (4) it serves as a means of instigating the sexual bond (female attracted by singing males).

From the foregoing definitions and explanations of the function of territory it is evident that there are many divergent views on this concept. However, there is so much diversity in animal behavior that it is often difficult to establish clearcut definitions. Perhaps the idea of territory is best explained by presenting some selected examples.

## BIRDS

In most passerine birds song is one of the most conspicuous manifestations of territory. It is a means of announcing possession and as such is given along the boundary lines and usually from conspicuous perches. Many birds that live in the short-grass prairies of western North America, where conspicuous perches are largely absent, sing in the air. This is characteristic of lark buntings (*Calamospiza melanocorys*), chestnut-collared longspurs (*Calcarius ornatus*), and McCown's longspurs (*Rhynchophanes mccownii*).

Nonpasserine birds may make use of visual communication as well as various types of vocalization other than song to ward off aggressors from their territory. Detailed behavioral studies on gulls have shown that certain calls serve to inhibit an intruder, but if such a call is given by the intruder it stimulates aggressive behavior by the resident. Other calls may not stimulate aggression by either. Postures play a major role in territorial behavior in these colonial birds. The head is important in aggressive display as well as in courtship. Experiments in breeding colonies with models of glaucous-winged gulls (*Larus glaucescens*) with adjustable stuffed heads have shown that when the head and neck are in an upright (threat) position they are subject to more attack by territorial residents than when the head is lowered. The latter posture in some animals is a submissive one or at least represents a lower level of aggressiveness.

Territoriality has been studied in many different kinds of birds, including resident, migratory, colonial, and noncolonial species. In some species the area occupied and defended may be large, as, for example, among birds of prey. In colonial species it may be a matter of feet or even inches. Large territories generally include feeding areas as well as nesting sites, while in most colonial species, such as marine birds or swallows, the nesting site alone is the territory.

An excellent example of a resident, territorial bird is the wrentit (*Chamaea fasciata*), which was the subject of a detailed study by Erickson (1938) (Fig. 9.1). This is a species that is restricted to the Pacific coast of North America, ranging from the vicinity of the Columbia River, in western Oregon, south to northern Baja California. It belongs to the monotypic family Chamaeidae, thought by some to be closely related to the Timaliidae or babbling thrushes of the Old World.

Wrentits are inhabitants of chaparral or brush and rarely venture even a few feet into the open. Once a pair has established a territory which averages about eight-tenths of an acre but may vary from one half to two and one-half acres, the members remain there for life. They are, therefore, monogamous and strictly resident. Song is indulged in throughout the year, principally by the male. It serves to

FIGURE 9.1.  Wrentit (*Chamaea fasciata*).

announce territory to other wrentits and to indicate location to the other member of the pair if the male and female become separated. Both members of the pair move about the territory and defend it against intruders.

There are colonial species also that are resident and maintain territory throughout the year. The smooth-billed ani (*Crotophaga ani*), a member of the cuckoo family noted for its parasitic habits, presents an example of this (Fig. 9.2). The smooth-billed ani occurs in the West Indies, in Central America, and in tropical parts of South America. Davis (1940) made an interesting study of this species in Cuba. Here he found that flocks usually are composed of about seven birds, although the number may range from a pair to as many as 24 individuals.

Each flock possesses its own territory which it defends against intruders. The territory contains a nesting and a feeding area. The former is dominated by some bushy trees and the latter is essentially savanna. The members of the flock stay in the feeding area during the day and sleep close together in densely foliaged trees at night. The birds frequently preen one another and seem quite solicitous over each other's welfare. Both males and females participate in the defense of territory. This may be done by chasing the intruder or

**FIGURE 9.2.**   Smooth-billed ani (*Crotophaga ani*).

"rushing." When rushing, the birds fly as a group from tree to tree uttering a peculiar call. Fighting may ensue if the intruder does not leave. Sometimes a stranger may be accepted in the colony after being chased for several days. Although anis have a number of call notes they have no song to announce their territories. While territorial boundaries are strictly adhered to, adjacent flocks usually get along with little fighting.

As can be deduced from Davis' study of the smooth-billed ani, this species is definitely territorial, but its territory does not fit in with most of the earlier definitions of this concept. Territory here is a piece of land which is defended by a community of birds, composed of both sexes, against other members of the same species of either sex throughout the year.

An interesting variation in territorial behavior is shown by the scissor-tailed flycatcher (*Muscivora forficata*), which breeds in south western United States and winters in Central America (Fitch, 1950). During the day the male is strictly territorial and defends his nesting territory against intrusion by other males. In the evening, however all the males in one region join the unmated females and young of the year in a communal roost which consists of a specific tree. As many as 250 individuals have been observed coming to a single roost before dark. In the morning the males return to their respective territories and defend them throughout the day.

Emlen (1954) pointed out that two changes in behavior appear to be involved in the establishment of territory. These are the localization of activity and increased intolerance of associates. In many colonial birds only the former change is apparent. This is especially true of species such as albatrosses, certain terns and gulls, gannets, and some other marine birds. Many birds of this type range widely outside of the breeding season, since they have a plentiful supply of food in the sea, but they are limited to particular islands or parts of shore lines for nesting sites (Fig. 9.3). As a consequence they become colonial at such times. This, as pointed out by Summers-Smith (1954), has both advantages and disadvantages. A breeding colony is more easily located by predators, but all members of the colony may participate in the defense and drive the enemy away.

In cliff swallows (*Petrochelidon pyrrhonota*), which are highly colonial, the territory selected by the male is the nest site (Fig. 9.5). After the nest is built, its mud walls offer protection on five sides and only the sixth or open end needs to be defended. According to Emlen (*op. cit.*), "the extent of the defended territory around the nest was apparently determined by the reach of the bill from the nest

**FIGURE 9.3.** Colonial nesting sea birds, such as Brandt cormorants (*Phalacrocorax penicillatus*) shown above, tend to space themselves rather evenly over the limited habitat available. (Photograph taken on the Farallon Islands, California, by O. J. Heinemann. Courtesy of the California Academy of Sciences.)

**FIGURE 9.4.** Nesting elegant terns (*Thalasseus elegans*) space themselves so that each is just out of reach of its neighbor's bill. (Photographed on Isla Raza, Sea of Cortez, Mexico.)

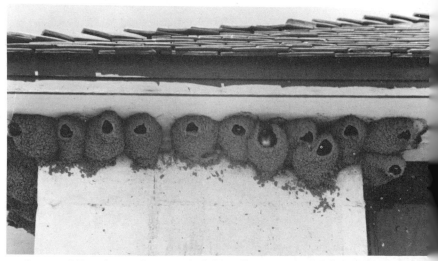

**FIGURE 9.5.** The mud nests of cliff swallows (*Petrochelidon pyrrhonota*) are frequently located under the eaves of buildings. The heads of some of the occupants may be seen in the entrances in the above picture. (Photography by Allan Cruickshank.)

rim, a distance of about four inches." In this species as in many other species, the members of the colony share the foraging area, which may extend several miles from the nesting site.

One of the major contributions to our knowledge of avian territory was made by Mrs. Margaret M. Nice (1943) as a result of her studies on the song sparrow (*Melospiza melodia*) in Ohio. The song sparrow is a species that is widespread over North America from Alaska to Mexico (Fig. 9.6). Its behavior varies in different parts of its range, to some extent at least. Some of the northern populations are migratory, moving south to temperate regions in the winter. In the southern part of the range of the species, song sparrows in general are resident. At Columbus, Ohio, where Mrs. Nice made her studies, some song sparrows remained resident and noncolonial throughout the year, whereas others became migratory and flocking in the fall. Changing day length and climate are considered as most important in regulating song sparrow behavior.

These birds were found to be highly territorial for seven or eight months. Territorial defense, however, is entirely the responsibility of the male, as is defense of his mate and the nest. The female is concerned with nest building, incubating, and caring for the young

FIGURE 9.6.   Song sparrow (*Melospiza melodia*).

although the male assists her in the latter effort to some extent. About four-fifths of the breeding females at Columbus migrate in winter, while only half of the males do so. The resident males tend to stay on or near their established territory although they do not defend it in winter. Weather, however, plays an important part. Cold weather in the fall stimulates migration. It also stimulates flocking in fall and winter on the part of the nonmigrating birds. Warm weather has the opposite effect.

The snow bunting (*Plectrophenax nivalis*), studied in Greenland by Tinbergen (1939*b*), represents a species that is territorial in summer only and flocking and migratory outside of the breeding season (Fig. 9.7). In eastern Greenland male snow buntings, still in the process of assuming summer plumage, arrive in the latter part of March in flocks. By the latter part of April the flocks break up. The males become solitary and establish territory. The males announce their territories by means of song from such singing posts as rocks that command a view of the surrounding terrain. Each male sleeps in his territory, which is largely covered with snow at this time of year, and attacks any other bunting that intrudes. Defending males sing in flight when going after intruders. The females do not arrive until the end of April and at first fly in flocks. Subsequently, pairing takes place and nesting activities get under way.

As was pointed out previously, Tinbergen considers territory as the area in which sexual fighting occurs. In the snow bunting the female builds the nest, but it is not always within the territory defended by the male. Food is often gathered outside the territory. Sometimes this is on neutral ground, but occasionally food is secured

FIGURE 9.7.   Snow bunting (*Plectrophenax nivalis*).

on the territory of another male where the intruding bird may be tolerated if it is not singing.

In the various examples of territoriality among open-nesting birds given so far, the selection of territory, whether it be a matter of acres or only a few inches, precedes the building of the nest. In birds that are primarily hole-nesting species, as pointed out by von Haartmann (1957), the selection of the nest site comes first. After the nesting hole has been chosen, territorial defense manifests itself.

Birds have been considered first in discussing territory for several reasons. More is known about their behavior than is true of other vertebrate groups. They are most suitable for field study because they live above ground and water, are largely diurnal and usually conspicuous in their environment, and generally announce their presence by vocal utterances, frequently in the form of song.

## MAMMALS

Since the majority of small mammals are nocturnal and secretive in their habits it is difficult in most instances to determine whether they exhibit territorial behavior in the sense that this implies defense of a definite area. Studies made on the opossum (*Didelphis marsupialis*) by Fitch and Sandidge (1953) in Kansas failed to show any evidence of territorial behavior. Numbers of individuals of both sexes and diverse ages were found to occur together in the same area

In certain groups of rodents, notably members of the Heteromyidae (kangaroo rats, pocket mice, and their allies) and the Geomyidae (pocket gophers), many species at least appear to be solitary for much of the year. Whether they maintain territories comparable to those of some birds is a subject for further investigation.

Linsdale and Tevis (1951), in their study of the dusky-footed woodrat (*Neotoma fuscipes*), have shown that these animals depend for their existence on their houses, which consist of conical piles of sticks situated on the ground or in trees. Each rat guards its house throughout most of its life and will defend it by fighting. Fitch (1958) found this also to be true of the eastern woodrat (*Neotoma floridana*) and states that "The house itself may be thought of as constituting a small territory." Although the foraging area about the house is also defended, there do not appear to be any sharply defined territorial boundaries.

Very interesting observations made by Altmann (1959) and others on the howling monkeys (*Alouatta palliata*) of the American tropics have shown that small and essentially permanent groups of these animals occupy definite territories. The defense consists primarily of vocalizations by the males when intruders approach the territorial boundaries.

There are certain mammals which exhibit very definite territorial behavior. This is true of some of the pinnipeds, notably the northern fur seal (*Callorhinus ursinus*). In North America these animals breed on St. Paul and St. George Islands in the Pribilof group in the Bering Sea. Outside of the breeding season fur seals are pelagic in occurrence and range widely over the North Pacific. In April or May the adult bulls migrate to the Pribilof shoreline rookeries and stake out territories which range in diameter from 10 to more than 30 feet.

Once the bulls come to shore and establish territories they remain there for a considerable period of time and are referred to as harem masters. Since the females do not arrive until sometime between early June and mid-July, a harem master may have to defend his territory for a month or more preceding the arrival of the cows. Although much of the defense consists in rushing from the center of the territory to the periphery, severe fighting may occur. Bulls with adjacent territories may even join together to force an intruder away.

As the females begin to arrive the bulls nearest the water are the first to acquire a harem; the average harem consists of from 40 to 50 cows. Each cow gives birth to a pup a few days after arriving and is bred shortly thereafter.

**FIGURE 9.8.** Northern fur seal (*Callorhinus ursinus*) beachmasters with their harems of cows and pups. (Photographed on St. Paul Island, Bering Sea, Alaska.)

**FIGURE 9.9.** A Steller sea lion (*Eumetopias jubata*) bull, slightly left of center, with a harem of cows, some of which have newborn pups. This species is found in the North Pacific.

**FIGURE 9.10.** A northern elephant seal (*Mirounga angustirostris*) with his harem of females and young. The bull is trumpeting as a challenge to the presence of human beings, and the female in the foreground has just thrown sand in the air with her right front flipper. This species breeds on offshore islands from central California south to northwestern Mexico.

Since the bulls have to establish and hold their territories before the arrival of the females and then later maintain and defend their harems, they must go without food and water for many days, living on accumulated fat. The average period of duty for each harem master is 31 days although some remain on constant guard for more than twice this length of time.

There may be a number of bulls that are unsuccessful in securing a territory. These idle individuals congregate about the edge of the rookeries and frequently attempt to take harems away from established bulls. They become more successful as the season progresses and the harem masters become weakened and thin from many weeks of vigilance. By August the territorial defense wanes and bulls without territories and subadults may take over late arriving cows.

Somewhat similar territories which consist primarily of harems of females are maintained during the breeding season by some other pinnipeds, especially the Steller sea lion (*Eumetopias jubata*) and the California sea lion (*Zalophus californianus*) in North America.

The behavior of the northern elephant seal (*Mirounga angustirostris*) during the breeding season differs from that of most other Northern Hemisphere pinnipeds. In this species the dominant males do not maintain a spatial territory but have a social hierarchy, essentially like the peck order in domestic fowl, whereby the highest ranking males stay with the breeding females. Specific sites are not defended, but the dominant male moves to wherever the greatest number of females are aggregated and permits no other male to encroach on his harem. As a result most of the females are bred by a relatively small number of the total available males.

Studies made recently by Dr. Roger S. Payne of Rockefeller University indicate that whales may have marine territories just like certain terrestrial animals. He made recordings of the underwater sounds of the humpback whale (*Megaptera novaeangliae*) and found a definite pattern in the vocalization. A single "song" may last for five to thirty minutes and be repeated by the hour. It is suggested that this may be a territorial announcement just like the song of birds. Song believed to be made by the same whale was recorded in the same locality off Bermuda two years in a row.

## REPTILES AND AMPHIBIANS

Relatively little information is available on the subject of territory among reptiles and amphibians. Certain lizards that have been studied, notably the Cuban lizard (*Anolis sagrei*) and some North American members of the genus *Sceloporus*, aggressively defend certain areas in the breeding season.

Fitch (1940) in his studies on the fence lizard (*Sceloporus occidentalis*) marked 366 individuals so that each could subsequently be recognized. He found that there was usually but one male to a stump, boulder, or log although the territories were not sharply defined. Each male defended his basking territory but the defense became progressively weaker as the distance away from the basking area increased.

Martof's (1953) interesting observations on the green frog (*Rana clamitans*) in the vicinity of Ann Arbor, Michigan, indicate that a type of territory seems to exist among the males of this species during the breeding season. Frogs were captured and marked by a system of toe cutting. During the breeding season the adult male green frogs moved from streamside habitats to ponds and lakes or to large pools in quiet streams. In the breeding ponds the males spread out and remained rather uniformly spaced about 2 to 3 meters apart. They vocalized at night and remained in these breeding clusters for about two months. Although groups might move more than 100 meters from one pond to another during a season, individuals tended to maintain their same positions relative to one another. Even the movement of many of the frogs appeared to be synchronized. It is suggested that this spacing represents a type of territorial behavior on the part of the males which aids in conserving their energies and in more easily detecting females that come to the breeding ponds for short periods of time.

Some evidence of territoriality is exhibited by certain caudate amphibians. Grant (1955) found that when two-lined salamanders (*Eurycea bislineata*) were placed in an aquarium in which there was a simulated natural habitat of moss, bark, and forest litter, each individual wandered from 5 to 6 inches away from a centrally located shelter such as a piece of bark. However, when another individual came within 2 inches of the shelter site, the occupant came and placed its snout against that of the intruder. This method of intimidation was the usual defense, although on occasions the resident bit the intruder on the snout or tail.

Similar observations made on the four-toed salamander (*Hemidactylum scutatum*) in an aquarium showed this species to have a defended territory of about the same size as that of the two-lined salamander. Defense, however, was seldom by bodily contact but rather by intimidation; the resident merely advanced toward the intruder.

## FISHES

Territorial behavior is much more marked in fishes than in other kinds of poikilotherms (Fig. 9.11). Anyone who has kept Siamese

FIGURE 9.11. The male Mozambique mouth breeder (*Tilapia mossambica*) "blows" out a depression in the sand which he defends against other males. The female deposits her eggs in this depression, and after they are fertilized she takes them into her mouth, where they remain until the young hatch in about two weeks. The male in the foreground is seen "blowing" out sand to form a depression, while the male in the background is guarding his depression.

fighting fish (*Betta*) is aware that no more than one male can be kept in a tank. Although the males are compatible with females, they viciously defend their territories against other males. Female salmon zealously guard their eggs for a considerable time after spawning. Steelhead trout, on the other hand, show very little territorial behavior.

One of the most interesting types of territorial behavior is exhibited by the bitterling (*Rhodeus amarus*), which was introduced locally into eastern United States about 1925. This is a species that deposits its eggs in the gills of a freshwater mussel. During the breeding season each male selects a mussel and fights any other male that comes near. The mussel, since it is capable of locomotion, becomes in a sense a moving territory.

The male jewel fish (*Hemichromis bimaculatus*), a tropical cichlid from Africa which is commonly kept in aquaria, behaves in certain respects like many birds. The male assumes a nuptial color at the onset of the breeding season and then selects a territory which he defends. The territory consists of a suitable substratum for the deposition of the eggs. Any other fishes that approach within 8 to 15 inches of this site are attacked.

Interesting geographic variation in territorial response appears to be exhibited by the black-nosed dace (*Rhinichthys atratulus*) of central and eastern United States. Raney (1940*a*), in his study of *R. a. meleagris* in New York State, found that in the breeding season males may guard a potential spawning area about 2 feet or less in diameter. The area was frequently defended by fighting, although only intruding males were attacked. Schwartz (1958), however, found no indication of territorial defense in *Rhinichthys atratulus obtusus*, a southern subspecies of the black-nosed dace.

Members of the stickleback family (Gasterosteidae) are very territorial during the breeding season. The males select territories where there is adequate vegetation to provide nesting material as well as shelter against currents and enemies. The nest material is woven together with a threadlike mucus produced by special kidney tubules. The male guards the nest both before he secures a female that will deposit her eggs in it and afterwards until the young are ready to leave. The sight of another male serves as a visual stimulus to aggressive behavior. If there are rocks or other objects that limit visibility, territories may be quite close together.

In his studies on the darters (family Percidae), Winn (1958) found that males had a specific territory which they defended during the daytime but not at night. Females entered these territories only to spawn.

Many other examples of territorial behavior among fishes are known. Some of these will be considered under "Reproduction."

## HOME RANGE

The idea of home range repeatedly has been confused with the territorial concept by many persons. This probably stems from the fact that the two are interrelated and, in some species, may be synonymous. However, home range as defined by Burt (1943) "is the area, usually around a home site, over which the animal normally travels in search of food." It has also been defined as an area about its established home or nest which is traversed by an animal in its normal activities of food-gathering, mating, and caring for the young. There is no implication that the area must be defended as is necessary in the case of territory.

Although all vertebrates are capable of locomotion and move about in search of food, the term home range implies a certain degree of permanence. It cannot, therefore, aptly be applied to nomadic or wandering animals. Most vertebrates, however, appear to limit their activities to within a definite area, at least during the reproductive season.

There are a few species, such as the wrentit previously discussed, in which the individual generally remains permanently within its defended territory once this has been established. Food is
secured within this rather strictly circumscribed area the year
around. In such an instance the home range is the same as the
territory. Most territorial species, however, venture outside the territorial boundaries even in the reproductive season. Thus the home
range is generally larger than the territory.

The home range varies greatly in size, depending on the species
involved. An adult male mountain lion has been known to range over
100 square miles, although this is much greater than average for this
species. Pine marten may wander as much as 15 miles in search of
food.

Leopold *et al.* (1951), in their study of deer on the west slope of
the Sierra Nevada in California, found that the home range of each
deer in summer comprised an area averaging about ½ to ¾ mile.
Pregnant does or those with very young fawns appeared to have
home ranges somewhat smaller than average, while some bucks had
ranges that were larger. The home range had to have food, water,
cover, and bedding areas. In winter the home range of these deer
was less than half the size of the summer home range.

A great many studies have been made on the home ranges of
small mammals. Such undertakings involve the repeated live trapping and marking of individuals and the recording of the exact
locality where each animal is trapped. The latter is usually accomplished by using the grid or quadrant method (Fig. 9.12). To illustrate this let us suppose we wish to study the home range of meadow
voles. A suitable area inhabited by these animals is selected and it is
marked off in squares of 50 feet. Live traps are placed either in the
center of each square or at the corners once a week throughout the
year. Each animal is given a distinguishing mark the first time it is
captured so that it may be recognized in subsequent captures. When
the various points of capture of any one individual that has been
caught repeatedly are plotted on paper and connected together they
present a polygon which might be considered as the minimum home
range. This is perhaps an oversimplification of this method of studying home range, but it illustrates the basic principles involved.
Various students have developed different methods of interpreting
data acquired in this manner. Marginal records may often merely
represent the occasional wanderings of an animal outside its ordinarily traversed range. Conversely, an animal is more likely to be
trapped in the proximity of its den or nest than on the periphery of
its range. Hayne (1949) presents some interesting views on this
subject to which the reader is referred.

Adams (1959), in his study of varying hares (*Lepus americanus*)

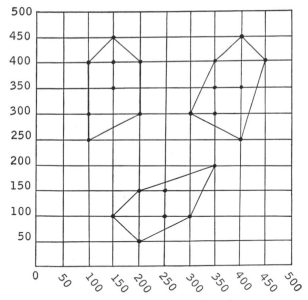

**FIGURE 9.12.**   Diagram illustrating one method of plotting home ranges of small mammals. The black dots represent sites of capture of individuals as a result of re- peated live trapping with the traps placed 50 feet apart in an area of 250,000 square feet. The polygons represent lines connecting the peripheral points of capture of three different individuals.

in northwestern Montana, found, by means of live trapping and ear marking with tags, that the minimum home range for males of this species was 25 acres. For females it was 19 acres and for immature hares 14 acres.

Blair (1940*a, b, c, d*) made detailed studies of the home ranges of several species of small mammals in southern Michigan. He found that the average size of the home range of the female meadow vole (*Microtus pennsylvanicus*) in both moist and dry grassland was from ⅕ to ¼ acre. The home range of males, on the other hand, was slightly less than ⅓ acre in moist grassland and slightly less than ½ acre in dry grassland. There was considerable overlap in these home ranges. No difference in the size of the home ranges between males and females was observed in the jumping mouse (*Zapus hudsonius hud- sonius*) or in the deer mouse (*Peromyscus maniculatus bairdii*). In the former species the average home range was about ⁹/₁₀ acre, in the latter slightly more than ½ acre. The short-tailed shrew (*Blarina brevicauda talpoides*) was found to have a home range of about 1 acre, but there is a marked difference in the distance wandered by males and females. The largest female range was .88 acre, whereas the largest range for males was 4.43 acres.

In the systematic live trapping of small mammals to determine home ranges, many other important life history data may be secured, especially pertaining to sex ratio, growth rate, reproduction, movements of young, external parasites, molt, and longevity.

The home ranges of many carnivores such as foxes, coyotes, and wolves can frequently be determined with great accuracy by a study of tracks. This is especially true in winter where the ground is covered with snow.

Studies have been made on the home ranges of a number of kinds of cold-blooded vertebrates. Stickel (1950), in her studies of the box turtle (*Terrapene c. carolina*) at the Patuxent Research Refuge in Maryland, found that adult turtles occupied and maintained the same home ranges from year to year. These ranges were retained even when they were inundated by floods. The average home range for adult males was 330 feet and for adult females 370 feet. The difference between the size of male and female home ranges was not considered statistically significant. One of the interesting methods employed in following the daily activities of these turtles involved the attachment of trailers to their backs. A trailer consisted of a housing containing a spool of white thread which unwound as the turtle moved along, thus revealing its exact route.

The home range of the desert tortoise (*Gopherus agassizi*) of southwestern United States was found by Woodbury and Hardy (1948) to vary from about 10 to 100 acres.

A number of species of both reptiles and amphibians kept under observation by Fitch (1958) on the University of Kansas Natural

FIGURE 9.13.    The pika (*Ochotona princeps*) of the Nearctic inhabits talus slopes. Each individual appears to deposit its droppings in a definite place among the rocks.

History Reservation proved to have definite home ranges, but there were other species that seemed to wander indefinitely. The latter was true of the leopard frog (*Rana pipiens*) and the timber rattlesnake (*Crotalus horridus*).

A comprehensive study of plethodontid salamanders (*Ensatina eschscholtzi*) in California by Stebbins (1954) indicated that the home range of males was from 32 to 135 feet in width and that of females from 20 to 75 feet.

One is frequently inclined to visualize a home range as circular or at least ameboid in shape. Very often, however, it is linear or rectilinear, as Pearson (1955) found to be true for the spadefoot toad (*Scaphiopus h. holbrooki*) of eastern North America. In this species the average home range is slightly more than 100 square feet.

The study of home ranges in fishes, especially marine species, presents many more difficulties than in the case of terrestrial vertebrates. Many kinds have been studied, nevertheless, and seem to have very definite foraging and resting areas in which they usually limit their daily activities. Winn (1958), in his work on the behavior of darters (family Percidae), uses the terms food range and escape range in addition to territory or reproductive range. Male darters were found to move into the feeding range when there were no other males nearby. The escape range is shared commonly by males and females when danger threatens.

## METHODS OF MARKING VERTEBRATES

In any study on territory or home range it is necessary to identify the individual in some way so that it may be recognized. Ornithologists have long been using serially numbered aluminum bands which are attached to the legs of birds to study migratory movements. Such bands, however, are not practical when daily observations must be made on the activities and local movements of individuals, since the bird must be captured each time in order to determine the identity of the individual. As a consequence it has been found practical to attach colored bands, in various combinations, to the legs of birds whose territorial behavior or home ranges are being studied. These bands are usually made of colored celluloid or plastic and are of sufficient size to be visible to the observer from a distance.

Other methods that are occasionally employed in marking individual birds for subsequent recognition in the field consist in attaching colored feathers to the tail and in sewing colored nylon ribbons to the tarsometatarsus, leaving a free end trailing.

Since mammals are more diverse than birds in structure, size, and habits, a great many marking techniques have been developed. Most of these have been summarized by Taber (1956) and Fitch and Sandidge (1953). Some kinds of mammals can be so marked that they are easily recognized at a distance, but for secretive or nocturnal species the capture and examination of marked individuals is necessary for identification.

The principal methods employed in marking mammals are by mutilation, tagging, or coloring. The first of these involves the permanent or semipermanent mutilation of a part or parts of the body. Small mammals are usually marked by clipping the toes in various combinations or by cutting distinctive marks in the ears or by both. In marking the ears, they may be notched or punched. Other types of mutilation involve fur clipping, tattooing, and tail docking. Branding has been used successfully on seals and as a means of marking the tails of beavers.

Tagging involves the attachment of metal or plastic bands or strips or even of nylon ribbons to various parts of the body. Aluminum bands, similar to those used by bird banders, with serial numbers inscribed on them, are attached to the forearms of bats as a means of identifying individuals. These bands can also be attached to the legs of small terrestrial mammals. Some investigators have even incorporated radioactive substances in these bands so that the movements of the banded individuals can be followed with the use of a Geiger counter. Kaye (1960) described a method of using 20 gauge gold − 198 wires, 10 mm. long, with an activity of 0.7 to 4.5 mc. in studying the movements of eastern harvest mice (*Reithrodontomys humilis*). The wires were inserted under the skin of the abdomen through a hypodermic needle.

A very practical method of tracking certain kinds of mammals involves the use of small radio transmitters which are attached to the animals. Such a device, complete with batteries, may be given an epoxy-resin coating and thus be impervious to water or damage as a result of the wearer rubbing against brush, trees, or rocks. Small transmitters strapped to the body or attached to a collar which is placed around the neck have proved most useful in following the movements of deer, foxes, raccoons, skunks, and rabbits. Portable directional-type receivers enable the operator not only to record activity but locate resting places, dens, and burrows of the animals to which the transmitters are attached.

Strap tags and plastic disks are often attached to the ears of larger mammals. Such tags may be clipped on, sewn on with nylon thread, or riveted on to the ear. Nylon ribbons in various colors may be attached to the ears of rabbits to mark them individually. Metal straps or box car seals attached to the antlers of deer are used to identify these structures after they are shed. Even whales have been marked by shooting stainless steel tubes into the blubber. If these animals are captured later the tubes will be found when the blubber is being removed from the body.

Some biologists have used various kinds of dyes, including commercial fur dye, to mark both small and large mammals. In the case of small mammals the dye is generally applied to the animal by hand after it is captured. For large mammals various devices have been developed whereby the animal triggers a mechanism placed along a trail so that dye is fired at or dropped upon it. The use of small pellets of dye inserted subcutaneously has proved useful in following the movements of cottontail rabbits in the snow. Brown (1961) used certain dyes in this manner that would color the urine for periods of four to seven days. Some dyes will also color the feces.

Since many of the methods used to study the movements of small mammals may possibly influence the activities of the animals themselves, another technique, recently developed, has considerable merit. This involves the preliminary live trapping of a population and marking each individual by removing one or more toes in combination before release. Subsequent data on the movement of each individual is obtained by use of smoked cards which are widely

FIGURE 9.14. Schematic drawing of feet to show system of assigning serial numbers. (From Martof, B. S.: Ecology, Vol. 34, 1953.)

distributed over the study area. Each card is placed in a small shelter and checked daily for tracks. If any are found the card is replaced by a new one. In this way accurate data on the daily movements of small mammals may be obtained without any further trapping or disturbance.

Most lizards as well as frogs, toads, and salamanders can be marked by some system of toe clipping, in which the distal part of one or more toes is removed. Martof (1953) worked out a satisfactory method for marking frogs, as illustrated in Figure 9.14. The toes on the left hind foot are numbered 1 to 5, those on the right hind foot 10 to 50, those on the left front 100 to 400, and those on the right front foot 800 to 3200. By removing no more than two toes on each foot one can mark 6399 individuals in series. The latter number is obtained by removing the two outer toes on each foot.

FIGURE 9.15.    Petersen disk tag used in tagging surgeon fish in the Hawaiian Islands. The ends of the monofilament nylon connecting the two disks have been melted into knobs. The heating caused a blackening of the one knob that is visible. (From Randall, J. E.: California Fish and Game, Vol. 42, Jan., 1956.)

Snakes as well as lizards may be marked by excising specific scales or patches of scales in various combinations. Another method of marking involves the attachment of colored threads to the body.

Fishes are marked both by mutilation and by tagging. Removal of part or all of a fin is one method employed. This, however, does not permit the individual recognition of large numbers of fish. Natural mutilation must also be reckoned with and in certain fish, such as salmon, this is very high.

Tagging is most satisfactory and is accomplished in several ways. Strap tags of monel metal may be attached to the jaw, the preopercle, or the operculum. Steamer tags can be attached to various parts of the body by wire or filament. Recently the threading of a hollow plastic tube through the bodies of marine fishes has proved very satisfactory.

Many studies have been made on the use of the Petersen disk in marking fishes (Fig. 9.15). These disks are generally made of cellulose nitrate cut in the form of a circle. A disk is placed on each side of the upper base of the caudal fin, and these disks are attached to each other by a wire or filament run through the bony base of the fin. Randall (1956) has recommended the use of monofilament nylon leader to connect the disks through the body. The ends are not knotted but melted by means of a flame into a small ball which holds each disk tightly against the body. The necessary data are inscribed on the tags.

## References Recommended

Adams, L. 1959. An Analysis of a Population of Snowshoe Hares in Northwestern Montana. Ecol. Monog. 29:141-170.

Altmann, S. A. 1959. Field Observations on a Howling Monkey Society. J. Mamm. 40:317-330.

Baldwin, P. H. 1953. Annual Cycles, Environment and Evolution in the Hawaiian Honeycreepers (Aves, Drepaniidae). Univ. Calif. Publ. Zool. 52:285-398.

Beer, J. R., Frenzel, L. D., and Hansen, N. 1956. Minimum Space Requirements of Some Nesting Passerine Birds. Wilson Bull. 68:200-209.

Bellrose, F. C. 1967. Radar in Orientation Research. Proc. XIV Internat. Ornith. Congr.:281-309.

Bellrose, F. C., and Graber, R. R. 1963. A Radar Study of the Flight Directions of Nocturnal Migrants. Proc. XIII Internat. Ornith. Congr.:362-389.

Blair, W. F. 1940a. Home Ranges and Populations of the Jumping Mouse. Am. Mid. Nat. 23:244-250.

Blair, W. F. 1940b. A Study of Prairie Deer Mouse Populations in Southern Michigan. Am. Mid. Nat. 24:273-305.

Blair, W. F. 1940c. Notes on Home Ranges and Populations of the Short-tailed Shrew. Ecology 21:285-288.

Blair, W. F. 1940d. Home Ranges and Populations of the Meadow Vole in Southern Michigan. J. Wildlife Mgmt. 4:149-161.

Blair, W. F. 1951. Population Structure, Social Behavior, and Environmental Relations in a Natural Population of the Beach Mouse (*Peromyscus polionotus leucocephalus*). Univ. Mich., Contrib. Lab. Vert. Biol. 48:1-47.

Bogert, C. M. 1947. A Field Study of Homing in the Carolina Toad. Am. Mus. Novitates 1355:1-24.

Bole, B. P. 1939. The Quadrat Method of Studying Small Mammal Populations. Scientific Publ. Cleveland Mus. Nat. Hist. 5:15-77.

Briggs, J. C. 1955. Behavior Pattern in Migratory Fishes. Science 122:240.

Brown, L. N. 1961. Excreted Dyes Used to Determine Movements of Cottontail Rabbits. J. Wildlife Mgmt. 25:199-202.

Burt, W. H. 1940. Territorial Behavior and Populations of Some Small Mammals in Southern Michigan. Misc. Publ., Mus. Zool., Univ. Mich. 45:1-58.

Burt, W. H. 1943. Territoriality and Home Range Concepts as Applied to Mammals. J. Mamm. 24:346-352.

Burt, W. H. 1949. Territoriality. J. Mamm. 30:25-27.

Carr, A. 1965. The Navigation of the Green Turtle. Sci. Amer. 212(5):78–86.

Carr, A. 1967. Adaptive Aspects of the Scheduled Travel of Chelonia, pp. 35–55. In Storm, R. M. [ed.], Animal Orientation and Navigation. Proc. 27th Annual Biol. Coll., Oregon State Univ. Press, Corvallis.

Cochran, W. W., and Lord, R. D., Jr. 1963. A Radio-tracking System for Wild Animals. J. Wildlife Mgmt. 27:9-24.

Collias, N. E. 1944. Aggressive Behavior Among Vertebrate Animals. Phys. Zool. 17:83-123.

Davis, D. E. 1940. Social Nesting Habits of the Smooth-billed Ani. Auk 57:179-218.

Emlen, J. T., Jr. 1954. Territory, Nest Building, and Pair Formation in the Cliff Swallow. Auk 71:16-35.

Erickson, M. M. 1938. Territory, Annual Cycle, and Numbers in a Population of Wrentits (Chamaea fasciata). Univ. Calif. Publ. Zool. 42:247-334.

Estes, R. D. 1969. Territorial Behavior of the Wildebeest (Connochaetes taurinus Burchell, 1823). Zeitschrift für Tierpsychologie, 26:284-370.

Fisler, G. F. 1962. Homing in the California Vole, Microtus californicus. Am. Mid. Nat. 68:357-368.

Fitch, F. W., Jr. 1950. Life History and Ecology of the Scissor-tailed Flycatcher, Muscivora forficata. Auk 67:145-167.

Fitch, H. S. 1940. A Field Study of the Behavior of the Fence Lizard. Univ. Calif. Publ. Zool. 44:151-172.

Fitch, H. S. 1958. Home Ranges, Territories and Seasonal Movements of Vertebrates of the Natural History Reservation. Univ. Kansas Publ., Mus. Nat. Hist. 11:63-326.

Fitch, H. S., and Sandidge, L. L. 1953. Ecology of the Opossum on a Natural Area in Northeastern Kansas. Univ. Kansas Publ., Mus. Nat. Hist. 7:305-338.

Friedmann, H. 1933. Size and Measurement of Territory in Birds. Bird-Banding 4:41-45.

Garcea, R., and Gorman, G. 1968. A Difference in Male Territorial Display Behavior in Two Sibling Species of Anolis. Copeia 1968:419-420.

Gordon, R. E. 1952. A Contribution to the Life History and Ecology of the Plethodontid Salamander Aneides aeneus (Cope and Packard). Am. Mid. Nat. 47:666-701.

Grant, W. C., Jr. 1955. Territorialism in Two Species of Salamanders. Science 121:137-138.

Gullion, G. W. 1953. Territorial Behavior of the American Coot. Condor 55:169-186.

Hayne, D. W. 1949. Calculation of Size of Home Range. J. Mamm. 30:1-18.

Hinde, R. A. 1956. The Biological Significance of the Territories of Birds. Ibis 98:340-369.

Howard, E. H. 1920. Territory in Bird Life. London, John Murray.

Howard, E. H. 1929. An Introduction to the Study of Bird Behaviour. London, Cambridge University Press.

Justice, K. E. 1961. A New Method for Measuring Home Ranges of Small Mammals. J. Mamm. 42:462-470.

Kaye, S. V. 1960. Gold-198 Wires Used to Study Movements of Small Mammals. Science 131:824.

Kenyon, K. W., and Scheffer, V. B. 1954. A Population Study of the Alaska Fur-seal Herd. U.S. Dept. Interior, Fish and Wildlife Service, Special Scientific Report-Wildlife, No. 12.

Leopold, A. S., Riney, T., McCain, R., and Tevis, L., Jr. 1951. The Jawbone Deer Herd. Calif. Div. Fish and Game, Game Bull. *4*:1-139.

Linsdale, J. M., and Tevis, L. P., Jr. 1951. The Dusky-footed Wood Rat. Berkeley, University of California Press.

Marler, P. 1959. Developments in the Study of Animal Communication, pp. 150-206. *In* Bell, P. R. [ed.], Darwin's Biological Work. Cambridge [Eng.] University Press.

Martof, B. S. 1953. Territoriality in the Green Frog, *Rana clamitans*. Ecology *34*:165-174.

Mayr, E. 1935. Bernard Altum and the Territory Theory. Proc. Linn. Soc. N.Y., Nos. 45 and 46, pp. 24-38.

Nice, M. M. 1933. The Theory of Territorialism and Its Development. *In* Fifty Years Progress of American Ornithology. American Ornithologists' Union, pp. 89-100.

Nice, M. M. 1937. Studies in the Life History of the Song Sparrow. I. A Population Study of the Song Sparrow. Trans. Linn. Soc. N.Y., Vol. 4.

Nice, M. M. 1941. The Role of Territory in Bird Life. Am. Mid. Nat. *26*:441-487.

Nice, M. M. 1943. Studies in the Life History of the Song Sparrow. II. The Behavior of the Song Sparrow and Other Passerines. Trans. Linn. Soc. N.Y., Vol. 6.

Noble, G. K. 1939. The Role of Dominance in the Social Life of Birds. Auk. *56*:263-273.

Odum, E. P., and Kuenzler, E. J. 1955. Measurement of Territory and Home Range Size in Birds. Auk *72*:128-137.

Orr, R. T. 1965. Interspecific Behavior Among Pinnipeds. Zeitschrift für Saügetierkunde *30*:163-171.

Pearson, P. G. 1955. Population Ecology of the Spadefoot Toad, *Scaphiopus h. holbrooki* (Harlan). Ecol. Monog. *25*:233-267.

Pettingill, O. S., Jr. 1946. A Laboratory and Field Manual of Ornithology. Minneapolis, Burgess Publishing Co.

Pitelka, F. A. 1942. Territoriality and Related Problems in North American Hummingbirds. Condor *44*:189-204.

Randall, J. E. 1950. A New Method of Attaching Petersen Disk Tags with Monofila ment Nylon. Calif. Fish and Game *42*·63 67.

Raney, E. C. 1940*a*. Comparison of the Breeding Habits of Two Subspecies of Blacknosed Dace, *Rhinichthys atratulus* (Hermann). Am. Mid. Nat. *23*:399-403.

Raney, E. C. 1940*b*. Summer Movements of the Bullfrog, *Rana catesbeiana* Shaw, as Determined by the Jaw-tag Method. Am. Mid. Nat. *23*:733-745.

Schwartz, F. J. 1958. The Breeding Behavior of the Southern Black-nosed Dace, *Rhinichthys atratulus obtusus* Agassiz. Copeia *1958*:141-143.

Sheppe, W. A. 1965. Characteristics and Uses of *Peromyscus* Tracking Data. Ecology *45*:630-634.

Sock, D. 1940. Pair-formation in Birds. Condor *42*:269-286.

Stebbins, R. C. 1954. Natural History of the Salamanders of the Plethodontid Genus *Ensatina*. Univ. Calif. Publ. Zool. *54*:47-124.

Stickel, L. F. 1950. Populations and Home Range Relationships of the Box Turtle, *Terrapene c. carolina* (Linnaeus). Ecol. Monog. *20*:351-378.

Stickel, L. F. 1954. A Comparison of Certain Methods of Measuring Ranges of Small Mammals. J. Mamm. *35*:1-15.

Storm, G. L. 1965. Movements and Activities of Foxes as Determined by Radiotracking. J. Wildlife Mgmt. *29*:1-13.

Stout, J. F., Wilcox, C. R., and Creitz, L. E. 1969. Aggressive Communication by *Larus glaucescens*. Part I. Sound Communication. Behaviour *34*:29-41.

Stout, J. F., and Brass, M. E. 1969. Aggressive Communication by *Larus glaucescens*. Part II. Visual Communication. Behaviour *34*:42-54.

Summers-Smith, D. 1954. Colonial Behavior in the House Sparrow. British Birds *47*:249-265.

Taber, R. D. 1956. Marking of Mammals; Standard Methods and New Developments. Ecology *37*:681-685.

Tinbergen, N. 1936. The Function of Sexual Fighting in Birds; and the Problem of the Origin of "Territory." Bird-Banding *7*:1-8.

Tinbergen, N. 1939*a*. On the Analysis of Social Organization Among Vertebrates, with Special Reference to Birds. Am. Mid. Nat. *21*:210-234.

Tinbergen, N. 1939*b*. The Behavior of the Snow Bunting in Spring. Trans. Linn. Soc. N.Y. 5:1-94.

Tinbergen, N. 1951. The Study of Instinct. Oxford, Clarendon Press.

Tinbergen, N. 1953. The Herring Gull's World. London, Collins.

Verts, B. J. 1963. Equipment and Techniques for Radio-tracking Striped Skunks. J. Wildlife Mgmt. 27:325-339.

von Haartman, L. 1957. Adaptation in Hole-nesting Birds. Evolution *11*:339-347.

Winn, H. E. 1958. Comparative Reproductive Behavior and Ecology of Fourteen Species of Darters (Pisces-Percidae). Ecol. Monog. 28:155-191.

Woodbury, A. M. 1955. Ecology of Disease Transmission in Native Animals. Semi-annual Report, 1 June to 30 November, 1954. Ecological Research, University of Utah.

Woodbury, A. M., and Hardy, R. 1948. Studies of the Desert Tortoise, *Gopherus agassizii*. Ecol. Monog. *18*:145-200.

# Chapter Ten

# POPULATION MOVEMENTS

The movements of vertebrate populations fall into several categories. Some are seasonal or annual. Other movements are engaged in only once in a lifetime. Some are of such irregular occurrence that they may not take place within the lifetime of many individuals of the species involved. These movements may be influenced by environmental factors, by physiological factors, by age, and by sex. In general, population movements are known as migrations, irruptions, dispersal movements, and nomadism.

## MIGRATION

The term migration in the biological sense refers to the periodic movements of animals away from a region and their subsequent return to the same region. This implies a round trip which may be accomplished annually, as is true of the migrations of many kinds of birds and mammals, or may require a lifetime, as is true of some salmonid fishes and freshwater eels. The distance traveled varies with the species. It is generally conceded that the arctic tern (*Sterna paradisaea*), whose annual circuit may nearly equal the circumference of the earth, takes top honors in this regard.

Other types of population movements lack the regular periodicity of migration. Irruptions or invasions are of sporadic occurrence and dependent either upon unusual environmental conditions or cyclic changes of several years or more. Furthermore, the organisms engaging in these kinds of movements usually do not return to the point of origin. Dispersal movements, likewise, are of a permanent nature and imply emigration or exodus from the place of birth or hatching. Nomadism is irregular and unpredictable wandering, usually in search of food, and involves no return to a home territory.

True migration is engaged in by many kinds of vertebrates, from jawless fishes to higher mammals. Many fishes have annual movements to feeding and breeding areas which may be some distance apart. For some species the movement is not annual but is completed once in the lifetime of the individual. This is true of some *diadromous* fishes — those that migrate between fresh and salt water. Breeding occurs in one environment and growth to maturity in the other. Species reproducing in freshwater systems and then migrating to the sea to mature are referred to as *anadromous*. Good examples are the Pacific salmon of the genus *Oncorhynchus*. *Catadromous* species, on the other hand, are those that reproduce in the sea, like the freshwater eels of the genus *Anguilla*, and then move into rivers and streams to grow to maturity.

Most amphibians engage in annual movements to breeding ponds or streams because they require a wet or moist environment in which to deposit their eggs. Subsequently there is a dispersal to other areas. The young upon metamorphosing generally move away from the hatching site, not to return until maturity.

The most significant reptilian migrants are the sea turtles, of which there are two families: the Dermochelidae, which contains the leatherback turtle (*Dermochelys coriacea*), and the Cheloniidae, representing the true sea turtles of the genera *Lepidochelys*, the ridley; *Caretta*, the loggerhead; *Eretmochelys*, the tortoise shell or hawksbill; and *Chelonia*, the green turtle. All appear to have migratory movements between feeding and breeding areas. The most extensive studies have been made on the green turtle, which may migrate more than 1000 miles to remote oceanic islands to lay its eggs. Interestingly, it appears that such migratory movements are not made annually but every two or three years by the adults.

The best known migrants are birds, especially those of the Northern Hemisphere where marked seasonal changes in the environment occur. Most of the migrants are water birds, birds of prey, and small insectivorous passerine species. The northern summer provides a suitable environment for nesting and rearing the young, but the winter is inhospitable and food is lacking. Movement southward to a more favorable habitat is necessary for survival.

Mammalian migrants consist principally of certain kinds of bats, cetaceans, pinnipeds, and large herbivorous hoofed animals. Mammals belonging to these groups are most capable of traveling long distances and therefore can move closer to the equator to winter or, in the case of many herbivores, from tundra to forest or from high mountain pastures to lower valleys until the following spring.

## CAUSATIVE FACTORS

Migration is an adaptive type of behavior which provides a greater distributional range, a more constant food supply, and a suitable reproductive environment for many kinds of animals. The proximal causes of migration can be considered as alimental, gametic, and climatic. There are few species, however, in which only one of these factors is related to the migratory pattern.

The alimental or food requirements of many species are such that seasonal migration is decidedly advantageous and results in a favorable food supply throughout the year. This does not apply to all vertebrates. Many mammals are active all winter, and certain species store food for this season. Some birds such as jays, nuthatches and chickadees find sufficient animal and plant food in winter, when most other species have left, that they can survive in the north without migrating. A few, such as the acorn-storing woodpecker (*Melanerpes formicivorus*), Lewis's woodpecker (*Asyndesmus lewis*), Clark's nutcracker (*Nucifraga columbiana*), and the gray jay (*Perisoreus canadensis*), store food for winter use, but this is not a common avian trait. Most amphibians and reptiles in cold regions have solved the problem of winter food by becoming dormant at this season. Freshwater fishes, too, may become relatively inactive in winter so as to reduce metabolic activity. Few cold-blooded vertebrates have the capability for extended movements anyway. Some do, however: certain fishes regularly migrate between breeding area and feeding grounds. The distance may involve several hundred miles or more. Many salmon along the Pacific coast of North America travel several thousand miles from their hatching sites in the headwaters of river systems to the Gulf of Alaska, where they spend two or more years feeding before attaining sexual maturity. Green turtles may migrate hundreds of miles from the oceanic islands where they lay their eggs to sea pastures where food is abundant.

The greatest migrants are the birds of northern Eurasia and northern North America. Water birds, whether they be ducks, geese or waders, cannot feed on frozen lakes and mudflats. Birds such as flycatchers, swallows, thrushes, warblers, and vireos have no access to insect food in a cold winter environment. Areas that provide sufficient food in summer become barren six months later. Likewise, predatory birds, with a few exceptions, must migrate southward with their prey. Mountain quail (*Oreortyx pictus*) in western North America move to lower elevations in the autumn. They are ground feeders and their food source at higher elevations is covered with snow in winter. The reverse is true of blue grouse (*Dendragapus obscurus*). These gallinaceous birds spend the winter in thick fir clumps on the higher mountain ridges above the usual breeding area. The buds and needles of these conifers comprise their principal food at that season.

The movement of whales to polar areas, where they spend the arctic and antarctic summers after winter breeding in tropical and subtropical waters, is basically alimental. The cold polar seas are rich in food, especially krill (euphausiids) upon which baleen whales largely feed. Some porpoises and dolphins seem to follow the local seasonal migrations of their food fishes. Many northern insectivorous bats hibernate in winter, but a few, such as the red bat (*Lasiurus borealis*), the hoary bat (*Lasiurus cinereus*), the silver-haired bat (*Lasionycteris noctivagans*), and the guano bat (*Tadarida braziliensis*), do not. Instead they make extensive migrations to areas where there is a supply of flying insects in winter. Even the movement of certain big herbivores to lower elevations in the fall of the year seems to be in search of food.

There are many examples of migration for gametic reasons. The return of salmon or steelhead trout from the sea to the streams in which they hatched is solely for the purpose of reproduction. Pacific salmon of the genus *Oncorhynchus* do not feed once they have returned to fresh water, and after spawning they die. The migration of freshwater eels (genus *Anguilla*) from the freshwater streams of

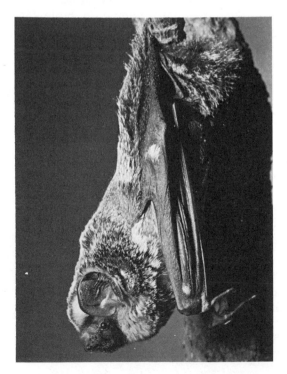

**FIGURE 10.1.** The hoary bat (*Lasiurus cinereus*), a species that is migratory, at least in parts of North America.

eastern North America and Europe to the region of the Sargasso Sea in the western Atlantic Ocean is likewise purely gametic. They too die after spawning. Similar gametic movements are those of sea turtles to oceanic islands far from their feeding grounds, where they deposit their eggs, and the migration of many whales from the cold polar seas to equatorial regions to calve. Examination of the stomach of these whales fails to indicate that there is any significant feeding during the many months they are away from their summering areas. The migrations of many pinnipeds to rookery areas to breed is gametic, while the movement away to feeding grounds after the breeding season is alimental.

Climate obviously is an important factor in the migratory movement of many vertebrates. It is difficult to separate this from alimental and gametic factors, since climate controls food production, and increased day length in the spring is a definite stimulus to gonadal development in many species. Shorter days and lowered temperatures in autumn tend to induce southward migration. It is easy to see why many terrestrial vertebrates avoid arctic and antarctic winters, whether they be shorebirds, waterfowl, penguins, or whales.

## ENVIRONMENTAL FACTORS

Since the migration of many kinds of animals is a seasonal phenomenon, it is logical that changes in the environment that are associated with the different seasons play an important part. These include such obvious factors as the changing length of day, climate, food availability, and even lunar periodicity.

Away from the equator the one constant change that occurs each day throughout the year is the length of the period of daylight. This decreases from its maximum at the time of the summer solstice to a minimum at the winter solstice, and the degree of difference is directly proportional to the distance from the equator. Among migratory endotherms, such as certain kinds of birds, bats, whales, and even some large herbivores, decreasing day length in late summer and fall in the Northern Hemisphere is correlated with a southward movement and, conversely, increased day length in spring is associated with the northward movement of individuals of the same species.

The daily amount of light received, called the *photoperiod*, was first investigated by the late Professor William Rowan in Alberta, Canada. Through his experiments he discovered that when certain male birds, captured before they migrated south in autumn, were exposed to artificially increased daily periods of light in winter, their gonads increased in size. When these birds were released they left the winter area, whereas controls that were not subjected to such light stimulation did not. Most of Professor Rowan's early experiments were with slate-colored juncos (*Junco hyemalis*) and crows (*Corvus brachyrhynchos*). Other investigators subsequently carried on studies which proved that in many birds of the Northern Hemisphere, increase in the length of daylight in spring is directly associated with increased gonadal activity, higher metabolic rate, fat deposition, and migratory restlessness. It is believed that the longer period of wakefulness, resulting from the increase in the photoperiod, increases hypothalamic activity which in turn stimulates the production of gonadotropic hormones. The latter cause gonadal *recrudescence* or activity and may also be responsible for an increase in lipid deposition.

Despite the seeming correlation between amount of light, gonadal activity, and migratory movement in many species of birds, a number of kinds do not exactly conform to this pattern. Certain shorebirds that nest in the Arctic begin their southward migration in the latter part of June, when maximum daily amounts of light prevail. Southbound migrant male rufous hummingbirds (*Selasphorus rufus*), which breed from Alaska south to extreme northern California, are numerous in central California by the last week in June. Early seasonal reproductive activity in such species, however, may result in an early *refractory period*, which is a time of physiological rest and gonadal regression. Other birds of the Northern Hemisphere move south in autumn with decreasing day length, but when they arrive in their wintering areas in the Southern Hemisphere it is spring there and the period of day length is increasing. This, however, does not induce an immediate northward movement. The bobolink (*Dolichonyx oryzivorus*) is a good example of this. Bobolinks breed across the grasslands of temperate North America from southwestern Canada to California and east to Nova Scotia and Pennsylvania, but winter in Bolivia, western Brazil, Paraguay, and northern Argentina. Their northward movement does not occur until the period of day length is decreasing in the Southern Hemisphere, at which time it is spring in the Northern Hemisphere. The same is true of birds such as certain species of shearwaters (genus *Puffinus*) that breed in the Southern Hemisphere and spend the southern winter in the North Pacific. Their movement back to the nesting rookeries occurs in August and September when the northern days are growing shorter.

Climate has a decided effect on migratory movements and is thought to be one of the major factors influencing the evolution of this type of behavior. Extreme heat or cold usually produces an unfavorable environment which must be avoided by species that do not go into torpidity at such times. Migration provides a solution to this problem. Margaret M. Nice, on the basis of her study on song sparrows (*Melospiza melodia*) near Columbus, Ohio, concluded that spring migration in this species was dependent on both increasing day length and rising temperature. In the fall it appeared that cold or inclement weather stimulated migration in some individuals which under more favorable conditions would remain resident. Similar observations have been made on other species.

Temperature is very important in the migration of some fishes. Rainbow trout (*Salmo gairdnerii*) and cutthroat trout (*Salmo clarki*) are freshwater species that move to spawning areas in the spring or early summer when the water temperature is rising. The Dolly Varden trout (*Salvelinus malma*) of the Pacific coast and the brook trout (*Salvelinus fontinalis*) of eastern North America are species that spawn in autumn when stream and lake temperatures are declining. Similarly, seasonal changes in the temperature of ocean water along continental shores is correlated with the movements of many kinds of marine fishes. The cooling of shallow waters in winter causes some species to move away from shore. In summer when the temperature of coastal waters rises, these fish return.

In parts of the country the local movements of certain amphibians from above-ground situations to suitable estivating sites for summer is stimulated, at least in part, by decreased humidity and high temperature; conversely, their return in autumn is correlated with the arrival of the rainy season and cooler weather. The movement of many amphibians and reptiles to their winter hibernacula is associated with reduced environmental temperatures.

Food availability is obviously a significant factor influencing migration in many vertebrates. If the food supply of a species disappears in winter, the individuals of that species must either starve, hibernate, or move to more favorable regions where food can be obtained. Many small mammals, especially rodents, are active all winter beneath the snow in high mountains and on the arctic tundra. Their food requirements are met by available bark, roots, lichens, and other kinds of vegetation, and sometimes even by caches stored away during the summer and fall. These mammals in turn provide food for certain carnivores so that migration is unnecessary for either predator or prey. Other species, however, whose food requirements are greater or different must move out of these regions before the onset of winter.

Woodbury (1941) has suggested that migration is comparable to certain other biological rhythms. According to the theory of periodic response, the seasonal movements of some animals arose first as conditioned behavior to environmental changes. In time natural selection favored those individuals moving out of situations that were temporarily hostile, and migration became hereditary in nature.

Just as length of day, climate, and food availability seem to

**FIGURE 10.2.** In winter in the higher mountains of western North America, pocket gophers (*Thomomys*) develop their burrow systems up into the snow. This enables them at this season to secure food, such as bark, that is unavailable to a fossorial animal during the summer months. When the snow melts, cores of earth remain as evidence of winter activity. Pocket gophers neither migrate nor hibernate.

govern the migrations of some animals, others are influenced by the moon. Lunar periodicity is associated with the breeding activities of a number of invertebrates, particularly marine worms and crustaceans. It also seems to govern the movements of certain fishes of the genus *Leuresthes*. Along the coast of southern California the grunion (*L. tenuis*), a small, smeltlike fish belonging to the silversides family, come to shore in large masses to spawn on the first three or four nights following either the full moon or the new moon from late February to early September. The actual spawning is limited to a period of one to three hours following high tide. A closely related species, *L. sardinia,* that occurs in the Gulf of California, has very similar breeding habits except that spawning may occur in the daytime.

## MIGRATIONS OF COLD-BLOODED VERTEBRATES

Migratory patterns have developed in many kinds of fishes, a number of amphibians and a few reptiles. Among fishes migratory movements may involve thousands of miles of travel or merely the distance from feeding areas in a lake to a particular spawning locality near shore which may be a few hundred yards to several miles away.

Some of the greatest marine migrations are made by tunas, pelagic fishes that breed in tropical seas. The young may travel thousands of miles from the reproductive area in search of food before attaining maturity and returning to breed. Individuals tagged off the Pacific coast of North America have been taken two years later near Japan and Iwo Jima. Smelts and herrings also engage in very definite migrations from pelagic to shallow coastal waters to spawn.

The best known migratory fishes are salmonids belonging to the genera *Oncorhynchus* and *Salmo*. The former is confined to the North Pacific Ocean where five species, all of commercial importance, occur. These are the sockeye salmon (*O. nerka*), the silver or coho salmon (*O. kisutch*), the pink or humpback salmon (*O. gorbuscha*), the chum salmon (*O. keta*), and the king or chinook salmon (*O. tshawytscha*). These anadromous fishes breed in freshwater streams or lakes and the young migrate to sea when they are a few weeks to two years of age, depending upon the species. They reach sexual maturity in the sea in from two to four years and then return to the home stream where they were hatched to spawn and die. Species that breed at the headwaters of major rivers like the Yukon and the Columbia may travel many thousands of miles, while others merely move from estuarine situations to nearby coastal waters where they remain until it is time to return. Some landlocked populations of the sockeye are exceptions in that they remain in freshwater lakes or streams all their lives.

The steelhead (*Salmo gairdnerii*), a seagoing form of the rainbow trout along the Pacific coast of North America, and the Atlantic salmon (*Salmo salar*) of eastern North America spawn in fresh water and mature in the sea; but, unlike the Pacific salmon, they do not necessarily die after spawning. They may return to the sea and spawn again in a subsequent year.

Lengthy migrations are performed by freshwater eels (see Fig. 10.13). Two rather closely related catadromous species are found on either side of the Atlantic. The European species, *Anguilla vulgaris*, inhabits the coastal streams of the British Isles and continental Europe from Norway south to Portugal, while the American eel (*A. rostrata*) lives in streams along the east coast of North America. Both spawn in the vicinity of the Sargasso Sea. Many other fishes also migrate regularly to spawning areas, but their movements are less spectacular than those of salmon and eels.

The migrations of amphibians are primarily to and from breeding sites. Since members of this class must lay their eggs either in water or in moist situations, this entails travel to ponds and streams for many species. Here the sexes gather together in the spring to reproduce. Among frogs and toads the vocalization of the males serves to attract females and guide them to a specific location. The movement to and from such areas is usually overland for most anurans and caudate amphibians, although some remain in water all year round and the migration is either within a single water system or from one pond to another nearby.

Although a few terrestrial reptiles show some migratory behavior, such as the movements of rattlesnakes to wintering dens, the principal migrants are the sea turtles. These marine reptiles generally deposit their eggs on remote islands. The most striking migration is that of the green turtle in the South Atlantic. Females tagged as they came ashore to lay on tiny Ascension Island have been recovered along the Brazilian coast 1400 miles to the west.

## AVIAN MIGRANTS

The greatest migrants are birds, and more is known about their movements than those of other kinds of animals. As previously pointed out, the principal avian migrants of the Northern Hemisphere are insectivorous passerine species, water birds, and birds of prey. This does not hold true for the Southern Hemisphere, where the majority of migrants are nonpasserines such as cuckoos, nightjars, and swifts.

In North America as well as in Europe and Asia most species of hawks move southward in autumn. Since these birds of prey depend upon thermals and updrafts in their soaring and gliding type of flight, they are diurnal migrants. Many species follow ridges bordering valleys since updrafts are most often to be found near the crests of these ridges. One of the most famous in North America is Kittatinny Ridge in eastern Pennsylvania where hundreds of raptors may be seen passing daily in the autumn. These birds also follow shorelines but avoid flight over water because of the absence of thermals or updrafts over such bodies. Most of the birds of prey moving south from Europe and Asia to winter in North Africa avoid crossing the Mediterranean. Their passage is either along its eastern shores or down the Iberian Peninsula to Gibraltar and across the narrow strait to Morocco. The same is true of the white stork (*Ciconia ciconia*) of Europe.

Many kinds of water birds migrate. In the Antarctic most species of penguins move northward in the autumn so as to stay in open water north of the pack ice. Their migrations are by swimming and drifting on rafts of ice. Some species, such as the Adélie penguin, may travel as much as 400 miles each way. Many petrels and shearwaters have extensive migrations. The slender-billed shearwater

(*Puffinus tenuirostris*), or mutton bird as it is known in Australia and New Zealand, breeds in huge colonies along the shores and on islands in Bass Strait between Tasmania and southern Australia. The nesting season extends from late October until about March. By then the young are ready to fly. Their migration takes them northeast to the New Hebrides, after which they veer to the northwest and pass by the east coast of Japan and Korea. From here they make an arc across the North Pacific to Alaska and by late summer in the Northern Hemisphere pass down the Pacific coast of North America. Then they fly diagonally across the Pacific so as to arrive at the nesting grounds in Bass Strait by late October. The young do not return to the nesting area until they are several years old. A somewhat similar migration is made by the sooty shearwater (*Puffinus griseus*).

The most conspicuous avian migrants are duck and geese. In North America the principal breeding grounds for these waterfowl are the northern marshlands of Alaska and Canada and the prairie potholes of Canada and northern United States. Although no two

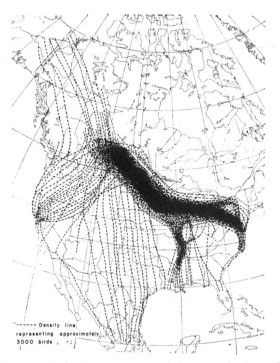

**FIGURE 10.3.** Principal migratory routes of the canvasback (*Aythya valisneria*). (From Stewart, R. E., Geis, A. D., and Evans, C. D.: J. Wildlife Mgmt. 22:353, 1958.)

species have the same migration route, most waterfowl use the four major continental flyways, the *Atlantic Coast*, the *Mississippi*, the *Central*, and the *Pacific Coast*. The Mississippi Flyway is most heavily used by dabbling ducks. The number of individuals estimated to fly southward, east of the Rocky Mountains, is about 17,500,000 each year, with mallards (*Anas platyrhynchos*) comprising about 40 per cent of the total. Diving ducks make greater use of the Pacific and Atlantic flyways.

Some species of waterfowl, such as the Ross goose (*Chen rossii*) in North America, have migration routes that are most distinctive. This goose nests in the Perry River drainage, which flows into Queen Maud Gulf in the Northwest Territories, and winters in the Central Valley of California.

The greatest migrants are to be found among the shorebirds, many of which nest in the Arctic and winter in the Southern Hemisphere. The route taken by the American golden plover (*Pluvialis dominica*) is exceeded in distance only by that of the arctic tern (*Sterna paradisea*), which is said to be about 22,000 miles. There are two populations of this species of plover. One nests from Point Barrow, Alaska, across the Canadian tundra to Baffin Land, while a more western population is found in summer in eastern Siberia and western Alaska. The eastern golden plovers migrate across Canada to the Atlantic Ocean, then fly south over what is termed the Atlantic Ocean Flyway to southeastern South America to winter. On their return they take a very different route, passing northwest to Central America and then up the Mississippi Flyway; from here they continue north to the arctic coast. This involves a round trip of at least 16,000 miles. The western breeding population of American golden plover moves south in migration over what is called the Pacific Ocean Flyway to various Pacific islands as far away as New Zealand.

A number of the smaller sandpipers nest on the arctic tundra of North America, Asia, and Europe and winter in South America, Australia, and southern Africa. The white-rumped sandpiper (*Erolia fuscicollis*) of the American Arctic winters as far south as Tierra del Fuego and South Georgia Island. Baird's sandpiper (*Erolia bairdii*) may travel nearly as far. The sanderling (*Crocethia alba*), a Holarctic species as regards breeding range, winters in South America, the Pacific islands, Australia, and southern Africa.

Although the general pattern of migration in the Northern Hemisphere is from north to south in autumn and the reverse in spring, not all species of birds conform to this. The great majority of elegant terns (*Thalasseus elegans*) and Heermann's gulls (*Larus heermanni*) breed on a small desert island called Isla Raza in the Gulf of California, Mexico. Postbreeding movements by members of both species take them north and south along the Pacific coast. The elegant tern

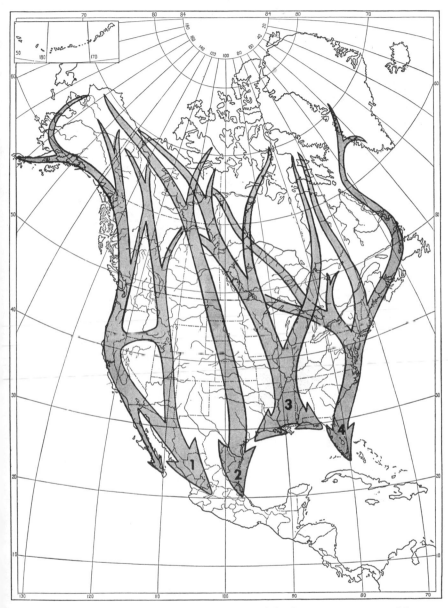

**FIGURE 10.4.** A very schematic illustration of the four major continental flyways used by birds in North America. *1,* Pacific Flyway; *2,* Central Flyway; *3,* Mississippi Flyway; *4,* Atlantic Flyway. It is probable that no two species follow exactly the same lines of flight, and even within a single species different populations have different routes.

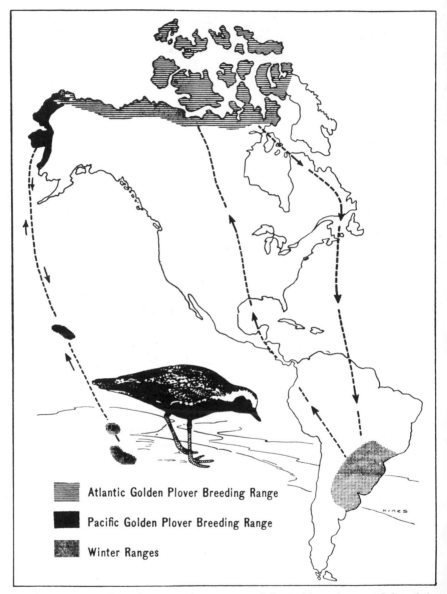

**FIGURE 10.5.** Distribution and migration of the golden plover. Adults of the eastern form migrate across northeastern Canada and then by a nonstop flight reach South America. In spring they return by way of the Mississippi Valley. Their entire route is therefore in the form of a great ellipse with a major axis of 8000 miles and a minor axis of about 2000 miles. The Pacific golden plover, which breeds in Alaska, apparently makes a nonstop flight across the ocean to Hawaii, the Marquesas Islands, and the Low Archipelago, returning in spring over the same route. (Lincoln, F. C., Biologist, and Hines, B., Illustrator, in Migration of Birds. Circular 16, U.S. Government Printing Office, Washington, 1950.)

winters from central California to the coast of Chile, while the Heermann gull ranges from Guatemala to southern British Columbia.

The order Passeriformes contains the greatest number of avian migrants. In Europe and Asia the principal groups making seasonal movements are the swallows, thrushes, Old World warblers, Old World flycatchers, pipits, and wagtails. Most are insectivorous and must leave northern Eurasia before winter. Those from Europe and western Asia migrate to central or southern Africa. A few from eastern Asia make this extensive trip, but the majority of species breeding in Siberia, Kamchatka, Manchuria, Korea, Japan, and northern China winter in Malaysia and southern India.

In North America the principal passerine migrants are tyrant flycatchers, swallows, thrushes, vireos, wood warblers, and tanagers. Fewer finches and sparrows engage in extensive migrations, although some, including the fox sparrow (*Passerella iliaca*), white-crowned sparrow (*Zonotrichia leucophrys*), and golden-crowned sparrow (*Zonotrichia atricapilla*), may winter hundreds or even thousands of miles south of the northern part of the breeding grounds. Most New World flycatchers and swallows winter from Mexico to northern South America. A few, such as the cliff swallow (*Petrochelidon pyrrhonota*), which summers widely over much of North Amer-

**FIGURE 10.6.** Most Heermann gulls (*Larus heermanni*) migrate to Isla Raza, a small desert island in the Sea of Cortez, to nest. In former years commercial eggers cleared the flatter areas of rocks, which they piled in stacks, in order to provide more surface for nesting.

ica, migrate as far south as central Argentina. Most thrushes and wood warblers that nest in Canada and the United States winter in Mexico and Central America.

One of the most extensive migrants among New World passerines is the bobolink (*Dolichonyx oryzivorus*), a member of the family Icteridae. Bobolinks nest in southern Canada and northern United States, from British Columbia to Nova Scotia and from eastern California to Pennsylvania. In migration the great majority moves to southern Florida, then across the Caribbean to northern South America. From there the bobolinks move down to Bolivia, Brazil, Paraguay, and northern Argentina, wintering about 6000 miles south of the northern part of the breeding range.

## MAMMALIAN MIGRANTS

Mammals that migrate are bats, cetaceans, pinnipeds, and large herbivores. Most insectivorous bats living in temperate and north temperate regions hibernate. The movement from summer colonies to hibernacula such as caves may entail only local movements, but sometimes considerably greater distances are involved. Indiana bats (*Myotis sodalis*) have been recorded migrating over 300 miles, and in Europe the noctule (*Nyctalus noctula*) has been found in summer 470 miles from its wintering ground. A pipistrelle (*Pipistrellus pipistrellus*) banded in the U.S.S.R. was recovered in Bulgaria, 720 miles distant. In North America, guano bats (*Tadarida braziliensis*) banded in summer in Carlsbad Cavern, New Mexico, have been recaptured in winter 810 miles south in central Mexico.

Lasiurine bats are nonhibernators that make extensive north-south migrations between their summer and winter homes. Flights of both hoary bats (*Lasiurus cinereus*) and red bats (*L. borealis*) have been reported in Bermuda, 600 miles off the eastern coast of the United States. These two species as well as the silver-haired bats (*Lasionycteris noctivagans*) regularly move southward or coastwise from Canada and northern United States to a more temperate winter environment.

The greatest mammalian migrants are to be found in the order Cetacea. Most of the large baleen whales of the world summer in polar or sub-polar waters where there is an abundance of food, particularly small shrimplike crustaceans called krill. After amassing a thick layer of blubber, which not only insulates against heat loss but provides a great resource for energy, the whales move toward equatorial waters. The young are born in tropical or subtropical seas, where they grow rapidly on the rich milk of their mothers and are able to travel by spring to arctic or antarctic waters.

One of the best known cetacean migrations is that of the gray whale (*Eschrichtius gibbosus*). This is a species that summers in the Okhotsk Sea, Bering Sea and Arctic Ocean, feeding principally on krill for about four months. In autumn a small population moves down the Asiatic coast to Korean waters, but the majority follows the Pacific coast of North America to the shallow lagoons along the west coast of Baja California, Mexico. They arrive there by January, have their young, breed, and by March have started north again to arctic waters. They make a round trip of over 6000 miles and, as far as known, take no food during the entire eight months that they are away from their summering grounds.

Some pinnipeds are resident, but a number of sea lions, some fur seals, and elephant seals make extensive migrations. The best known as well as the greatest migrant in the order is the northern fur seal (*Callorhinus ursinus*). Its principal rookery area is the Pribilof Islands in the Bering Sea, with small populations also occurring on the Commander Islands off Kamchatka, Robben Island near Sakhalin, and some of the Kurile islands. Females and immatures winter in the ocean as far south as coastal southern California and Japan. The adult males, or bulls, stay in the Gulf of Alaska in winter but move to the breeding rookeries in spring a few weeks before the females arrive. The females come in June and July and join the various harems guarded by the dominant and successful bulls, or beachmasters, who have fought each other for their territories. A day or so after arrival the females bear their single pup and one week later are bred by the harem bull. Immature and subadult males too young to command a harem are permitted to gather together in bachelor groups. The harems break up in late summer and the bulls leave, having gone without food and water for weeks during the vigil. The females and young of the year do not leave until late October and November.

A somewhat similar migratory pattern is seen in the Steller sea lion (*Eumetopias jubata*), whose breeding range extends from islands in the Bering Sea south to California. In this species the males, at least in the southern part of the range, migrate northward after the breeding season. They return to the rookeries in May and establish territories before the females come in June. Shortly after a female joins the harem she gives birth to her pup and is bred about two weeks later. The bachelors or nonbreeding males are isolated from the harem areas but leave the rookery islands in late July and August when the bulls leave. The females remain with their young for some months, a few remaining all year.

In the North Atlantic region the harp seal (*Phoca groenlandica*) engages in fairly extensive migrations. Summers are spent in high latitudes where food is abundant. In autumn these seals move southward to the breeding areas in the vicinity of Newfoundland, around

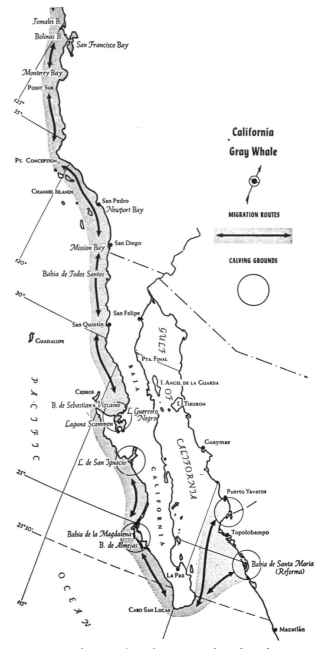

FIGURE 10.7.   Map showing the calving grounds and southern part of the migration route of the California gray whale (*Eschrichtius gibbosus*). (From Gilmore, R. M., and Ewing, G. C.: Pacific Discovery, Vol. 2, No. 3, 1954.)

FIGURE 10.8.   Aerial view of a gray whale (*Eschrichtius gibbosus*) cow and calf in Scammons Lagoon, Baja California, Mexico. (Courtesy of George E. Lindsay.)

Jan Mayan in the Norwegian Sea, and in the White Sea. There the young are born between January and April.

Many other pinnipeds in both Northern and Southern hemispheres engage in extensive migrations between feeding areas and breeding rookeries.

Marked migratory patterns are found in a number of hoofed animals. In North America, caribou (*Rangifer tarandus*), moose (*Alces alces*), elk (*Cervus elaphus*), and mule deer (*Odocoileus hemionus*) move from their summer ranges in autumn to more favorable wintering areas and then return the following spring. In Alaska and northern Canada the distance between summer and winter ranges of the caribou may be as much as 300 miles.

The migrations of moose, elk, and deer are usually altitudinal. Summer range in higher parkland and subalpine forests provides an abundance of food until the first heavy snow comes. By November there is a downward movement to valley floors.

In parts of Africa various kinds of antelopes make seasonal movements that are seemingly migratory. The wildebeest (*Gorgon taurinus*) moves in herds of hundreds of thousands on the Serengeti Plain to calving grounds where favorable food and water can be found. The route may involve a round trip of 200 miles.

FIGURE 10.9.  In East Africa, wildebeests (*Gorgon taurinus*) migrate annually to calving grounds where favorable food conditions prevail. (Photograph courtesy of George E. Lindsay.)

## POPULATION, AGE, AND SEX DIFFERENCES

Variances in migratory behavior sometimes occur between individuals of the same species and even the same population. In southwestern Alaska some arctic lampreys (*Lampetra japonica*) migrate to the sea, whereas others that appear morphologically identical spend their entire lives in fresh water. Nearly all Pacific salmon are anadromous, but one form of the sockeye known as the kokanee (*Oncorhynchus nerka kennerlyi*) is landlocked and matures in fresh water.

In some kinds of birds the northern populations are migratory while more southerly members of the same species are resident. Good examples of this are the song sparrows (*Melospiza melodia*) and white-crowned sparrows (*Zonotrichia leucophrys*) of North America. As a result, wintering northern populations and more southerly residents may mingle together at that season of the year. On the Pacific coast of North America there are a number of recognizable subspecies of the fox sparrow, *Passerella iliaca*. All the races are migratory, at least to some extent, but those that breed the farthest north migrate the farthest south (Fig. 10.10). There are some who have attributed these differences in the intensity of migration to hereditary variation, which may be true. However, the fact remains that the areas occupied by members of the more northerly races are uninhabitable in winter, and these birds in their southward movement in autumn are forced to pass over suitable territory which is already occupied by other fox sparrows. The latter behavior is, in a sense,

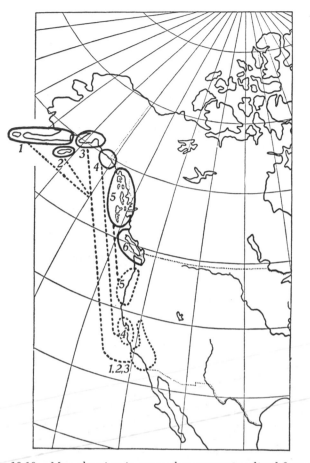

FIGURE 10.10. Map showing in somewhat conventionalized form the nature of the migrations performed by some northern subspecies of fox sparrows: *1, unalaschcensis; 2, insularis; 3, sinuosa; 4, annectens; 5, townsendi; 6, fuliginosa.* A solid line surrounds the summer habitat; a dotted line surrounds the main winter habitat. The broken lines connecting summer and winter habitats do not necessarily indicate migration routes; they are inserted to emphasize the overleaping features in the seasonal movements. The subspecies occupying the northernmost summer habitats (*unalaschcensis, insularis,* and *sinuosa*) travel the farthest south. The subspecies occupying the southernmost summer habitat (*fuliginosa*) is practically resident there. (Swarth, H. S., in Univ. Calif. Publ. Zool., Vol. 21, No. 4, 1920.)

somewhat analogous to that observed in certain populations of migratory fishes that spawn in fresh water. The first individuals arriving move to the uppermost parts of the spawning areas, while the late arrivals are restricted to areas farther down the streams. This is referred to as the "phenomenon of differential distribution." In Europe the robins (*Erithacus rubecula*) of Germany may migrate to North Africa to winter while those of France are essentially sedentary. Similarly, the common heron (*Ardea cinerea*) is resident in Great Britain but migratory in most of continental Europe.

Age is often a determining factor in migratory behavior. In many birds the young migrate before the adults. This has been observed in diverse species ranging from Adélie penguins (*Pygoscelis adeliae*) in the Antarctic to passerine migrants in the Northern Hemisphere. Banding carried on over many years in Wisconsin has shown that there are two peaks to the southward movement of sharp-shinned hawks (*Accipiter striatus*) in autumn. The first of these is in mid-September, when the migrants are mostly immatures. The second peak occurs in mid-October and consists largely of adults.

The young are not always first to migrate. In northern Alaska the males of the king eider (*Somateria spectabilis*) and the common eider (*S. mollissima*) are the first to start their migration from the Beaufort Sea to the Chukchi Sea. This begins early in July. By the middle of August the females begin to outnumber the males in

**FIGURE 10.11.** A large post-breeding aggregation of male California sea lions (*Zalophus californianus*) that have just arrived on Año Nuevo Island, which is about 250 miles north of the northernmost breeding rookery for this species along the coast of California.

FIGURE 10.12. In the spring, after the adult northern elephant seals (*Mioungu angustirostris*) leave the rookery islands, there is an influx of immature animals such as seen in the above picture taken in May on Ano Nuevo Island, California.

migration, and in September the young of the year start their migration.

Many other kinds of vertebrates differ in times of migration with both sex and age. This is especially noticeable in marine mammals, as has been indicated in describing the migratory movements of northern fur seals and Steller sea lions. In the California sea lion (*Zalophus californianus*), which breeds on islands along the west coast of Baja California and southern California, the males leave the rookery areas and move north in late July and August. They arrive in large numbers along the coast of central California by the first of September and many continue north to the waters of Oregon, Washington, and British Columbia. The females are largely resident. At least 80 per cent of the migratory males are adults.

The seasonal movements of the northern elephant seal (*Mirounga angustirostris*) vary with sex and age. Adult males arrive on breeding islands off the coast of Baja California and California in December, followed shortly by the females. The young are born in January and February, and the adults leave the islands in March and go to sea. Later that month and during April there is an influx of immature animals to the islands, and in May the young of the year

leave. By June or July most of the immatures leave and the adult and
subadult males return to molt. They remain until late summer and
are again replaced by immatures, who remain until the adult males
return in December.

The southward movement of gray whales (*Eschrichtius gibbosus*) from their summering area in arctic and subarctic waters to the
calving grounds of Baja California is led by pregnant females, followed by nonpregnant females, adult males, and immatures. On the
return trip, beginning in February and March, the newly pregnant
females lead the way, followed by the males and immatures. The
females with new calves are the last to start north.

## PHYSIOLOGICAL CHANGES

Many physiological changes are associated with migration. Some
of these are associated with reproduction. The reproductive cycles of
migratory fishes, amphibians, marine turtles, whales, pinnipeds, and
many birds are closely correlated with the annual movement to their
breeding areas.

In many vertebrates migration is preceded by an increase in
body fat. Herrings, salmon, eels, and other fishes that engage in long
migrations accumulate large amounts of fat before starting to the
spawning area. So, too, do whales and pinnipeds. More studies have
been made on the increase in lipid content in migratory birds than in
any other vertebrates. In some long-distance migrants such as bobolinks (*Dolichonyx oryzivorus*) and scarlet tanagers (*Piranga olivacea*)
the lipid content may amount to 50 per cent of the total body weight.
This fat is mostly subcutaneous and interperitoneal. It has been
estimated that any small bird whose fat content amounts to 27 per
cent of the wet (undesiccated) body weight can make a nonstop flight
of 600 miles under average flying conditions. The tiny ruby-throated
hummingbird (*Archilochus colubris*) acquires an average of two
grams of premigratory fat before it makes its nonstop flight of 500
miles across the Gulf of Mexico, between its summer range in eastern United States and Canada and its wintering grounds in Central
America. An individual of this species could travel 800 miles on the
energy provided by 2.1 grams.

Some avian species acquire migratory fat during the early part of
their migration rather than before it. Most of the warblers that nest in
northeastern England start south in autumn without a noticeable
increase in body weight. However, they stop over in southern England and parts of continental Europe where the fruit is ripe and
there build up their fat reserves before moving on to North Africa.

There appears to be a seasonal change in the function of the thyroid gland in some vertebrates that is correlated with their migratory patterns. Young coho or silver salmon (*Oncorhynchus kisutch*) usually remain a year or longer in fresh water before migrating to the sea. When they become smolts and are ready to migrate to salt water there is an increase in thyroid activity, which is also associated with increased day length. After arrival in the ocean thyroid activity decreases. In some migratory birds whose thyroid cycle has been studied there is a similar rise in the thyroxin level immediately prior to migration. It may be that the thyroxin stimulates an increase in the utilization of fat at the time when it is needed most.

In birds the annual molt is closely related to the fall migration. Since molting utilizes a great deal of energy, it generally precedes the premigratory fat deposition and also, in nocturnal migrants, the development of nocturnal restlessness. All are associated with a decrease in the daily photoperiod. In those species in which the molt is completed in late summer or early autumn, prior to migration, there is no conflict in energy requirements. This is very important for long-distance migrants. In species that travel a relatively short distance from summer to winter range, molting often continues during and after the migration. Some species, notably long-distance migrants such as the slender-billed shearwater (*Puffinus tenuirostris*), Sabine's gull (*Xema sabini*), the northern phalarope (*Lobipes lobatus*), the barn swallow (*Hirundo rustica*), and a few others, do not molt their wing feathers until they arrive at the wintering quarters.

The young of most migratory birds make their first autumn migration with their juvenal wing feathers.

# ORIENTATION

One of the most remarkable aspects of migration is the ability of organisms to return to specific sites. Marking has shown that salmon come back from the sea to spawn in the same stream in which they were hatched, and banding has proved that wintering birds, such as song sparrows (*Melospiza melodia*) and golden-crowned sparrows (*Zonotrichia atricapilla*), often come back to the same garden to winter year after year. Other vertebrates show the same pinpoint accuracy in their seasonal movements. How this is accomplished has been the subject of much research in recent years.

*Fishes.* Studies on salmon have shown that olfaction is a major factor in orientation. The small fry seemingly have the odor of the home stream imprinted in them at an early age. This odor is the result of chemicals in the water that seep into the stream from the soil and vegetation of the drainage system and differ slightly from those of other streams. After several years at sea, sometimes more than a thousand miles away from where they hatched, the memory of that odor is still retained by these fish and serves to guide them up great rivers like the Yukon and the Columbia to where their progenitors spawned.

This remarkable guiding system was demonstrated some years ago by Wisby and Hasler (1954). Silver salmon (*Oncorhynchus kisutch*) were captured in two forks of a stream in the state of Washington as they were ascending to spawn. Each fish was marked and the olfactory pits of about one-half of the trapped fish were occluded. All the salmon were then released one-half mile below the fork. Those with unoccluded olfactory pits selected the same fork to ascend when they came to the junction, while the others were unable to differentiate at this point and showed a random distribution.

Experiments on the kokanee salmon (*Oncorhynchus nerka kennerlyi*), a form of the sockeye that is not anadromous, have shown that even it makes use of olfaction in its local movements. Migrating kokanee leave the lakes in which they are landlocked and move into their home streams to spawn in August and September, following the shore line until they reach the mouth of their native stream. Here they pause until ready to ascend to the spawning ground. When kokanee are removed as they ascend the home stream and released with the nasal apparatus unharmed in the lake near the mouth of this tributary, a fair percentage return again to the spawning area. However, if the released kokanee have their nasal receptors destroyed they are seemingly unable to find their way back. Studies in olfactory perception in eels have shown that they can detect dilutions of alcohol equivalent to one-half a teaspoonful in a mass of water equal to the Lake of Constance, which has an area of 207 square miles.

Other factors play a part in the orientation of young salmon migrating to the sea. Some Pacific salmon, such as the chum (*Oncorhynchus keta*) and the pink (*O. gorbuscha*), spawn in the lower parts of streams, generally within 100 miles of the ocean. The young of these species, after hatching, are attracted both by light and by swift current. Although they may stabilize their positions by daylight hours, they drift with the current at night and thereby reach the sea at an early age. The behavior of other kinds of Pacific salmon is quite different. The king or chinook (*O. tshawytscha*), the silver or coho (*O. kisutch*), and the sockeye (*O. nerka*) spawn considerably farther inland. This is also true of the Atlantic salmon (*Salmo salar*). As in the chum and pink salmon, the young tend to stay hidden during the daylight hours, avoiding the light, and then drift with the current at night. This continues until they reach a lake. There they remain a year or longer, until they attain what is called the smolt stage. Careful studies on their movements through these lakes, which may be very long and complicated, indicate that they make use of sun-compass orientation in order to arrive at the outlet. Furthermore, they are able to adapt themselves to seasonal changes in the position of the sun. This was proved by maintaining captured young in containers exposed to the sun and recording their directional orientation as the season progressed.

There is some evidence, as a result of studies on sockeye smolts, that young salmon make use of polarized light from the sky. Most of their movements occur at twilight when neither sun nor stars are visible, but when the polarization of light is most pronounced. When a polaroid filter is placed at twilight over a tank containing smolts, they tend to orient directly to the polarized light.

Work on the white bass (*Roccus chrysops*) by Dr. Arthur D. Hasler and his associates in Lake Mendota, Wisconsin, has shown that migration by this species to the two spawning areas at the north end of the lake is accomplished by means of sun-compass orientation. Individual adult fish captured at these spawning sites in May and early June were marked and released 2.4 km. away. Each had a float attached to the posterior part of the back with a long nylon thread so that the course of the fish could be traced. After being released the bass oriented in a northerly direction if their eyes were not covered and the sky was not covered with clouds. If either of the latter conditions prevailed there was no directional orientation in their movement. Sun visibility appeared to be essential to migratory orientation in this species. Visual orientation in movement to the spawning area occurs in other kinds of freshwater fishes, although it is not necessarily dependent on the sun.

Most remarkable migrations are performed by freshwater eels (Fig. 10.13). Two rather closely related species are found on either side of the Atlantic. The European species, *Anguilla vulgaris*, inhabits the coastal streams of the British Isles and continental Europe from Norway south to Portugal. It requires anywhere from five to more than 20 years to attain maturity, at which time it migrates down to the ocean and crosses a major part of the Atlantic to the region of the Sargasso Sea. Here, at great depths, the sexually mature eels spawn and die. The eggs hatch, and the young, known as *Leptocephali*, gradually rise toward the surface and begin to drift with the Gulf Stream during their first summer. However, it is not until the third autumn and winter that they approach the shores of Europe and develop into elvers. The following spring they move inland to ponds and lakes, where they remain for some years. Finally, when they mature, they start down the streams and rivers to the coast and begin their migration back across the Atlantic to the spawning area. Although years are involved, many thousands of miles are traveled by each successful eel.

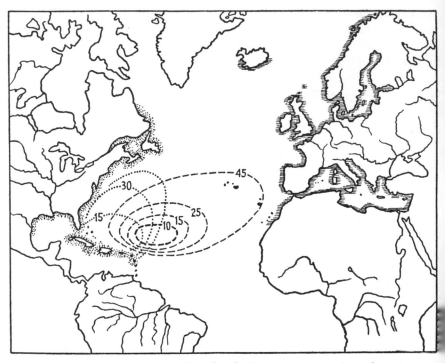

FIGURE 10.13. Migrations of the eels. The European species (A. *vulgaris*) occurs along the coasts outlined with lines, the American species (A. *rostrata*) where there are dots. The curved lines show where larvae of the lengths indicated (in millimeters) are taken. (After Norman, in Young, J. Z.: The Life of Vertebrates. Oxford, Clarendon Press, 1950.)

The American eels (*Anguilla rostrata*) spawn slightly to the south and west of the European eels, but their larval development into elvers requires only one year, at which time they have arrived at the mouths of various streams along the east coast of North America. After ascending these watercourses they, like their European relatives, require some years to reach maturity. When this occurs they too return to the ocean and go to the ancestral spawning grounds in the Sargasso Sea.

Currents seem to play a major part in orientation in both the European and the American eel. Adult European eels moving southward may follow the North Equatorial Current across the Atlantic to the Sargasso Sea, and the young reach their European watercourses by drifting with the Gulf Stream. Adult American eels follow the Labrador Current southward, and the young move north along the western edge of the Gulf Stream.

*Amphibians.* The movements of amphibians to their breeding

sites involve several different means of orientation. In anurans, vocalization is one of the ways it is accomplished. The nightly calls of male toads and frogs at breeding ponds guide females there, as well as other males. Visual clues also appear to be important. Experiments involving the displacement of tree toads (*Hyla regilla*) and giant toads (*Bufo marinus*), the latter at a time of year when they were silent, have shown that they can readily return to the original site of capture. Recognition of topographical features rather than celestial navigation is thought to be the important factor.

In some anurans there is definite celestial orientation. Studies by Ferguson and Landreth (1966) on Fowler's toad (*Bufo fowleri*), a species found in the Great Lakes region and the Mississippi Valley, have shown that these amphibians make use of sun-compass orientation. Young toads, recently metamorphosed, were placed in plastic pens 10 feet in diameter constructed so that only the sky and the walls of the container could be seen from the water. The pens were located about 75 yards from the shore of the lake when the toads hatched. In transporting the young toads to these sites they were kept in light-proof bags. After being released, when they could see the sun, they oriented in a direction that would correspond to a line bisecting the home shore at right angles. This is referred to as the Y-axis. Even toads displaced as much as 90 miles from the home shore displayed the same orientation. When the sun was obscured by a cover over the pen there was essentially no orientation. Similar sun orientation, as well as orientation on nights when the moon was full and on dark nights when only stars could be seen, has been observed in studies on both southern cricket frogs (*Acris gryllus*) and northern cricket frogs (*A. crepitans*).

The most extensive studies on orientation in caudate amphibians were carried on by the late Victor C. Twitty of Stanford University, using the red-bellied newt (*Taricha rivularis*), a species of the redwood belt of northwestern California. Newts were captured after the early autumn rains, when they emerged from their summer hiding places, and marked for future recognition. Some were displaced as much as two and one-half miles, either downstream or into other watercourses separated by a ridge from the home stream. The great majority of displaced individuals were retaken at the original site of capture during succeeding breeding seasons. The fact that some of those that returned had been blinded tended to rule out visual orientation and suggested that olfaction plays a major role in the movements of these amphibians. However, other experiments have shown that salamanders are also capable of celestial orientation, both by day and by night.

*Reptiles.* Much of our knowledge of migratory orientation in sea turtles is the result of studies made by Dr. Archie Carr of the

**FIGURE 10.14.** The rough-skinned newt (*Taricha granulosa*), like the red-bellied newt (*T. rivularis*), is believed to use celestial as well as olfactory clues in migratory orientation.

University of Florida and his associates. Major research was done on the green turtle (*Chelonia mydas*), a species found in tropical and subtropical oceans of the world. In this species the eggs are deposited on the sandy beaches of remote oceanic islands often very far from the feeding grounds. This entails an extensive migration across hundreds of miles of open sea where topographic guides are lacking. Females lay every two or three years.

One of Dr. Carr's most important projects involved the banding of turtles on tiny Ascension Island in the center of the South Atlantic. Ascension Island is 1400 miles east of Brazil, and no one knew where the turtles that layed there came from. The tagging project was started in 1960, in the laying season from February to July. In 1962 some of the turtles tagged in 1960 were recaptured on the island. In 1963 more were retaken during the banding operation, and in 1964 two turtles that had been banded in 1960 and recaptured in 1962 were again taken. These various recaptures showed a return of laying females to the beaches every two or three years. In 1965 however, nine green turtles that had been banded years earlier on Ascension Island were taken by fishermen along the coast of Brazil. The evidence strongly indicated a migratory route between the mid-Atlantic island and the South American coast.

Ascension Island is in the path of the South Equatorial Current which drifts toward the Brazilian coast at a rate of four and one-hal miles an hour. This would enable the young as well as post-laying adults to arrive at the sea pastures off South America where these turtles feed. The return trip is a different matter. There exists the remote possibility that olfactory guidance might be provided by chemicals oozing out of this island and being carried to the west, but this seems very unlikely. The current itself does provide a guide, since an adult turtle weighing up to 500 pounds must swim more than four and one-half miles an hour to make headway toward Ascension Island. It seems more likely that green turtles navigate by means of sun-compass orientation, but this has yet to be proved.

*Birds.* Many birds migrate along routes that parallel the axes of mountain ranges, valleys, or even river systems, as though these geographical features marked their way. Among flocking species in which adults and young migrate in compact groups, such as geese, ducks, or even certain passerine birds, the young may learn the route from the older birds. The training of homing pigeons shows that memory and familiarity, combined with the desire to return to a particular locality, are important in enabling these birds to come back to their homes. Young birds are acquainted with their environs, then taken to progressively greater distances each time they are released. Despite various theories to the contrary it appears that training and memory are significant factors involved here.

Training and memory, however, cannot account for the migrations of many other kinds of birds. In some species the young leave the nesting territory alone and are not accompanied by the adults. Other species are solitary migrants, and many kinds of birds migrate at night. A young bird flying alone at night, southward from the Arctic to the tropics for the first time, must rely on sources other than memory or the guidance of other members of the same species for orientation. The same appears true of pelagic migrants. There are no familiar landmarks on the open ocean.

Some have suggested that the pecten, a peculiar, fanlike structure that extends into the posterior chamber of the avian eye, may be of some significance in orientation. Whether it is an aid in determining direction in relation to the sun or whether its lines produce a grid-pattern which assists in navigation is not known. The fact remains, however, that a somewhat similar structure occurs in certain reptiles that do not do any extensive traveling.

Matthews (1955) concluded that some birds navigate by the sun. According to this theory the bird is familiar with the path of the sun through the sky and with the positions of the sun on this arc at different times. In migration over new localities the bird will have to reconstruct the arc from observations made on the movement of the sun over a small portion of the arc. This means that it must extrapolate to obtain the highest point. Knowledge of this point plus the angle of the horizontal and a knowledge of the same values of the previous day will give the change in latitude.

In recent years, work by Kramer, Sauer, and a number of other investigators has shown that some birds make use of the star pattern in the sky to navigate. For some time it has been known that certain kinds of migratory passerine birds, when kept in captivity, show definite restlessness at night or *Zugunruhe*. When these birds are kept out-of-doors so that they can observe the sky, this restlessness shows a definite orientation. Captive migrants in spring tend to face in a northerly direction at night, whereas in the autumn the direction is southerly in our latitudes. The direction in which the birds orient themselves at such times correponds to the direction in which they would ordinarily migrate if not confined. The fact that these passerine species, although diurnal throughout most of the year, do migrate at night suggested that the position of stars or constellations might be the navigational guide. To check this, experiments were carried on at the proper season in a planetarium where the celestial pattern which is projected onto a dome can be changed. The instruments used permit rotation of the position of stars and planets so that one might really look to the true south, but it would appear to be the north. Furthermore, the celestial pattern of any time of the night, past, present, or future, at any point on the earth can be presented by proper use of the projector. The birds that were observed under such artificial skies were found to orient themselves to the sky as they saw it, not as it actually was outside. That is to say, if under the real sky their nocturnal orientation was to the north and they were placed in a planetarium where the sky pattern was rotated 180°, they would orient themselves to the artificial north, which was really south. Significant also is the fact that cloudy or overcast skies at night interfere with the definite orientation of caged birds at such times.

Solar and celestial navigation is a new and interesting field in which many advances in our knowledge of avian orientation are being made. Dr. Frank C. Bellrose, studying mallards (*Anas platyrhynchos*) in Illinois, found that when they were displaced some miles from their home and released on clear nights they headed directly home, but on cloudy nights their movements appeared to be random. Canada geese (*Branta canadensis*) captured south of their breeding grounds on their northward migration in spring have been found to continue traveling north afoot after their wings were pinioned. One goose was recorded walking nearly 25 miles after being released.

Interesting experiments on orientation in crowned sparrows of the genus *Zonotrichia* have been performed by Dr. L. Richard Mewaldt of San Jose State College, California, and some of his associates. The forms studied included two races of the white-crowned

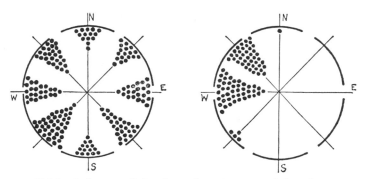

FIGURE 10.15.  Influence of clouds on the orientation ability of captive starlings.
*Left*, Dense clouds; the starling wanders at random. *Right*, Clearing; the starling
immediately gets its bearings in the direction in which it is flying. (After Kramer, in
Dorst, J.: The Migration of Birds. Houghton Mifflin Co., 1962.)

sparrow: *Z. leucophrys pugetensis,* a Pacific coastal subspecies that
breeds from British Columbia south to northwestern California, and
*Z. l. gambelii,* which breeds from Alaska east to Hudson Bay. The
golden-crowned sparrow (*Z. atricapillus*), which breeds in western
Canada and Alaska, and the white-throated sparrow (*Z. albicollis*),
which summers in Canada and northeastern United States, were also
used. The birds studied were captured in the vicinity of San Jose,
where populations of all these birds winter. The experimental cages
were circular and each had eight activity-sensitive perches arranged
around its periphery so as to form a sort of octagon. Each perch had a
microswitch connected with an Esterline-Angus recorder so that use
of the perch by a bird was indicated on a revolving drum. The cages
were used out-of-doors at night in order to determine the directional
response of crowned sparrows during the spring and fall periods of
migration. All four populations showed a strong northward orientation in spring and southward orientation in autumn, as indicated by
their choice in perch use.

Further experiments were made on the ability of golden-
crowned sparrows and the two kinds of white-crowned sparrows to
return to their wintering grounds in San Jose after extensive geographical displacement. Four hundred and twelve birds were transported
by plane to Baton Rouge, Louisiana, in the winter of 1961–62 and
released. In the winter of 1962–63, 26 of these displaced birds were
recaptured in San Jose. Again during the winter of 1962-63 a total of
360 birds trapped and banded at San Jose, including 22 of those that
had returned from Baton Rouge, were taken to Laurel, Maryland, and
released. The following winter 15 were recaptured at San Jose,
including six of the 22 that had previously been displaced to Louisiana and had returned to San Jose.

For the past hundred years or so there has existed the recurring theory that birds have a sixth sense that enables them to detect the polarity of the earth. More recently the idea has developed that they are aware of the magnetic fields of the earth. Little positive evidence has been brought forth to back this theory. An account by Knorr (1954), however, tells of observations made during World War II on the effect of radar on flocks of scoters and scaup. Repeated experiments in which a radar beam was directed at flocks of these birds flying parallel to the coastline resulted in individuals falling out of formation and seemingly becoming bewildered. As soon as the beam was diverted, the birds regrouped and continued on their way in an orderly manner. This poses the possibility that these ducks were affected by magnetic radiation. The author suggests that "one cannot help but wonder if the behavior described above does not support the theory that birds indeed perceive the earth's magnetic field. In flight, the crossing of these lines of force may result in the production of phosphenes, or perhaps the answer lies in the setting up of tiny oscillating currents somewhere in the animals' central nervous system." Similar observations of the effect of radar on birds were previously recorded in Germany (Drost, 1949).

Other experiments along this line have provided negative results. However, it is known that radar beams can affect man adversely. In all probability the reactions that have been observed in some avian species are of the same nature and do not support the theory that birds can perceive magnetic fields of force.

*Mammals.* There are far fewer mammalian than avian migrants, but even so less is known of their means of orientation during their seasonal movements. It appears likely, however, that olfactory, visual, and auditory clues are employed. We know that bats, cetaceans, and pinnipeds make use of echolocation in their daily lives, but the use of such a sonar sensory system in navigation has yet to be proved. It seems likely that such a mechanism would be of great advantage to migrating whales in determining depth as well as conformation of the sea bottom, shorelines, and the proximity of other migrants of the same species. Whales cannot make use of olfaction nor can visual perception be of much use under water. Some species do, however, raise their heads out of water periodically. This is a characteristic of the gray whale, which migrates fairly close to the shoreline, and is believed to be a means of locating familiar headlands.

There have been occasional observations on the migration of bats of the genus *Lasiurus* during daylight hours, but to the best of our knowledge most of the bats that migrate do so at night. Because of the high frequency of sound pulses they emit it seems somewhat questionable that these are of much use in long-distance travel although they are a vital navigational aid as far as gaining knowledge of the immediate environment is concerned. There is some indication that familiarity with major topographical features such as river systems and valleys may be of major importance in migratory orientation.

Numerous displacement experiments have been made on bats in both North America and Europe to determine their homing capabilities. In one such instance 155 brown bats (*Eptesicus fuscus*) taken in Cincinnati, Ohio, on July 20 were released 450 miles to the north at Pilgrim, Michigan, the next evening. Three of the displaced bats were found at the home roost on August 24 and four more were discovered there on October 26. All were adult females although most of the displaced bats were juveniles. In Europe there have been returns of mouse-eared bats (*Myotis myotis*) from distances up to 165 miles. It has generally been found that the return of displaced bats to their home roost is greater in migratory species than in those that move about only locally.

Temperature and ocean currents may be guiding factors in the movements of marine mammals. There are marked differences in ocean temperature between the summering and wintering areas occupied by whales as well as by certain pinnipeds. Even within the same species sexual differences in the temperature of the water inhabited may be marked. Northern fur seal bulls, for example, come no farther south than the Gulf of Alaska in winter, while females and young range south to the ocean off southern California.

Olfaction as well as vision is important in the migrations of large herbivores. Most of these animals have well-developed interdigital and tarsal glands whose secretion produces a scent that is long-lasting on game trails. Spring and autumn movements of certain kinds of deer as well as elk and caribou are along definite trails that are used year after year.

## MECHANICS OF MIGRATION

Apart from methods of orientation, there are other features of migration that must be considered under the heading of mechanics. Vertebrates migrate by swimming, drifting with the current, walking, and flying. Some species are solitary migrants whereas others move in groups, small or large. Some travel rapidly while others make a leisurely trip. Migration may occur at night or by day. Flying species may travel at considerable height or may fly low. They may soar, glide, or flap.

The migrations of fishes and marine mammals is by means of swimming. Penguins also use this method, but in addition they make

**FIGURE 10.16.** Aerial view of elk in migration. (Photograph courtesy of Idaho Fish and Game Department.)

use of drifting ice floes to carry them north at the end of the antarctic summer. It has already been pointed out that currents play a major role in the movements of young salmon in streams and of freshwater eels and even young sea turtles in the ocean. Land mammals depend entirely on walking in migration, as do certain gallinaceous birds like quail and grouse. Most birds, like all bats, migrate by means of flight

Group movement occurs in most migratory species of fishes but is not characteristic of amphibians and reptiles. Many mammals are gregarious during migration. Elk, caribou, and wildebeests are examples of mammals that travel in herds ranging from hundreds to hundreds of thousands.

Birds show great variation in solitary versus social movements. In general most diurnal migrants as well as species that travel either by day or night, like waterfowl, fly in flocks. Some flocks are compact, like those of waxwings and blackbirds, while others are loose as with swallows, where individuals may spread widely apart in flight yet move as a group.

Birds of prey tend to move in loose groups often composed of several species. This pattern is largely determined by air current and rising thermals. The day must be sufficiently advanced before these thermals develop. Then migrating hawks soar from one rising thermal to another, depending upon them for gaining altitude. Thi avoids the enormous expenditure of energy that wing flapping would require and explains why most soaring and gliding species avoid travel over large bodies of water like the Great Lakes of North America or the Mediterranean Sea.

Most nocturnal migrants, apart from shorebirds and waterfowl, seem to fly in small groups in which the individuals are spaced well apart. This has been shown by radar observations as well as by watching migrants through telescopes directed at the full moon. Such groups seem to maintain their entity by means of call notes or "chips." There appear to be decided advantages to flocking. The chances of success for a single bird making its first migratory flight alone would seem to be far less than if it were a member of a flock where others have had previous experience in traversing the route.

Among birds the time of day that migration occurs is closely associated with the food habits of the species. Birds of prey migrate during daylight hours when they can secure food as they travel. This is also true of swifts and swallows that feed on the wing. Warblers and thrushes must stop to feed, since their food is largely associated with vegetation and must be secured from leaves, branches, or the ground during daylight hours. This necessitates nocturnal migration in these birds that at other times of the year are strictly diurnal. For most nocturnal migrants travel begins at dusk, reaches its peak shortly before midnight, and terminates between dawn and sunrise.

FIGURE 10.17. Adélie penguins (*Pygoscelis adeliae*) drifting northward on ice floes after leaving their breeding area near Cape Hallett Station, Antarctica. (Photograph courtesy of Dietland Müller-Schwartze.)

Salmon and trout tend to move into the spawning streams under cover of darkness. Young salmon in their downstream movements to the sea migrate principally in the evening and early morning and hide from various predators during the daytime.

The total time involved in any one migration as well as the speed of travel is subject to great variation. Some species of amphibians move to their breeding ponds or streams in a matter of hours, while some whales take up to eight months to complete their round trip from arctic to subtropical waters. Arctic terns are almost continually in migration between the north and south polar areas except for the brief summer breeding season. Young European freshwater eels take nearly three years to travel from the site of their hatching to the freshwater streams where they will grow to maturity. The albacore tuna of the North Pacific is thought to move between the continental waters of the United States and Japan. One tuna tagged off the California coast was recovered three years later near Tokyo. Another that was tagged about 1300 miles north of Hawaii was captured 471 days later near Japan, 2370 miles distant.

The majority of vertebrates that engage in extensive migrations do so in a leisurely manner. Many land birds are estimated to travel no more than 1000 miles a month during migration. This, of course, does not necessarily imply that they move at a relatively uniform rate each day. After a prolonged flight they may rest for a few days and build up a fat reserve sufficient to provide energy for the next flight. Recent studies on the migratory movements of white-crowned sparrows (*Zonotrichia leucophrys*) of North America suggest a migration rate of about 100 miles a day in spring. This appears to be more rapid than the southward movement in fall. The three shortest recovery times were 310 miles in one day, 175 miles in two days, and 520 miles in six days.

Gray whales moving southward to their breeding grounds have been found to travel around 5 miles per hour. Eels moving southward along the coast of northern Europe have been recorded moving at the rate of 9 miles per day. There are some migrations, however, that are made rapidly and some that are of necessity nonstop. The blue geese (*Chen caerulescens*) that winter in the marshes along the coast of Louisiana fly north in late March or early April to James Bay in Canada. For the most part this is a nonstop flight, at least throughout the length of the United States. After their arrival at James Bay they rest for several weeks before moving north to Baffin and Southampton islands. Land birds crossing extensive bodies of water must necessarily do so without stopping. By attaching a tiny radio transmitter weighing 2.5 grams to the back of a gray-cheeked thrush (*Hylocichla minima*), Dr. Richard R. Graber was able to follow the night flight of this bird in its northward migration. The bird was captured at noon on May 25, 1965 near Urbana, Illinois, and released at 7:55 that evening. By following the radio signals with a plane, the thrush was found to fly directly to Chicago and to the northwest shore of Lake Michigan; it was eight hours in the air and flew close to 400 miles, more than half of this over water. It had a tailwind, so while its ground speed was 50 miles per hour its air speed was only 33 miles per hour.

In general the ground speed of most small nocturnal avian migrants is close to 40 miles per hour while that of birds of prey is more often between 45 and 50 miles per hour. Bats are much slower flyers than birds. The guano bat (*Tadarida brasiliensis*) is estimated to make its southward migration in western North America at a rate of slightly more than 20 miles a night.

Sockeye salmon are said to migrate in the sea at an average speed of 30 miles per day. Pink salmon travel at 24 miles a day and chum salmon, 16 miles a day.

The height at which migrating birds fly depends on the type of flight and the time of day as well as the terrain below. Most diurnal passerine migrants fly considerably lower than those that travel at night. Radar studies on nocturnal migrants show that most fly between 1500 and 2500 feet altitude with the peak probably a little more than 2000 feet. Many diurnal migrants fly within several

**FIGURE 10.18.** Marbled godwits (*Limosa fedoa*) resting on a gravel bar during spring migration.

hundred feet of the ground, but the median altitude for most is 1800 feet. Storks and birds of prey generally migrate at higher altitudes than most land birds because they depend on rising air, although they do lose altitude between successive thermals. Most sea birds, such as loons, grebes, and shearwaters, fly low over the sea when migrating.

## HAZARDS OF MIGRATION

Travel is hazardous. Whether you are a member of the human species driving from New York to San Francisco or a bird flying from Alaska to Patagonia, the element of danger is always present. Many of these hazards to migrants are natural, but a number are man made. Natural enemies are prone to take advantage of migrating animals. Attrition at this time by predators is often at its greatest. An example of this is seen in the habits of Eleonora's falcon (*Falco eleonorae*) of the Mediterranean that has its nesting period on the islands so timed as to coincide with the fall migration of passerines from Europe to Africa. During this period it is estimated to capture nearly one million small birds in flight. This of course represents only a small percentage of the one billion passerine migrants that move across the Mediterranean at this season.

One can often locate schools of marine fishes moving into spawning areas by the aggregations of gulls, cormorants, and other sea birds voraciously feeding on the migrants or their eggs. Salmon moving upstream to the spawning areas are often subject to concentrated predation by eagles, by bears, and, of course, by man. In certain streams salmon, steelhead, and lampreys have always had to contend with waterfalls and fight their way through rapids. However, dams, pumping stations, and pollution are more recent dangers that now confront them in many waterways.

Many birds are blown off their course by storms, succumb to sudden drops in temperature, or become lost in fog. In coastal areas it is a common event during the migration period to observe small land birds a considerable distance at sea. Such individuals will frequently alight in an exhausted condition on passing ships. Sudden temperature drops resulting in snow and ice have been known to produce appalling results. In 1907, as a result of a heavy snowfall, an estimated 750,000 Lapland longspurs (*Calcarius lapponicus*) were found dead on the ice on two small lakes in Minnesota. This probably represented a small percentage of the mortality to this species on the particular occasion, as dead birds were reported over an area of more than 1500 square miles. Catastrophes of this sort, although not usually of this magnitude, are not uncommon.

Nocturnal migrants are frequently attracted to lighthouses or to lights in tall buildings and towers and, as a result, crash into these objects and die. The Washington Monument in Washington, D.C., as well as a number of the skyscrapers in cities along the east coast of the United States, have been responsible for the death of a vast number of birds. Most casualties occur when there are low clouds and rainfall that force the birds to lower elevations. More recently a new hazard to migrant birds has developed in the form of airport ceilometers. These instruments direct a beam of light upward at night to determine the height of the cloud ceiling. The lights, however, under certain climatic conditions may confuse nocturnal migrants and cause them to collide with buildings or with each other or to crash to the ground. The first ceilometer casualties were reported in the fall of 1948. The greatest toll occurred during three successive nights early in October 1954, in the eastern half of the United States. It so happened that the southward movement of a cold front, with consequent low clouds and rain, coincided with the southward migration of many kinds of small birds, including thrushes, vireos, warblers, tanagers, and sparrows. Each night more and more of the southbound birds became confused in the vicinity of airport ceilometers with fatal results. On the last night alone 75,000 birds were killed, and the total casualties from this one type of hazard for the three nights was known to have exceeded 100,000. Since then, through the cooperation of many individuals and agencies, efforts have been made to use filters that produce a blue or purple light, which apparently the birds see either very poorly or not at all.

## METHODS OF STUDYING MIGRATION

The migration routes of many vertebrates have been accurately determined both by observation and by marking individuals. The latter method, employing bands, has been used successfully for a number of years in the study of bird migration in the United States and in many other countries. Since 1920, in the United States, use of this method has been under the direction of the U.S. Biological Survey and its successor, the U.S. Fish and Wildlife Service, in cooperation with many other federal, state, and private agencies as well as thousands of private individuals.

Bird banding involves the capturing of wild birds and the placing of a serially numbered metal band of the proper size on the leg of each bird, after which it is released. Several methods are employed in capturing birds. These generally involve the use of either wire traps of various types or nets. In recent years Japanese mist nets have become very popular with many bird banders. These nets are made of a single fine mesh and are almost invisible at any distance. Bergstrom and Drury (1956) have presented an excellent summary of the current techniques used in this work.

The band, in addition to bearing a number, requests that anyone finding a bird so tagged advise the U.S. Fish and Wildlife Service, Washington, D.C. (Fig. 10.19). The finder should send to this organization the number of the band as well as the date and locality where the banded bird was captured or, if the bird is dead, send the tag itself. The returns from such banding activities carried on over many parts of the country have provided a vast amount of data which have made it possible to trace the migration routes of a great many species and to determine general flyway patterns.

The banding and marking of individuals to determine migration movements and routes has by no means been restricted to birds. In recent years many persons have been engaged in banding bats to study the population movements of various species. The bands used are principally bird bands which are generally attached to the forearm of the bat in such a way that they do not injure the wing membrane. Various methods of marking have also been employed in studying the movements of certain big game mammals. These include ear cutting, the attaching of plastic discs to the ears, and the use of serially numbered boxcar seals attached to the antlers. The Petersen disc (see Fig. 9.15) has been used extensively in the marking of fish to determine their migration routes. The migrations of

**FIGURE 10.19.** Bands of various sizes, made of soft, lightweight metal, are provided by the U.S. Fish and Wildlife Service for bird banding in order that the migratory movements of various species may be determined.

FIGURE 10.20 An electric brand being applied to a subadult elephant seal (*Mirounga angustirostris*). (Photograph courtesy of Richard Jennings.)

other animals, such as caribou and whales, have been determined primarily through observation without the use of special marking devices.

Methods involving minor mutilation, such as toe clipping for amphibians, the removal of certain scales on reptiles and fishes, or excising part or all of a fin in fishes, have been described in the previous chapter. Branding is another means of marking that has been used successfully in mammals.

Recently, the use of small radio transmitters which may be attached to migrants has proved extremely useful, especially in following big game animals. To do this it is generally necessary to tranquilize the animal before the transmitter is attached. Tiny transmitters have been developed which can be attached to small birds, and others have been shot into whales. The use of radio transmitters in following whales, however, presents certain problems. Not only must such instruments be able to withstand the corrosive effects of salt water, but they must withstand great pressure changes and be able to transmit during the very brief time that a cetacean surfaces.

Sonar is now being used in following the migration of certain fishes, and aural recordings are employed in research on avian migration. The latter system, employing a parabolic reflector with a microphone in the center to pick up sound, is very useful in detecting relatively low-flying migrants on dark nights that vocalize in flight. It also permits specific identification in many instances.

For some years the appearance of blips or echoes on radar screens were referred to as "angels." Later it was discovered that these were birds. Radar tracking is now used extensively in following migratory birds. Flight density and height can be recorded, but not specific identity.

Although it was known for years that if one looked at the full

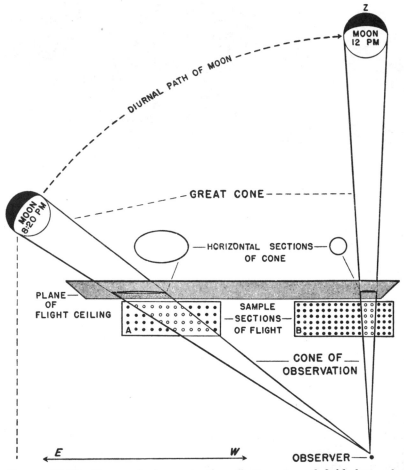

**Figure 10.21.** Temporal change in the effective size of field during lunar observation. The sample sections, A and B, represent the theoretical densities of flight at 8:20 and 12:00 P.M., respectively. Though twice as many birds are assumed to be in the air at midnight when the moon is on its zenith (Z) as there were at the earlier hour, only half as many are visible because of the decrease in size of the cone of observation. (From Lowery, G. H., Jr., Univ. of Kansas Publications, Museum of Natural History, 3:365, 1951.)

moon through a low-powered telescope, birds would occasionally be seen flying across the field of vision, it remained for Dr. George H. Lowery, Jr. and Dr. Robert J. Newman of Louisiana State University to make use of this knowledge in studying the migration of birds. The technique they developed involves the use of a 20-power telescope during the four or five nights of the full moon in spring and fall. The moon is viewed as a clock and the direction and line of movement of all migrants across its face is recorded. Naturally as the moon rises the distance between the observer and the plane of flight as viewed decreases, as does the cone of observation. By use of mathematical formulae the data derived from such observations give an index to the number of migrants passing over a region and the directional trends of their flight.

## IRRUPTIONS, INVASIONS, AND DISPERSAL

As has already been pointed out, migration is a rhythmic phenomenon involving population movements that are synchronized in some species with various stages of development and in others with season of the year. Also generally implicated with this type of behavior is the desire on the part of the individual organism to return to the region of its birth or origin for purposes of reproduction.

There are other types of population movements, however, that are not rhythmic, at least as far as the individual is concerned, and that do not include a return. One of these which involves the sudden movement of a large part of an animal population out of its normal range is generally referred to as an irruption. Irruptions are mostly associated with areas of climatic extremes such as arctic or desert regions. In temperate or tropical areas conditions are more uniform and cyclical fluctuations are rarely noted. These differences are due to the few plants and animals in northern ecosystems as contrasted with those farther south. In arctic communities a change in one species may affect many others, whereas in the tropics the number of plant and animal species is so great that a change in one has little or no effect on the many others present. This is well illustrated by the behavior of the snowy owl (*Nyctea scandiaca*), a species that breeds on the arctic tundra of North America and regularly winters as far south as southern Canada. Periodically, however, snowy owls move much farther southward and invade many parts of the United States (Fig. 10.23). They have even been recorded in Bermuda. These southern invasions seem to occur at times when food in the form of small mammals such as lemmings and hares is scarce in the north. They are not of annual occurrence as are the migrations of many other arctic breeding birds and may not take place during the lifetime of many individual owls.

**FIGURE 10.22.** Snowy owls (*Nyctea scandiaca*), which are circumpolar, nest on the tundra and may lay their eggs before the snow is gone. (Photograph by John Koranda.)

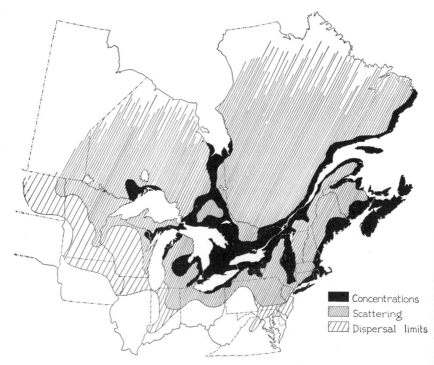

■ Concentrations
▨ Scattering
▨ Dispersal limits

**FIGURE 10.23.** The great "flight" of the snowy owl (*Nyctea scandiaca*) into northern United States in 1941–1942. (Courtesy of L. L. Snyder.)

There are other birds, such as red crossbills (*Loxia curvirostra*), that wander far out of their regular range in years when food is scarce. Sometimes they will remain in newly invaded regions for a year or so. Thick-billed parrots (*Rhynchopsitta pachyrhyncha*) from the tableland of Mexico occasionally invade the mountains of southern Arizona. They have been known to remain there during the breeding season, although they have not been found nesting under such circumstances. These nomadic movements, probably stimulated by scarcity of food within the normal range of the species involved, are not migrations in the ordinary sense of the word, nor are they of the permanent nature that is generally associated with emigration.

In the Old World the sporadic invasions of the Pallas sandgrouse (*Syrrhaptes paradoxus*) are well known. This is a species that normally inhabits the steppes of central Asia. In May and June of 1863, however, great numbers of these birds became established sufficiently to breed in northern Europe. Another even greater invasion occurred in 1888, when nearly 2000 were estimated to have reached Scotland alone. The last invasion of Europe by sandgrouse took place in 1908.

Irruptions of this sort have been attributed to population pressure from within, combined with the need for a greater food supply for survival. Many species of vertebrates maintain fairly well balanced numbers, fluctuating to some extent from year to year, but generally not to the point where their numbers exceed the food supply. Others, however, do not. Perhaps the best known example of a vertebrate that is subject to marked cyclic fluctuations that result at times in emigrations is the Norwegian lemming (*Lemmus lemmus*). These small rodents reach a population peak about every four years. If their numbers are sufficiently high in northern Europe they may, locally, undergo mass movements away from the home grounds. Many individuals perish at such times as a result of predation, disease, accident, and starvation, and a few fail to participate in these movements. If all remained the species might be in danger of complete disaster from starvation and disease. However, with only a few remaining and capable of surviving on the limited food supply, there is still a nucleus for perpetuating the population. Furthermore, such periodic emigrations provide a means for possible expansion or change in the distribution of the species if new, suitable, previously unoccupied habitats are encountered. Although the lemmings of North America are subject to somewhat similar cyclic fluctuations, they have not been observed undergoing emigrations like those of northern Europe.

The Clark's nutcracker (*Nucifraga columbiana*) is a member of the crow family that is usually resident in the higher montane regions of western North America. Its winter food consists principally of the seeds of conifers. These seeds are stored for winter use in the fall of the year, when the cones open. In certain years, however, marked irruptions of nutcrackers occur and large scale movements to the lowlands take place. At least six such irruptions have occurred in western United States between 1898 and 1961. Such movements do not appear to be local. Simultaneous movements have been recorded in California, Arizona, and New Mexico, with wandering birds reported as far east as Missouri. Considerable evidence indicates that Clark nutcracker irruptions coincide with low cone crops. In California, where accurate records have been kept of the cone crop in recent years, each nutcracker irruption occurred during such a winter. However, irruptions have not always occurred when cone production was low, possibly because the nutcracker populations were also low at such times, and therefore the food supply, although reduced, was still adequate.

The Old World nutcracker of Asia, *Nucifraga caryocatactes*, also has occasional population irruptions when, after some years of increase correlated with a buildup in cone production by the Arolla pine (*Pinus cembra*), the main food of this bird, there is a drastic decline in production. At such times nutcrackers invade Europe, even reaching the British Isles. Such irruptions are also characteristic of the Bohemian waxwing (*Bombycilla garrulus*), a Holarctic species.

In Australia the budgerigar (*Melopsittacus undulatus*), a desert inhabitant that occurs in large groups, may invade coastal areas when the food supply becomes low.

Invasions represent slower types of population movements that may bring about changes in the distribution of species and are often correlated with environmental changes of various sorts. Some of these may even be man made. A very good example of this is demonstrated by the invasion of the sea lamprey (*Petromyzon marinus*) into the Great Lakes region. Lampreys always had access to Lake Ontario by way of the St. Lawrence River. They could not extend their range beyond this, however, because of the barrier presented by Niagara Falls. In 1829, with the completion of the Welland Canal, these cyclostomes had access to Lake Erie and ultimately all of the other Great Lakes. Lake Erie, because it is shallow and warm, does not present a favorable environment for lampreys, nor are there suitable spawning streams flowing into the lake. Consequently, it was nearly 100 years before they reached Lake Huron. Once this was accomplished, however, the emigration of these animals proceeded with speed. They were well established in Lake Huron by the end of the 1930's. Within a few years they had moved into Lake Michigan and by the middle of the 1950's were in Lake Superior. The speed of their invasion is evidenced by the startling decline in the commercial catch of lake trout. The annual catch of trout in Lake Huron in 1939 amounted to 1,372,000 pounds, whereas in 1951 it had dropped to 25 pounds. The annual catch in Lake Michigan was 5,650,000 pounds in 1945 and only 400 pounds in 1954.

The recent expansion of the cattle egret (*Bubulcus ibis*) from Africa to South America (about 1930) and thence to North America (in 1948) is attributed in part to expansion of the cattle industry, which provided suitable situations for these birds. Careful study on the expansion of the range of this species in Florida has shown that individuals emigrate northward with other groups of migrating herons. Rookeries of the latter appear to provide the necessary stimulus for small numbers of the cattle egrets to breed and soon become established. At present cattle egrets have reached the west coast of North America and have even been recorded on Cocos and Clipperton islands in the eastern Pacific. They have also invaded New Guinea and Australia from the Moluccas.

While many species of vertebrates were brought to extinction in the Pleistocene, a number of kinds were able to reach North America by land or ice bridges from Asia. Included are such mammals as caribou, moose, elk, and grizzly bears. Conversely, camels invaded the Old World. It is not known how much time was involved in such movements, but it is likely that the process was long and slow and correlated with gradual changes in the environment. Even at present we see slight but continuing changes in ranges of many species in our own country. Within the past half century there has been a considerable northward extension of the ranges of the mockingbird (*Minus polyglottos*) and the hooded oriole (*Icterus cucullatus*) in California. Similarly the gray fox (*Urocyon cinereoargenteus*) has moved north in the Great Plains region in recent years, invading North Dakota, and has even been recorded in Alberta. Many other examples could be cited. Gradual northward invasions of this sort may in part be associated with the onset of a more arid and warmer climate, although changes brought about by man are no doubt significant factors.

Many of the gradual invasions made by vertebrate species into newly developed or previously unoccupied territory are the result of dispersal of the young and their selection of breeding territories for the first time. Among most noncolonial, resident species of birds and mammals, the young, when they have attained a certain age, either move out of the home territory of their own accord or are forced to do so by one or both parents. This is a natural means of preventing overcrowding and keeping the population density compatible with the food supply.

Mention has been made in Chapter 9 of the voluntary movement of wrentits from the parental territory when they are nine or 10 weeks old. Studies made on young pocket gophers (genus *Thomomys*) indicate that they are driven from the burrow of the mother at a relatively early age. In captivity female pocket gophers have been found to develop a pugnacious attitude toward their offspring when the latter are about six weeks old. The young in turn become intolerant of each other at this same time. Observations made on deer mice, voles, chipmunks, and other small mammals have shown that the young after reaching a certain age tend to wander. These movements are in large measure responsible for the invasion of seral communities by vertebrate faunas. The destruction of a climax forest by fire may result in a grassland community in several years. Though temporary in nature and merely representing the first seral stage leading to the redevelopment of the climax forest, it will be invaded by grassland-frequenting species of small mammals. This invasion will not be effected primarily by the movement of adults with previously established territories, but rather by successive waves of young individuals searching for suitable unoccupied homesites.

This seemingly innate desire on the part of the young to wander when they reach a certain age not only prevents overcrowding, but it performs other services that benefit the species. Territories that are unoccupied because of mortality are not long left unused. Areas that ordinarily serve as barriers preventing population expansion may be traversed at such times. Furthermore, the possibility of inbreeding, which might prove detrimental, is greatly reduced.

### References Recommended

Allen, G. M. 1939. Bats. Cambridge, Mass., Harvard University Press.

Bellrose, F. C. 1958a. The Orientation of Displaced Waterfowl in Migration. Wilson Bull. 70:20-40.

Bellrose, F. C. 1958b. Celestial Orientation by Wild Mallards. Bird-Banding 29:75-90.

Bellrose, F. C. 1967. Radar in Orientation Research. Proc. XIV Internat. Ornith. Congr.:281-309.

Bellrose, F. C. 1968. Waterfowl Migration Corridors East of the Rocky Mountains in the United States. Biol. Notes 61, Illinois Nat. Hist. Surv.

Bergstrom, E. A., and Drury, W. H., Jr., 1956. Migration Sampling by Trapping: A Brief Review. Bird-Banding 27:107-120.

Brattstrom, B. H. 1962. Homing in the Giant Toad, *Bufo marinus*. Herpetologica 18:176-180.

Briggs, J. C. 1955. Behavior Pattern in Migratory Fishes. Science 122:240.

Burt, W. H. 1940. Territorial Behavior and Populations of Some Small Mammals in Southern Michigan. Misc. Publ. Mus. Zool., University of Michigan, No. 45.

Carr, A. 1962. Guideposts of Animal Navigation. Amer. Inst. Biol. Sci. BSCS Pam. 1.

Carr, A. 1965. The Navigation of the Green Turtle. Sci. Amer. 212(5):78-86.

Carr, A. 1967. Adaptive Aspects of the Scheduled Travel of *Chelonia*, pp. 35–55. *In* R. M. Storm [ed.], Animal Orientation and Navigation. Proc. 27th Ann. Biol. Colloq., Oregon State University Press, Corvallis.

Casteret, N. 1938. Observations sur une Colonie de Chauves-Souris Migratrices. Mammalia 2·29-34.

Cockrum, E. L. 1956. Homing, Movements, and Longevity of Bats. J. Mamm. 37:48-57.

Cortopassi, A. J., and Mewaldt, L. R. 1965. The Circumannual Distribution of White-crowned Sparrows. Bird-Banding 36:141-169.

Davis, J., and Williams, L. 1964. The 1961 Irruption of the Clark's Nutcracker in California. Wilson Bull. 76:10-18.

Delvingt, W., and Leclercq, J. 1963. The Sense of Direction in Migratory Birds. Endeavor 22(85):27-30.

Dennis, J. W. 1954. Meteorological Analysis of Occurrence of Grounded Migrants at Smith Point, Texas, April 17–May 17, 1951. Wilson Bull. 66:102–111.

Devlin, J. M. 1954. Effects of Weather on Nocturnal Migration as Seen from One Observation Point at Philadelphia. Wilson Bull. 66:93-101.

Dorst, J. 1962. The Migration of Birds. Boston, Houghton Mifflin Co.

Drost, R. 1949. Zugvogel perzipieren ultrakurzwellen. Vogelwarte 15:57-59.

Drury, W. H., and Nisbet, I. C. T. 1964. Radar Studies of Orientation of Songbird Migrants in Southeastern New England. Bird-Banding 35:69-119.

Eastwood, E. 1967. Radar Ornithology. London, Methuen.

Eastwood, E., and Rider, G. C. 1964. The Influence of Radio Waves upon Birds. British Birds 57:445-458.

Eastwood, E., and Rider, G. C. 1965. Some Radar Measurements of the Altitude of Bird Flight. British Birds 58:393-426.

Eisentraut, M. 1943. Zehn Jahre Fledermausberingung. Zool. Anz. 143:20-32.

Emlen, S. T. 1969. Bird Migration: Influence of Physiological State upon Celestial Orientation. Science 165:716-718.

Engels, W. E. 1962. Day-length and Termination of Photorefractoriness in the Annual Testicular Cycle of the Transequatorial Migrant Dolichonyx (the Bobolink). Biol. Bull. 123:94-104.

Evans, P. R. 1966. Migration and Orientation of Passerine Night Migrants in Northeast England. J. Zool., London 150:319-369.

Farner, D. S. 1955. The Annual Stimulus for Migration: Experimental and Physiological Aspects, pp. 198–237. In A. Wolfson [ed.], Recent Studies in Avian Biology. Urbana, Ill., University of Illinois Press.

Ferguson, D. E., and Landreth, H. F. 1966. Celestial Orientation of Fowler's Toad, Bufo fowleri. Behaviour 26:105-123.

Ferguson, D. E., Landreth, H. F., and Turnipseed, M. R. 1965. Astronomical Orientation of the Southern Cricket Frog, Acris gryllus. Copeia 1965:58-66.

Gilmore, R. M. 1954. Calving of the California Grays. Pacific Discovery 7(3): 13, 15, 30.

Graber, R. R. 1965. Night Flight with a Thrush. Audubon 67:368-374.

Graber, R. R. 1968. Nocturnal Migration in Illinois – Different Points of View. Wilson Bull. 80:36-71.

Graber, R. R., and Cochran, W. W. 1959. An Audio Technique for the Study of Nocturnal Migration of Birds. Wilson Bull. 71:220-236.

Graber, R. R., and Cochran, W. W. 1960. Evaluation of an Aural Record of Nocturnal Migration. Wilson Bull. 72:253-273.

Grant, D., Anderson, O., and Twitty, V. 1968. Homing Orientation by Olfaction in Newts (Taricha rivularis). Science 160:1354-1356.

Griffin, D. R. 1945. Travels of Banded Cave Bats. J. Mamm. 26:15-23.

Griffin, D. R. 1952. Bird Navigation. Biol. Rev. 27:359-393.

Griffin, D. R. 1955. Bird Navigation, pp. 154-197. In A. Wolfson [ed.], Recent Studies in Avian Biology. Urbana, Ill., University of Illinois Press.

Griffin, D. R. 1958. Listening in the Dark. New Haven, Conn., Yale University Press.

Griffin, D. R. 1964. Bird Migration. Garden City, New York, Doubleday and Co., Inc.

Groot, C., and Wiley, W. L. 1965. Time-Lapse Photography of an ASDIC Echo-Sounder PPI-Scope as a Technique for Recording Fish Movements during Migration. J. Fish. Res. Bd. Canada 22:1025-1034.

Hager, W., Jr. 1953. Pacific Salmon. Hatchery Propagation and Its Role in Fishery Management. U.S. Department of Interior, Fish and Wildlife Service, Circular 24.

Hamilton, W. J., III, and Hammond, M. C. 1960. Oriented Overland Spring Migration of Pinioned Canada Geese. Wilson Bull. 72:385-391.

Hasler, A. D. 1960. Guideposts of Migrating Fishes. Science *132*:785-792.
Hasler, A. D. 1966. Underwater Guideposts. Madison, Wisc., University of Wisconsin Press.
Hasler, A. D., Horall, R. M., Wisby, W. J., and Braemer, W. 1958. Sun-orientation and Homing in Fishes. Limnology and Oceanography *3*:353-361.
Hassler, S. S., Graber, R. R., and Bellrose, F. C. 1963. Fall Migration and Weather. Wilson Bull. *75*:56-77.
Heape, W. 1931. Emigration, Migration and Nomadism. Cambridge [England], Cambridge University Press.
Johnson, W. E., and Groot, C. 1963. Observations on the Migration of Young Sockeye Salmon (*Oncorhynchus nerka*) through a Large, Complex Lake System. J. Fish. Res. Bd. Canada *20*:919-938.
Keeton, W. T. 1969. Orientation by Pigeons: Is the Sun Necessary? Science *165*:922-928.
Kemper, C. A. 1964. A Tower for TV: 30,000 Dead Birds. Audubon *66*:86-90.
Kenyon, K. W. 1960. Territorial Behavior and Homing in the Alaska Fur Seal. Mammalia *24*:431-444.
Kenyon, K. W., and Wilke, F. 1953. Migration of the Northern Fur Seal, *Callorhinus ursinus.* J. Mamm. *34*:86-98.
King, J. R., and Farner, D. S. 1963. The Relationship of Fat Deposition to *Zugunruhe* and Migration. Condor *65*:200-223.
King, J. R., Farner, D. S., and Mewaldt, L. R. 1965. Seasonal Sex and Age Ratios in Populations of the White-crowned Sparrows of the Race *gambelii.* Condor *67*:489-504.
Knorr, O. A. 1954. The Effect of Radar on Birds. Wilson Bull. *66*:264.
Kramer, G. 1957. Experiments on Bird Orientation and Their Interpretation. Ibis *99*:196-227.
Kramer, G. 1959. Recent Experiments in Bird Orientation. Ibis *101*:399-416.
Lack, D. 1954. The Natural Regulation of Animal Numbers. Oxford, Clarendon Press.
Lack, D. 1958. Watching Migrant Birds by Radar. Listener *60*(1544):691, 694.
Lack, D. 1962. Radar Evidence on Migratory Orientation. British Birds *55*:139-158.
Landreth, H. F., and Ferguson, D. E. 1967. Newts: Sun-Compass Orientation. Science *158*:1459-1461.
Lidicker, W. Z., Jr. 1962. Emigration as a Possible Mechanism Permitting the Regulation of Population Density Below Carrying Capacity. Amer. Nat. *96*:29-33.
Lincoln, F. C. 1939. The Migrations of American Birds. New York, Doubleday, Doran and Co., Inc.
Lincoln, F. C. 1950. Migration of Birds. U.S. Department of Interior, Fish and Wildlife Service, Circular 16.
Lorz, H. W., and Northcote, T. G. 1965. Factors Affecting Stream Location, and Timing and Intensity of Entry by Spawning Kokanee (*Oncorhynchus nerka*) into an Inlet of Nicola Lake, British Columbia. J. Fish. Res. Bd. Canada *22*:665-687.
Lowery, G. H., Jr. 1951. A Quantitative Study of the Nocturnal Migration of Birds. University of Kansas Publ., Mus. Nat. Hist. *3*:361-472.
Lowery, G. H., Jr., and Newman, R. J. 1955. Direct Studies of Nocturnal Bird Migration, pp. 238-263. *In* A. Wolfson [ed.], Recent Studies in Avian Biology. Urbana, Ill., University of Illinois Press.
Lowery, G. H., Jr., and Newman, R. J. 1966. A Continentwide View of Bird Migration on Four Nights in October. Auk *83*:547-586.
MacGinitie, G. E., and MacGinitie, N. 1949. Natural History of Marine Animals. New York, McGraw-Hill Book Co., Inc.
Matthews, G. V. T. 1955. Bird Navigation. Cambridge [England], Cambridge University Press.
Medway, Lord, and Nisbet, I. C. T. 1968. Bird Report: 1966. Malayan Nature J. *21*:34-50.
Mewaldt, L. R. 1964. California Sparrows Return from Displacement to Maryland. Science *146*:941-942.
Mewaldt, L. R., Kibby, S. S., and Morton, M. L. 1968. Comparative Biology of Pacific Coastal White-crowned Sparrows. Condor *70*:14-30.
Mewaldt, L. R., Morton, M. L., and Brown, I. L. 1964. Orientation of Migratory Restlessness in *Zonotrichia.* Condor *66*:377-417.

Mewaldt, L. R., and Rose, R. G. 1960. Orientation of Migratory Restlessness in the White-crowned Sparrow. Science *131*:105–106.

Miskimen, M. 1955. Meteorological and Social Factors in Autumnal Migration of Ducks. Condor 57:179-184.

Moreau, R. E. 1966. The Bird Faunas of Africa and Its Islands. New York, Academic Press, Inc.

Moreau, R. E. 1967. Water-Birds over the Sahara. Ibis *109*.232-259.

Mueller, H C , and Berger, D. D. 1967. Fall Migration of Sharp-shinned Hawks. Wilson Bull. 79.397-415

Murie, O. J. 1935. Alaska Yukon Caribou. U.S. Department of Agriculture, Bur. Biol. Surv., N. Am. Fauna, 54.

Murie, O. J. 1951. The Elk of North America. Harrisburg, Pa., The Stackpole Co., and Washington, D.C., Wildlife Management Institute.

Murray, B. G., Jr. 1965. On the Autumn Migration of the Blackpoll Warbler. Wilson Bull. 77:122-133.

Murray, B. G., Jr. 1966. Blackpoll Warbler Migration in Michigan. Jack-Pine Warbler *44*:23-29.

Murray, B. G., Jr. 1967. Dispersal in Vertebrates. Ecology *48*:975-978.

Myers, G. S. 1951. Fresh-water Fishes and East Indian Zoogeography. Stanford Ichthyol. Bull. *4*:11-21.

Neave, F., Ishida, T., and Murai, S. 1967. Salmon of the North Pacific Ocean. Part VII. Pink Salmon in Offshore Waters. Internat. N. Pac. Fish. Comm. Bull. *22*:1 39.

Nice, M. M. 1937. Studies in the Life History of the Song Sparrow. I. A Population Study of the Song Sparrow. Trans. Linn. Soc. New York *4*.1-247.

Nisbet, I. C. T., and Drury, W. H., Jr. 1967. Orientation of Spring Migrants Studied by Radar. Bird-Banding *38*:173-186.

Odum, E. P. 1960. Lipid Deposition in Nocturnal Migrant Birds. Proc. XII Internat. Ornith. Congr.:563–576.

Odum, E. P., and Connell, C. E. 1956. Lipid Levels in Migrating Birds. Science *123*:892-894.

Orr, R. T. 1070. Animals in Migration. New York, Macmillan Co.

Orr, R. T., and Poulter, T. C. 1965. The Pinniped Population of Año Nuevo Island, California. Proc. Calif. Acad. Sci. (4) *32*:377-404.

Pearson, O. P. 1961. Flight Speeds of Some Small Birds. Condor *63*:506-507.

Rice, D. W. 1956. Dynamics of Range Expansion of Cattle Egrets in Florida. Auk *73*:259-266.

Rowan, W. 1925. Relation of Light to Bird Migration and Developmental Changes. Nature *115*:494-495.

Rowan, W. 1926. On Photoperiodism, Reproductive Periodicity, and the Annual Migrations of Birds and Certain Fishes. Proc. Boston Soc. Nat. Hist. *38*:147-189.

Rowan, W. 1929. Experiments in Bird Migration. I Manipulation of the Reproductive Cycle: Seasonal Histological Changes in the Gonads. Proc. Boston Soc. Nat. Hist. *39*:151-208.

Rowan, W. 1932. Experiments in Bird Migration. III. The Effects of Artificial Light, Castration and Certain Extracts on the Autumn Movements of the American Crow (*Corvus brachyrhynchos*). Proc. Nat. Acad. Sci. *18*:639-654.

Ryder, J. P. 1967. The Breeding Biology of Ross' Goose in the Perry River Region, Northwest Territories. Canadian Wildl. Serv. Rept. Ser., 3, Ottawa.

Sauer, E. G. F. 1958. Celestial Navigation by Birds. Sci. Amer. *199*(2):42-47.

Sauer, E. G. F. 1963. Golden Plover Migration, Its Evolution and Orientation. Proc. XVI Internat. Congr. Zool. *4*:380-381.

Schevill, W. E., and Watkins, W. A. 1965. Underwater Calls of *Trichechus* (Manatee). Nature *205*:373-374.

Schmidt-Koenig, K. 1960. The Sun Azimuth Compass: One Factor in the Orientation of Homing Pigeons. Science *131*:826-828.

Smith, E., and Goodpaster, W. 1958. Homing in Nonmigratory Bats. Science *127*·644.

Stoddard, H. L., Sr. 1962. Bird Casualties at a Leon County, Florida TV Tower, 1955–1961. Tall Timber Research Sta., Tallahassee, Fla., Bull. 1.

Stoddard, H. L., and Norris, R. A. 1967. Bird Casualties at a Leon County, Florida TV Tower: An Eleven-Year Study. Tall Timbers Research Sta., Tallahassee, Fla., Bull. 8.

Swarth, H. S. 1920. Revision of the Avian Genus *Passerella*, with Special Reference to the Distribution and Migration of the Races in California. Univ. Calif. Publ. Zool. *21*:75-224.

Talbot, L. M., and Stewart, D. R. 1964. First Wildlife Census of the Entire Serengeti-Mara Region, East Africa. J. Wildlife Mgmt. *28*:815-827.

Tedd, J. G., and Lack, D. 1958. The Detection of Bird Migration by High-power Radar. Proc. Royal Soc. (B) *149*:503-510.

Terres, J. K. 1956. Death in the Night. Audubon *58*:18-20.

Twitty, V. C. 1959. Migration and Speciation in Newts. Science *130*:1735-1743.

Twitty, V., Grant, D., and Anderson, O. 1964. Long Distance Homing in the Newt *Taricha rivularis*. Proc. Nat. Acad. Sci. *51*:51-58.

Walter, H. 1968. Falcons of a Princess. Pacific Discovery *21*(3):2-9.

Wisby, W. J., and Hasler, A. D. 1954. Effect of Olfactory Occlusion on Migrating Silver Salmon (*O. kisutch*). J. Fish. Res. Bd. Canada *11*:472-478.

Wolfson, A. 1942. Regulation of Spring Migration in Juncos. Condor *44*:237-263.

Wolfson, A. 1952a. Day Length, Migration, and Breeding Cycles in Birds. Scientific Monthly *74*:191-200.

Wolfson, A. 1952b. The Occurrence and Regulation of the Refractory Period in the Gonadal and Fat Cycles of the Junco. J. Exper. Zool. *121*:311-325.

Woodbury, A. M. 1941. Animal Migration—Periodic Response Theory. Auk *58*:463-505.

Yeagley, H. L. 1951. A Preliminary Study of a Physical Basis of Bird Navigation. Part II. J. Appl. Physics *22*:746-760.

*Chapter Eleven*

# DORMANCY

## DEFINITION

The ability of some living things to slow down the normal processes of metabolism and growth under certain circumstances and thereby become dormant is a widespread phenomenon. In general, it is associated with unfavorable environmental conditions of a temporary or seasonal nature, such as cold, drought, or aridity, and therefore provides a means of survival for the organism. Among vertebrates dormancy has been observed in certain freshwater fishes, many amphibians and reptiles, two species of birds, and a number of mammals.

The fact that there are endotherms as well as ectotherms that periodically become dormant has led to a certain amount of confusion in terminology. For this reason the term *heterotherm* is frequently applied to those so-called "warm-blooded" vertebrates that at times behave like "cold-blooded" vertebrates and lose their powers of thermoregulation.

Since dormancy occurs most frequently in winter, it has commonly been referred to as hibernation at that season. Not too long ago this term would simply have been defined as "winter-sleep," which is a literal translation of the German equivalent *Winterschlaf.* With increased knowledge, however, problems in semantics have arisen and we are confronted with various interpretations of a word that at one time seemed to present no great problem.

For many generations the classical example of an animal that hibernated was the bear, because in many parts of the world where it is cold in winter these big carnivores are known to go into dens of one sort or another and sleep during that season. Physiologists, however, complicated things considerably by concluding that *hibernation* involves a state of torpor in which the metabolism is so greatly lowered that the body temperature approximates that of a cold environment. This sort of definition immediately excluded bears from the category of hibernating animals, as it was known that their body temperature did not drop markedly during their period of winter inactivity. Not only did such a definition exclude bears but it necessitated a re-evaluation of the word hibernation. The word itself comes from the Latin *hibernare*, which means to pass the winter. Some animals, however, retreat to cool places in the heat of summer, become torpid, and have a reduced body temperature and reduced metabolic rate, thus actually fulfilling all the necessary physiological qualifications for hibernation, although the term *estivation* is applied at this season of the year. Other animals, such as certain kinds of bats in temperate regions, might be said to partially hibernate many days during the summer, even though they be active at night. In the cool morning hours, after they have returned to their daytime roosts, they may become semitorpid with a greatly lowered respiration rate and a body temperature that is fairly close to that of their environment. There are, therefore, various kinds of torpidity, or dormancy.

In winter, when food is scarce and temperatures are low, the advantages of hibernation are obvious. In summer, high air temperatures on the surface of the ground and periods of food scarcity may be circumvented by summer sleep or estivation in cool underground burrows. With regard to bats, whose metabolic rate is very high during periods of activity, the daily lowering of the body temperature and respiration rate is certainly a means of conserving body energy.

Studies by Svihla and Bowman (1954) on a yearling bear cub in Alaska showed that it slept curled up from November to February, during which time the environmental temperature was low. The bear was sensitive to noise and, if disturbed, would slowly raise its head, but did not appear to see. Although its body temperature did not drop markedly (rectal, 95°–96° F.; oral, 95° F.), the respiration rate was reduced to two or three times per minute. No food was taken during these three months and there was complete cessation of urinary excretion and defecation. The loss in body weight over this period of 90 days amounted to about 25 per cent. The authors concluded, on the basis of their observations, that bears do hibernate, since their metabolic activities are reduced, even though they only partially lose their powers of sensibility and locomotion. They, therefore, suggested the advisability of recognizing two types of hibernation: "partial dormancy or torpor," as exemplified by bears, and "complete dormancy or torpor," such as that exhibited by certain ground squirrels.

One might suggest a somewhat different and more inclusive arrangement in which there are three major categories: winter dormancy (hibernation) which may be partial or complete, summer dormancy (estivation), and daily dormancy (diurnation).

## HIBERNACULA

Strictly speaking, the word *hibernaculum* refers to a winter resting place. However, for lack of a simple, more inclusive term it is being used here in a broader sense to cover those situations used by vertebrates during periods of dormancy. Since animals are at the mercy of their enemies at such times, it is necessary that most hibernacula afford suitable protection against predators and disturbances.

### FISHES

Although the blackfish (*Dallia pectoralis*), a freshwater inhabitant of arctic North America, has been the subject of much discussion, the fact remains that it passes many months at the bottom of bodies of water that are almost completely frozen. Although these fish may not be completely surrounded by a frozen medium, they are certainly provided with protection and freedom from disturbances by many feet of ice above.

Estivation in mud cells is well known in lungfishes in certain regions where periods of drought occur. There is some indication that the bowfin (*Amia calva*) of North America, under certain circumstances, may employ the same method to survive during dry spells. There is one record of a bowfin found in a spherical chamber about 1/4 mile from the Savannah River in a Georgia swamp. The area showed signs of having been flooded some time before, although the surface of the ground was dry and cracked. The chamber itself was 4 inches beneath the top of the ground and was 8 inches in diameter. Its walls were moist. The fish was active and moving when found. In fact the noise of its movements led to its discovery.

### AMPHIBIANS AND REPTILES

Since a great many North American amphibians and reptiles pass the winter in hibernation and a number of species estivate in summer, particularly in regions where it is hot and arid, it is not surprising that a wide variety of hibernacula are used. Many frogs hibernate in water, often in mud or debris at the bottom of ponds and lakes. Other species remain in pockets of soil beneath forest litter.

The green frog (*Rana clamitans*) has been observed in both situations in winter. Many toads and salamanders use deserted rodent burrows for this purpose. Spadefoot toads (*Scaphiopus*), in many instances at least, appear to dig their own burrows, making use of the hind feet and literally backing into the ground, where they may remain for some months.

In winter when ground temperatures are freezing, or in summer when the humidity is low, many salamanders move below the surface of the soil or into various situations which offer protection against either freezing or desiccation. Some make use of cracks or fissures in the soil or the burrows of other animals, some enter rotten logs or stay beneath bark, others have been found in caves and mine tunnels. Members of the plethodontid genus *Ensatina* frequently find refuge in or under nests of the dusky-footed wood rat (*Neotoma fuscipes*).

Animal burrows, rock crevices, caves, and mine tunnels, even deserted ant hills are often used by snakes during periods of quiescence in winter or summer. It has been suggested that the presence of available hibernacula may be a critical factor controlling the abundance of ectotherms in certain regions. The joint occupancy of such situations by a number of individuals is of fairly common occurrence.

Joint occupancy sometimes is not limited to members of one species or even one class. There is a record of 257 snakes consisting of eight *Thamnophis radix*, 101 *Storeria occipitomaculata*, and 148 *Opheodrys vernalis* found in a single ant hill in Manitoba. Another ant hill excavated in Michigan in February was found to contain 62 snakes representing seven species and 15 amphibians belonging to three species. During the course of excavating in Mercer County, Pennsylvania, early in January 1939, a hibernaculum was discovered containing 21 individuals representing seven species of reptiles and two species of amphibians. Extended studies over a period of years have shown that individual reptiles often return to the same den to hibernate year after year.

Many turtles hibernate in water. Recent studies on a marsh in Michigan have also revealed wintering garter snakes in crayfish burrows, where they appeared to be completely submerged.

## BIRDS

Although laboratory studies have shown that the trilling nighthawk (*Chordeiles acutipennis*) is capable of hibernation, only the poor-will (*Phalaenoptilus nuttallii*) has been found dormant in the wild. In one instance, a poor-will, presumably the same individual although it was not banded until the second year, was found for three successive years in a torpid condition in winter in the Chuckawalla Mountains of the Colorado Desert of California. The hibernaculum consisted of a crypt on the south-facing side of a granite boulder (Fig. 11.1). The lower portion of the cavity in which the bird stayed was exposed to sunlight for about three and one-half hours daily, although the rock was cooler than the air temperature during most of the day. Another hibernating poor-will was found in midwinter northwest of Tucson, Arizona, in a dormant condition. This bird was on the ground under a lower leaf of an agave plant. There are at least two other records of poor-wills that were found in a condition resembling dormancy. One was half buried beneath a pine limb and the other was in a small silt rivulet.

Hummingbirds are members of a New World family and are primarily tropical and subtropical in distribution. They have a very rapid wingbeat and a small body, which in a warm-blooded animal means a high metabolic rate. In equatorial and subequatorial regions

**FIGURE 11.1.** Poor-will (*Phalaenoptilus nuttallii*) in hibernation crypt in face of granite rock, Chuckawalla Mountains, Riverside County, California. (Photograph by Kenneth Middleham. From Jaeger, E. C.: Further Observations on the Hibernation of the Poor-will. Condor, *51*:106, 1949.)

where the ambient temperature is high, the problem of heat loss is small. Some species, however, range into temperate and north temperate regions in North America, and in South America a few species range up to high elevations in the Andes. These birds may reduce their body temperature where the nights are cold and go into a nocturnal torpidity which greatly reduces their energy outlay and prevents them from starving during overnight fasting. A species known as Estella's hummingbird (*Oreotrochilus estella*) occurs in the Altiplano of Peru, a high plateau in the Andes where elevations are mostly over 12,500 feet. To avoid the freezing night temperatures Estella's hummingbirds sleep in caves where the temperatures are fairly high and constant. Torpidity in a subfreezing environment would be fatal to a small bird.

## MAMMALS

Black bears frequently select hollows in the bases of trees, burned out logs, crevices under large rocks, or in piles of boulders to pass their period of partial winter dormancy. No special lining or bedding material is brought into these cavities.

Ground squirrels, on the other hand, usually have rather elaborate nests in which the period of dormancy is passed (Fig. 11.2). Detailed studies of the Columbian ground squirrel (*Citellus columbianus*) have shown that the shaft off which the hibernaculum is located is plugged off from the summer tunnel system by dirt. The den, a spherical chamber above the shaft, is carefully lined with dry grass and shredded material and a drain is excavated below the nest. The dirt from the latter is used to plug the entrance to the hibernaculum.

Woodchucks (*Marmota monax*) reportedly desert meadows in the autumn and move to burrows or other sites in hedgerows, woods, or stony inclines with a southern exposure. Some individuals have been observed hibernating in haystacks rather than dens.

Meadow jumping mice (*Zapus hudsonius*) seem to show a preference for elevated situations in which to winter. Hibernacula, consisting of round nests about the size of a baseball, have been found in hillocks of earth and, in several instances, in ash piles. The selection of these sites would seem to reduce the danger from flooding.

Eastern chipmunks (*Tamias striatus*) have been found curled up in nests of leaves in burrows underground or beneath boulders when dormant. Sometimes stored food may be found nearby. Since these animals do not become markedly fat during the fall of the year and yet do store up quantities of food at that season, it is possible that

**FIGURE 11.2.** Photograph of a captive hibernating golden-mantled ground squirrel (*Citellus lateralis*). Note how the head is tucked under the body and the tail is curled over the head. At the time this picture was taken the environmental temperature was 0° C. and the body temperature of the animal 2° C. The body temperature of these squirrels when active is about 37° C. (Photograph courtesy of Dr. E. T. Pengelley.)

even though they may become dormant in certain regions in winter they may also feed during periods of wakefulness.

There is relatively little information on the hibernating quarters of western chipmunks (genus *Eutamias*), although a few occupied winter nests have been found. The hibernacula of arboreal species have been found generally to be balls of dry grass, leaves, mosses, or lichens situated in holes in stumps. The torpid animals were curled up in the center of these masses of nest material.

Much has been learned in recent years regarding situations selected for hibernation by certain North American species of bats. In the eastern part of the continent most colonial bats aggregate in caves or mine tunnels in winter, although temperature and humidity requirements seem to vary with the species. The big brown bat (*Eptesicus fuscus*) is most tolerant to cold, dryness, and draft. In hibernating caves or tunnels these bats are frequently found clustered together, and then often in crevices close to the entrance where there is considerable air movement. Air temperature next to torpid big brown bats has been recorded as low as 23° F.

The least myotis (*Myotis subulatus*) is almost as tolerant of cold and dryness as the big brown bat, while the Keen myotis (*Myotis keenii*) shows preference for protected crevices where it is cool and moist. The little brown myotis (*Myotis lucifugus*) and the eastern pipistrel (*Pipistrellus subflavus*) both select situations where it is relatively warm, humid, and draftless. Most caves or tunnels show an increase in temperature and humidity as one proceeds from the entrance inward; consequently this frequently governs the relative positions of different species of hibernating bats. The fact that bats may periodically move to different parts of the same cave or to other caves in winter may be a result of changes in temperature and humidity.

Except for Townsend's big-eared bat (*Plecotus townsendii*), relatively little is known of the hibernacula of colonial bats in western North America. Scattered individuals or even small groups of most species have been found in caves, mine tunnels, crevices in rocks, cellars, and beneath loose boards or other objects hanging in abandoned buildings. In all such situations the temperature is usually low but above freezing and the atmosphere is not too dry. Warm, dry situations such as are often selected for summer roosts are rarely

FIGURE 11.3.  A group of Townsend's big-eared bats (*Plecotus townsendii*) in Jewel Cave, South Dakota. The thermometer indicates a temperature of 36° F. (Photograph courtesy of Charles E. Mohr.)

occupied. Hibernating big-eared bats have been found in aggregations numbering as high as 201 in a single cave in California. The temperature in caves occupied by this species has been observed to range from 28.5° F. to 55° F. Individuals of this species sometimes hang alone, but more often are found in clusters hanging by their feet from the ceiling or walls. The long ears are usually coiled when these animals are hanging alone, although they may be erect when in clusters.

No such large aggregations of other kinds of bats have yet been found in hibernation along the Pacific coast. From the evidence gathered to date, it appears likely that after the summer colonies break up in the fall in this region the members move singly or in small groups to scattered situations in which the winter is passed.

## Ectotherms

Some terrestrial ectotherms appear to have considerably more tolerance to lower temperatures than others, although on careful analysis this is largely a result of ability on the part of the particular organism to control its environment. Certain plethodontid salamanders remain active when the temperature is close to freezing. However, such species usually inhabit decaying organic litter, which would have a moderating influence on the environmental temperature and also house insect food. Some high mountain lizards that are able to move at remarkably low temperatures may appear in the early morning when it is still below freezing. By properly orienting themselves in relation to the sun, thus taking advantage of solar radiation and also reflected heat from the substratum, they may rapidly raise the body temperature many degrees above that of the air. In general, however, most terrestrial ectotherms are inactive when the environmental temperature approaches freezing.

Much remains to be found out concerning the minimum temperatures that ectotherms can withstand and yet live. There are many reports of blackfish, frozen in the fall of the year in Alaska and kept at low temperatures for weeks or months, that showed marked signs of life on being thawed out. L. M. Turner (1886) described the freezing of these fish for dog food by the natives and the amazing vitality of the fish to withstand such treatment. The present author has had reliable persons who have been residents of Alaska attest to the veracity of such reports. Several experimental studies, however, have failed to substantiate the ability of blackfish to withstand freezing, although the results for the most part are not completely conclusive because of the techniques used. Several important factors must be considered in this connection. Protoplasm has a lower freezing point than water, so that encasement in ice does not necessarily mean that a living organism is frozen. The author has seen leopard frogs (*Rana pipiens*) that have been left several days in a laboratory refrigerator survive encasement in a thin layer of ice. It must be remembered also that metabolic activity, even though greatly reduced in hibernating ectotherms, does not cease. Garter snakes have been known to withstand temperatures of −2° C. for prolonged periods, but the cloacal temperature of dormant snakes is higher than that of the environment. Another factor that must be considered is the necessity perhaps in many species of gradual physiological conditioning to low temperatures. This has even been demonstrated as necessary for endotherms that remain active in a cold winter environment. Experiments involving gulls have shown that birds removed from a warm environment and suddenly placed out-of-doors in freezing weather will suffer from frozen extremities, while those that are conditioned to cold will not.

Further studies will no doubt solve the perplexing problem presented by the interesting Alaskan fish, which has been a subject of argument from the camps of sourdoughs to scientific conclaves for several generations.

Summer heat and aridity cause many terrestrial ectotherms to estivate or at least become quiescent in underground hideouts. Spadefoot toads (*Scaphiopus couchi*) have recently been discovered on the Colorado Desert of California, where summer temperatures are up to 50° C. and the average rainfall is between 2 and 3 inches. During the dry parts of the year they are believed to remain buried in permanently moist layers of sand fairly close to potential pools in washes. The loss of body moisture in adults is partly reduced by the retention of old layers of skin on the body. Water loss is also reduced by the development of high osmotic concentrations of blood and lymph during dormancy in these amphibians.

Rattlesnakes may be fairly abundant in certain areas in the desert in spring, but as summer sets in and the daily temperatures rise fewer and fewer individuals can be found. This is true of many other snakes and certain species of lizards. The decrease in humidity and scarcity of food in summer, particularly in western North America, is also correlated with the disappearance of many salamanders from the surface of the ground. To avoid desiccation as well as starvation many species go down into the soil to a depth of several feet or take refuge beneath piles of debris or wood rat nests and do not appear again until the fall rains begin. During these periods of estivation there may be considerable loss of body water which appears to be rapidly replaced upon emergence in wet weather. A salamander of the genus *Ensatina* has been observed to increase its body weight about 40 per cent within a few hours as a result of absorption of water through the skin when placed in a moist, but not wet, environment.

There are certain vertebrates that can best be described as obligatory hibernators. For example, studies on the flat-tailed horned lizard (*Phrynosoma m'calli*), a desert reptile of southwestern United States, have shown that this species will become dormant in winter even under laboratory temperatures. In summer individuals of this species fail to go into dormancy even when the ambient temperature is lowered to 15°C. (Mayhew, 1965*b*). Certain squirrels have also been found to become partially dormant in winter under indoor laboratory temperatures.

Rising temperature in the late winter or spring unquestionably is important as a stimulus rousing many ectotherms from hibernation. There are other ectotherms in which changes in temperature seemingly have no effect on dormancy. Mayhew's (1963) studies on the granite spiny lizard (*Sceloporus orcutti*) of western North America have shown that once these lizards go into hibernation in the fall of the year they remain in this condition until time of emergence in the spring. This suggests that some other factor may trigger their return to an active state.

Water may also be of some significance in connection with the appearance of certain amphibians such as spadefoot toads in the spring. There is a direct correlation between the appearance of many salamanders and the first fall rains in western North America. Although some species may estivate several feet below the surface of the ground, a light rain that penetrates no more than half an inch will frequently be a sufficient stimulus to cause many individuals to come to the surface. It has been suggested that sensitivity to increased humidity may be responsible for these upward movements after the first light rain.

## HETEROTHERMS

Although environment alone does not provide a satisfactory reason for heterotherms becoming dormant, there is no question but that it acts as a trigger mechanism. However, it is impossible to assume that any one factor is the stimulus even within the same species. Under essentially identical external conditions some animals may become dormant while other members of the same species do not. This has been pointed out by Linsdale (1946) in his study on the California ground squirrel (*Citellus beecheyi*).

In view of the poorly developed heat regulating mechanism of heterotherms, it is obvious that a lowered environmental temperature must be considered of prime importance in initiating and maintaining dormancy. It is true that some animals go into this condition in the heat of the summer and that severe cold may either kill or awaken hibernators. Nevertheless, animals that estivate in summer usually go into burrows or dens where the environmental temperature is decidedly low as compared with that above ground. The temperature of burrows of dormant animals is rarely ever higher than about 70° F., which is low compared with average body temperatures of active heterotherms. Furthermore, it has been shown that mammals generally die if the body temperature gets below 32° F. It is obvious, therefore, that if the ambient temperature gets too low death will occur if the animal does not awaken.

Hibernation has been artificially induced in ground squirrels at all times in the year, but it is more difficult to induce in spring and early summer than at other seasons and the mortality is higher. In the fall of the year the gradual cooling of the ground seems to be an aid in preparing squirrels for hibernation.

There are a number of mammals, such as skunks and red squirrels, that are less active during periods of extreme cold, but it would appear that they merely sleep or rest until the climate becomes more favorable. In other groups, such as chipmunks, both *Tamias* and *Eutamias,* there is considerable variation with species and locality as regards hibernation. In regions where the climate is temperate chipmunks are active all winter. Even in certain areas where snow covers the ground some individuals may be observed on nice days. Where winter activity is not apparent above ground, excavation has shown that some individuals are wide awake and have stores of food in their dens, while others are completely torpid and no accumulated food has been found nearby.

Some investigators have attributed greater significance to lack of food than to temperature as a factor initiating dormancy. It is true that, for some species, summer estivation occurs when the heat has dried up much of the food and that hibernation commences in the fall or early winter when food again becomes scarce. There are other species though, like marmots, that become torpid when food is still abundant.

The role of food as a factor influencing dormancy appears to be meaningful only when the condition of the animal is taken into consideration. Fat is a predisposing factor and lack of surplus fat on the body has a delaying action. Most rodents and bats that hibernate show a marked increase in body weight, due to accumulated fat, in the fall of the year immediately preceding the period of winter dormancy. Individual ground squirrels that are active above ground in late fall, when other members of the same species have already entered hibernation, are generally thin and apparently trying to acquire the necessary surplus fat that is needed as a reserve for the long winter sleep.

As long ago as 1807 the possibility was suggested that confined air, that is, air in which there is an excess of carbon dioxide and a deficiency of oxygen, might be a factor in inducing dormancy. The fact is that a great many hibernators do live in burrows underground where there is a confined atmosphere and, as we shall see, the reduced metabolic activity of dormant animals greatly lowers the oxygen consumption.

Darkness is generally associated with the environment of torpid animals, but this probably has little effect on their physiology and merely results from the selection of safe refuges during such vulnerable periods. Experiments with dormant ground squirrels have shown that they do not awake when the illumination is increased. Certain insectivorous bats in captivity will become torpid during the day in a cool, undisturbed environment, irrespective of light, although they will try to select the darkest part of the cage. Disturbance and noise will tend to awaken dormant animals, but this rarely occurs in nature, since most animals select situations where they are not likely to be bothered by such environmental factors while dormant.

## PHYSIOLOGICAL FACTORS ASSOCIATED WITH DORMANCY

### TEMPERATURE

The most obvious physiological change in completely dormant endotherms is the lowering of the body temperature to essentially that of the environment. In other words they behave like ectotherms. This is true not only of insectivorous bats and rodents but also of the two species of birds known to become completely dormant, the poor-will and the trilling nighthawk. Furthermore, almost all warm-blooded vertebrates that do become completely dormant exhibit considerable daily variation in body temperatures outside the periods of extended dormancy.

The inadequacy in the thermoregulatory mechanism is most marked in certain microchiropteran bats in which the daily temperature when resting approximates the ambient temperature. Diurnation, as far as known, does not occur in other mammals, although the body temperatures of active squirrels have been found to range from 35° C. to 39° C. in a warm room and from 31° C. to 36° C. when exposed to cold. The extremes recorded in active woodchucks are 34.9° C. and 40° C. Fluctuations in body temperatures in these animals are produced not only by changes in the surrounding temperature but also by activity.

Among birds diurnation has been observed in certain kinds of hummingbirds. Both Anna (*Calypte anna*) and Allen (*Selasphorus sasin*) hummingbirds were shown by Pearson (1950) to have a markedly reduced body temperature and metabolic rate at night (Fig. 11.4). In captivity, recovery from nocturnal torpidity occurred before daybreak. Marshall (1955) found that resting poor-wills might have a temperature of 37.5° C. but after five minutes of flight it would rise to 40.5° C.

The degree to which the body temperature is lowered is correlated with the degree of torpidity. Bats in deep dormancy have body temperatures that range from 2° C. to 10° C., while those in light dormancy may range from 10° C. to 28° C.

The rate at which the body temperature of animals drops when they are going into dormancy is much slower than the rate of increase when they are awakening. Bats placed in a cold environment may require some hours before they become torpid and their body

**FIGURE 11.4.**  Hummingbirds have a high metabolic rate during periods of activity, but at night it drops to such a low level in some species that they become torpid. Photograph of a female Allen hummingbird (*Selasphorus sasin*) feeding young.

temperature approaches the ambient temperature. However, they will awaken and increase their body temperature to the point where they can fly in a matter of a few minutes. The body temperature of ground squirrels has been observed to drop at 0.07° C. per minute during entrance into hibernation. However, it will rise at four or five times this rate during the awakening process.

During the cooling process in ground squirrels the drop in body temperature is slower in the beginning than during the later stages of torpidity. On the other hand during the awakening process, whether undisturbed or initiated by some external disturbance, a rise in body temperature, heart rate, and respiration starts slowly and then increases after 40 to 60 minutes, either in a cold or moderately warm environment. At high environmental temperatures the body temperature increases more rapidly than the heart rate. Furthermore, large animals warm up more slowly than small animals.

Although it is generally said that the body temperature of completely dormant heterotherms is approximately that of the environment, the fact remains that it is usually a bit higher. In ground squirrels it generally ranges 1° C. to 3° C. above the ambient temperature. The body temperature of bats has been found to be 0.5° C. above an ambient temperature of 1.3° C., and 2° C. above an ambient temperature of 35.5° C.

## METABOLISM

Inadequate thermoregulation in heterotherms implies a deficiency in ability to regulate body metabolism so as to maintain a constant body temperature. At reduced external temperatures, therefore, such animals are unable either to raise or to maintain the metabolic rate at a level sufficient to compensate for the heat loss. This reduced metabolic rate is reflected in the reduction in oxygen consumption.

In hibernating bats it has been found that exhalation of carbon dioxide and intake of oxygen is a little over 1 per cent of the amounts recorded when the animals are active. Experiments performed on bats which were subjected to a reduction of the ambient temperature from 41.5° C. to 2.0° C. showed that the greatest decrease in oxygen consumption occurred between 20° C. and 10° C. However, as the ambient temperature approaches freezing the metabolic rate increases. Thus the oxygen consumption of hibernating bats at 0.5° C. is quadruple that at 2.0° C.

Heterotherms expend a comparatively large amount of energy in the process of awakening from dormancy. It has been found that ground squirrels expend 70 per cent more energy in the period of awakening from torpidity than they do in an equal time when awake. The rapid rise in temperature of awakening bats has generally been attributed to contractions of the heart and thoracic muscles and trembling movements.

The respiratory quotient ($CO_2/O_2$) of dormant mammals is very low, although some investigators have objected to the use of the values obtained as an accurate index of metabolic activity because of the enhanced solubility of the blood for carbon dioxide at low temperatures. However, experiments so conducted as to allow the blood gases to reach equilibrium have shown that the low quotients are not the result of enhanced solubility, since it is impossible for the blood to retain sufficient carbon dioxide to account for them. Since the enhanced solubility of the blood is not sufficient to compensate for the predicted extra amount of dissolved carbon dioxide, the reduced amount of the latter must be due to an actual decrease in production. A low respiratory quotient suggests partial oxidation, probably of fats since they contain less oxygen than carbohydrates.

The reduction of the rate of exchange of respiratory gases in dormant mammals is accomplished by a slowing and irregularity of the respiratory movements. Hibernating ground squirrels have been observed to respire at rates varying from 30 times per hour to as much as 83 times per hour. The latter was recorded in *Citellus lateralis* in captivity, when the ambient temperature was 4° C. and the animal had a rectal temperature of 5.2° C. Breathing in this instance was irregular. Dormant chipmunks of the species *Eutamias speciosus*, on the other hand, with a rectal temperature of 4.8° C. in an ambient temperature of 4° C. respired regularly at the rate of 138 times per hour. In dormant bats periods of apnea of up to four minutes have been recorded. However, the respiratory center remains sufficiently sensitive in its control over respiratory movements to maintain the blood concentration of carbon dioxide within bounds necessary for life. In torpid animals the increased solubility of the blood at lowered temperatures for oxygen and the decreased metabolic rate reduce the inspiration requirement of this gas. This is reduced to the extent that a remarkable ability to withstand oxygen deprivation or exposure to noxious gases has been observed. Hibernating bats that were awakening have been recorded as using up all the oxygen in a closed chamber without suffocating, while active individuals would become asphyxiated in an atmosphere of reduced oxygen. Similarly, it has been shown that some hibernating mammals can survive in atmospheres of pure carbon dioxide or pure nitrogen for short periods of time. It appears that under such circumstances there is sufficient dissolved oxygen in the blood to tide the animals over.

In most mammals that periodically become torpid there is an accumulation of fat in the fall of the year, principally under the skin, in the mesenteries, in the mesovaria, and in the mesorchia. This fat provides a source of energy for the body during its long period of quiescence. The amount of fat accumulated and the exact time are seemingly dependent upon the degree of hibernation and the time it is entered. In golden-mantled ground squirrels (*Citellus lateralis*) in the Sierra Nevada of California, there is a threefold increase in total body fat from August until just prior to hibernation in September. Chipmunks, both *Eutamias speciosus* and *E. amoenus*, living in the same region, show no increase in body fat until October, and it is proportionately less in relation to basic body weight than in the golden-mantled ground squirrels. Chipmunks enter hibernation later than the ground squirrels and also store food for winter. It is believed that they wake frequently and consume these stores, thus not requiring as much fat accumulation in the fall.

It has been calculated that ground squirrels may lose 30 to 45 per cent of their weight during five months of hibernation. The weight loss in woodchucks during four to five months' hibernation is estimated to amount to between one-third and one-half of the fall weight.

Pallid bats have been found to lose about 25 per cent of their weight during four months of dormancy. In all such animals the blood sugar level is considerably lower during torpidity than when they are awake. Furthermore, although it is likely that the fat is converted into glucose there is little evidence that it is stored as glycogen. Recent studies on hibernating arctic ground squirrels have shown that, except for cardiac muscle, most tissues of the body are characterized by a reduction in the glycogen content.

The fact that hypoglycemia is of common occurrence in most mammals during dormancy has led some investigators to suggest that lowered blood sugar plays an important role in the assumption and maintenance of torpidity. This seems doubtful, however, in view of the fact that injections of glucose into dormant squirrels have been found to have no more wakening effect than normal saline.

The hamster, which has been the subject of considerable experimentation, is peculiar among hibernating rodents that have been studied in that it exhibits an increase rather than a decrease in blood sugar during its short periods of dormancy. Unlike those rodents that accumulate sufficient surplus fat in the fall to tide them over the long period of winter dormancy, the hamster renews its energy sources every few days by awakening and eating stored food. In fact hibernating hamsters will die if deprived of food during their short waking intervals.

## CIRCULATORY SYSTEM

The reduced expenditure of energy during dormancy is accompanied by a slowing of the heart beat and therefore the circulation of the blood. The pulse rate of marmots has been noted to decrease from 90 beats per minute when awake at a room temperature of 19° C. to 10 to 12 beats per minute when torpid. The pulse of an actively struggling ground squirrel has been recorded at 506 beats per minute, while that of a dormant individual may be only 5.5 beats per minute.

The heart of dormant mammals appears to have a remarkable capacity for maintaining its beat. After decapitation the heart of a torpid marmot has been observed to continue beating for three hours, and in ground squirrels the heart has been found to continue beating an hour after the cessation of respiration.

Despite the slowing down of circulation in dormant heterotherms there is no complete cessation of the peripheral vascular flow. To prevent clotting as a result of the slow flow of blood there is a considerable decrease in the prothrombin content of the blood. This naturally prolongs the clotting time.

Several other marked changes occur in the blood of dormant mammals. There is a decrease in the plasma content and a concentration of red blood cells. In dormant ground squirrels the erythrocyte count has been found to be 12,000,000 per cubic centimeter as contrasted with 6,700,000 in active individuals. This is thought to account for the greater oxygen carrying capacity of the blood during dormancy. There also appears to be a prolongation of the life span of the erythrocytes. Many of the latter are stored in the spleen at such times, since this organ appears greatly enlarged and engorged with red blood cells during dormancy.

A reduction in the physiological activities of heterotherms during dormancy is also reflected in a reduction in the number of circulating leukocytes. Apparently there is a minimum need for these cells at such times.

Recent studies made by Dawe and Spurrier (1969) on thirteen-lined ground squirrels (*Citellus tridecemlineatus*) have shown that by transfusing blood from hibernating animals to nonhibernating individuals in spring and summer, the latter entered into hibernation in a room which was dark, quiet, and maintained at 7° C. This suggests that the "trigger" for hibernation in these ground squirrels may be carried in the blood.

## ENDOCRINE AND NERVOUS SYSTEMS

Because of the functional importance of the endocrine system in regulating the internal environment and metabolic activity, various attempts have been made to relate dormancy to certain observed changes in particular endocrine glands. Such changes as have been noted, however, appear to be a result of reduced bodily activities rather than a cause. Hypofunction of the thyroid has been suggested as a factor producing torpidity, yet thyroxin given orally or by intraperitoneal injection to ground squirrels has not been found to inhibit hibernation. Gonadal activity appears to inhibit dormancy in the breeding season whereas castration does not. This, however, does not appear significant when we consider that estivation and hibernation normally take place at those seasons of the year when there is a cessation of gonadal activity and that there are many animals that never become torpid at such times. Certain insectivorous bats have been found to copulate occasionally during brief periods of awakening during the winter, but spermatogenesis occurs prior to hibernation and ovulation generally after hibernation.

It is possible that the pituitary is of some importance in dormancy. A great reduction of cianofile cells and a reduction in the size of the anterior lobe have been observed in hibernating marmots. Injections of anterior pituitary extract have been shown to awaken hibernating marmots and to inhibit dormancy in ground squirrels. Hypofunction of the pituitary, therefore, appears to be correlated with torpidity in heterotherms. However, the fact that dormant animals can awaken and resume their activity within a very short period of time indicates that no profound structural changes occur in any of the endocrine glands even though their functional ability may be markedly reduced during periods of torpidity.

The physiological modifications involving the nervous system of dormant heterotherms thus far discovered would seem to indicate a predominance of spinal reflexes and parasympathetic activity with the consequent effects of loss of ability to perform coordinated movements and slowing of cardiac and respiratory action.

## References Recommended

Adolph, E. F., and Lawrow, J. W. 1951. Acclimatization to Cold Air: Hypothermia and Heat Production in the Golden Hamster. Am. J. Physiol. 166:62-74.

Alcorn, J. R. 1940. Life History Notes on the Piute Ground Squirrel. J. Mamm. 21:160-170.

Bailey, R. M. 1949. Temperature Toleration of Garter Snakes in Hibernation. Ecology 30:238-242.

Baldwin, F. M., and Johnson, K. L. 1941. Effects of Hibernation on the Rate of Oxygen Consumption in the Thirteen-lined Ground Squirrel. J. Mamm. 22:180-182.

Bartholomew, G. A., and Cade, T. J. 1957. Temperature Regulation, Hibernation, and Aestivation in the Little Pocket Mouse, Perognathus longimembris. J. Mamm. 38:60-72.

Bartholomew, G. A., and Hudson, J. W. 1960. Aestivation in the Mohave Ground Squirrel Citellus mohavensis. Bull. Mus. Comp. Zool. 124:193-208.

Bartholomew, G. A., and Hudson, J. W. 1962. Hibernation, Estivation, Temperature

Regulation, Evaporative Water Loss, and Heart Rate of the Pigmy Possum, *Cercaertus nanus.* Physiol. Zool. *35*:94–107.

Benson, S. B. 1947. Comments on Migration and Hibernation in *Tadarida mexicana.* J. Mamm. *28*:407–408.

Bogert, C. M. 1949. Thermoregulation in Reptiles, a Factor in Evolution. Evolution *3*:195–211.

Bohnsack, K. K. 1951. Temperature Data on the Terrestrial Hibernation of the Green Frog, *Rana clamitans.* Copeia *1951*:236–239.

Burbank, R. C., and Young, J. Z. 1934. Temperature Changes and Winter Sleep of Bats. J. Physiol. *82*:459–467.

Carpenter, C. C. 1953. A Study of Hibernacula and Hibernating Associations of Snakes and Amphibians in Michigan. Ecology *34*:74–80.

Chute, R. M. 1964. Hibernation and Parasitism: Recent Developments and Some Theoretical Considerations. Ann. Acad. Sci. Fennicae, Ser. A, IV, Biologica, *71*(7):115–122.

Criddle, S. 1937. Snakes from an Ant Hill. Copeia *1937*:142.

Dawe, A. R., and Spurrier, W. A. 1969. Hibernation Induced in Ground Squirrels by Blood Transfusion. Science *163*:298–299.

Dodgen, C. L., and Blood, F. R. 1953. Blood Sugar and Energy Relationships in the Bat During Hibernation. Federation of Am. Soc. Exper. Biol., Fed. Proc. *104*:34.

Eadie, W. R. 1949. Hibernating Meadow Jumping Mouse. J. Mamm. *30*:307–308.

Eisentraut, M. 1933. Winterstarre, Winterschlaf und Winterruhe. Mitt. Zool. Mus. Berlin *19*:48–63.

Evans, C. P. 1938. Observations on Hibernating Bats with Especial Reference to Reproduction and Splenic Adaptation. Am. Naturalist *72*:480–484.

Folk, G. E., Jr. 1940. Shift of Population Among Hibernating Bats. J. Mamm. *21*:306–315.

Gorer, P. A. 1930. The Physiology of Hibernation. Biol. Revs. Cambridge Philos. Soc. *5*:213–230.

Hamilton, W. J., Jr. 1934. The Life History of the Rufescent Woodchuck, *Marmota monax rufescens* Howell. Ann. Carnegie Mus. *23*:85–178.

Hendrickson, J. R. 1954. Ecology and Systematics of Salamanders of the Genus *Batrachoseps.* Univ. Calif. Publ. Zool. *54*:1–46.

Hitchcock, H. B. 1949. Hibernation of Bats in Southeastern Ontario and Adjacent Quebec. Canadian Field-Naturalist *63*:47–59.

Hock, R. J. 1951. The Metabolic Rates and Body Temperatures of Bats. Biol. Bull. *101*:289–299.

Howell, A. H. 1929. Revision of the American Chipmunks (Genera *Tamias* and *Eutamias*). U.S. Department of Agriculture, Bureau of Biologic Survey, North American Fauna, 52.

Howell, T. R. 1961. An Early Reference to Torpidity in a Tropical Swift. Condor *63*:505.

Hudson, J. W. 1962. The Role of Water in the Biology of the Antelope Ground Squirrel *Citellus leucurus.* Univ. Calif. Publ. Zool. *64*:1–56.

Hudson, J. W. 1964. Temperature Regulation in the Round-tailed Ground Squirrel, *Citellus tereticaudus.* Ann. Acad. Sci. Fennicae, Ser. A, IV, Biologica, *71* (15):219–233.

Hudson, J. W. 1965. Temperature Regulation and Torpidity in the Pigmy Mouse, *Baiomys taylori.* Physiol. Zool. *38*:243–254.

Irving, L. 1960. Birds of Anaktuvuk Pass, Kobuk, and Old Crow: A Study in Arctic Adaptation. U.S. Nat. Mus. Bull. 217.

Irving, L., Krog, H., and Monson, M. 1955. The Metabolism of Some Alaskan Animals in Winter and Summer. Physiol. Zool. *28*:173–185.

Irving, L., Schmidt-Nielsen, K., and Abrahamsen, D. N. 1957. On the Melting Points of Animal Fats in Cold Climates. Physiol. Zool. *30*:93–105.

Jaeger, E. C. 1949. Further Observations on the Hibernation of the Poor-will. Condor *51*:105–109.

Jameson, E. W., Jr. 1964. Patterns of Hibernation of Captive *Citellus lateralis* and *Eutamias speciosus.* J. Mamm. *45*:455–460.

Jameson, E. W., Jr., and Mead, R. A. 1964. Seasonal Changes in Body Fat, Water and Basic Weight in *Citellus lateralis, Eutamias speciosus* and *E. amoenus.* J. Mamm. 45:359-365.

Johnson, G. E. 1931. Hibernation in Mammals. Quart. Rev. Biol. 6:439-461.

Kayser, C. 1961. The Physiology of Natural Hibernation. South Bridge, Edinburgh, Scotland, James Thin.

Lachner, E. A. 1942. An Aggregation of Snakes and Salamanders During Hibernation. Copeia 142:262-263.

Lasiewski, R. C. 1963. Oxygen Consumption of Torpid, Resting, Active, and Flying Hummingbirds. Physiol. Zool. 36:122-140.

Linsdale, J. M. 1946. The California Ground Squirrel: A Record of Observations Made on The Hastings Natural History Reservation. Berkeley, University of California Press.

Lyman, C. P. 1948. The Oxygen Consumption and Temperature Regulation of Hibernating Hamsters. J. Exper. Zool. 109:55-78.

Lyman, C. P. 1951. Effect of Increased $CO_2$ on Respiration and Heart Rate of Hibernating Hamsters and Ground Squirrels. Am. J. Physiol. 167:638-643.

Lyman, C. P., and Chatfield, P. C. 1950. Mechanisms of Arousal in the Hibernating Hamster. J. Exper. Zool. 114:491-516.

Lyman, C. P., and Chatfield, P. C. 1953. Changes in Blood Sugar and Tissue Glycogen in the Hamster During Arousal from Hibernation. J. Cell. and Comp. Physiol. 41:471-491.

Lyman, C. P., and Dawe, A. R., Eds. 1960. Mammalian Hibernation. (Symposium.) Bull. Mus. Comp. Zool. 124:1-549.

Lyman, R. A., Jr. 1943. The Blood Sugar Concentration in Active and Hibernating Ground Squirrels. J. Mamm. 24:467-474.

Mann, F. C. 1916. The Ductless Glands and Hibernation. Am. J. Physiol. 41:173-188.

Mann, F. C., and Drips, D. 1917. The Spleen During Hibernation. J. Exper. Zool. 23:277-285.

Marshall, J. T., Jr. 1955. Hibernation in Captive Goatsuckers. Condor 57:129-134.

Mayhew, W. W. 1963. Temperature Preferences of *Sceloporus orcutti.* Herpetologica 18:217-233.

Mayhew, W. W. 1965a. Adaptations of the Amphibian, *Scaphiopus couchi,* to Desert Conditions. Am. Mid. Nat. 74:95-109.

Mayhew, W. W. 1965b. Hibernation in the Horned Lizard, *Phrynosoma m'calli.* Comp. Biochem. Physiol. 16:103-119.

McClanahan, L. 1964. Osmotic Tolerance of the Muscles of Two Desert-inhabiting Toads, *Bufo cognatus* and *Scaphiopus couchi.* Comp. Biochem. Physiol. 12:501-508.

Neill, W. T. 1950. An Estivating Bowfin. Copeia 1950:240.

Noble, G. K. 1931. The Biology of the Amphibia. New York and London, McGraw-Hill Book Co., Inc.

Orr, R. T. 1954. Natural History of the Pallid Bat, *Antrozous pallidus* (Le Conte). Proc. Calif. Acad. Sci., Ser. 4, 28:165-246.

Pearson, O. P. 1950. The Metabolism of Hummingbirds. Condor 52:145-152.

Pearson, O. P. 1953. Use of Caves by Hummingbirds and Other Species at High Altitudes in Peru. Condor 55:17-20.

Pearson, O. P. 1960. Torpidity in Birds. Bull. Mus. Comp. Zool. Harvard 124:93-103.

Pearson, O. P., Koford, M. R., and Pearson, A. K. 1952. Reproduction of the Lump-nosed Bat *(Corynorhinus rafinesquei)* in California. J. Mamm. 33:273–320.

Pengelley, E. T., and Fisher, K. C. 1961. Rhythmical Arousal from Hibernation in the Golden-mantled Ground Squirrel, *Citellus lateralis tescorum.* Canadian J. Zool. 39:105-120.

Rasmussen, A. T. 1915. The Oxygen and Carbon Dioxide Content of the Blood During Hibernation in the Woodchuck *(Marmota monax).* Am. J. Physiol. 39:20-30.

Rasmussen, A. T. 1916a. Theories of Hibernation. Am. Nat. 50:609-625.

Rasmussen, A. T. 1916b. A Further Study of the Blood Gases During Hibernation in the Woodchuck *(Marmota monax).* The Respiratory Capacity of the Blood. Am. J. Physiol. 41:162-172.

Rasmussen, A. T. 1916c. The Corpuscles, Hemoglobin Content, and Specific Gravity of the Blood During Hibernation in the Woodchuck (*Marmota monax*). Am. J. Physiol. *41*:464-482.

Rasmussen, A. T. 1918. Cyclic Changes in the Interstitial Cells of the Ovary and Testis in the Woodchuck (*Marmota monax*). Endocrinology *2*:353-404.

Rasmussen, A. T. 1921. The Hypophysis Cerebri of the Woodchuck (*Marmota monax*) with Special Reference to Hibernation and Inanition. Endrocrinology *5*:33-66.

Reeder, W. G. 1949. Hibernating Temperature of the Bat *Myotis californicus pallidus*. J. Mamm. *30*:51-53.

Reeder, W. G., and Cowles, R. B. 1951. Aspects of Thermoregulation in Bats. J. Mamm. *32*:389.

Scholander, P. F., Flagg, W., Hock, R. J., and Irving, L. 1953. Studies on the Physiology of Frozen Plants and Animals in the Arctic. J. Cell. Comp. Physiol. *42* (Suppl. 1):1-56.

Schwartz, C. W. 1951. A New Record of *Zapus hudsonius* in Missouri and Notes on Its Hibernation. J. Mamm. *32*:227-228.

Shaw, W. T. 1925*a*. Duration of the Aestivation and Hibernation of the Columbian Ground Squirrel (*Citellus columbianus*) and Sex Relation of the Same. Ecology *6*:75-81.

Shaw, W. T. 1925*b*. Observations on the Hibernation of Ground Squirrels. J. Agric. Research *31*:761-769.

Stebbins, R. C. 1954. Natural History of the Salamanders of the Plethodontid Genus *Ensatina*. Univ. Calif. Publ. Zool. *54*:47-124.

Svihla, A. 1951. Relation of Water to Dormancy in Mammals. Proc. Second Alaskan Science Conference, McKinley Nat. Park, Alaska, Sept. 5.

Svihla, A., and Bowman, H. 1952. Oxygen Carrying Capacity of the Blood of Dormant Ground Squirrels. Am. J. Physiol. *171*:479-481.

Svihla, A., and Bowman, H. 1954. Hibernation in the American Black Bear. Am. Mid. Nat. *52*:248-252.

Svihla, A., Bowman, H., and Ritenour, R. 1951. Prolongation of Clotting Time in Dormant Estivating Mammals. Science *114*:298.

Svihla, A., Bowman, H., and Ritenour, R. 1952. Relation of Prothrombin to the Prolongation of Clotting Time in Aestivating Ground Squirrels. Science *115*:306-307.

Thorburg, F. 1953. Another Hibernating Poor-will. Condor *55*:274.

Tinkle, D. W., and Milstead, W. W. 1960. Sex Ratios and Population Density in Hibernating *Myotis*. Am. Mid. Nat. *63*:327-334.

Tucker, V. A. 1962. Diurnal Torpidity in the California Pocket Mouse. Science *136*:380-381.

Turner, L. M. 1886. Contributions to the Natural History of Alaska. Arctic Ser. Publ., 2, Signal Service, U.S. Army, 226 pp.

Wade, O. 1930. The Behavior of Certain Spermophiles with Special Reference to Aestivation and Hibernation. J. Mamm. *2*:160-188.

Woodbury, A. M. 1951. Introduction—a Ten Year Study. *In* Symposium: A Snake Den in Tooele County, Utah. Herpetologica *7*:4-14.

# Chapter Twelve

# REPRODUCTION

Reproduction in vertebrates is a complicated phenomenon involving many factors, some of which will be considered in this chapter. In general, it is cyclic, and the cycle is so arranged that the young are born or the eggs hatch at the time of year that is most favorable for their survival.

## HORMONAL CONTROL

As will be seen, the reproductive cycle may be influenced by a number of different environmental stimuli, but, basically, it is under hormonal control (Fig. 12.1). The hormones involved are produced principally in the anterior lobe of the pituitary (*gonadotrophic hormones*) and in the gonads (*gonadal hormones*).

More is known about the endocrine control of the reproductive cycle in mammals than in other vertebrate groups, although the evidence presently available seems to indicate that similar hormones control this cycle in birds, reptiles, amphibians, and fishes. In the female placental mammal, the development of the ovum and the surrounding follicle in the ovary is instigated by the production of the *follicle stimulating hormone* (FSH) in the pituitary. During the course of its development, the follicle itself produces *estrogen*. Following ovulation the ruptured follicle, under the stimulation of the *luteinizing hormone* (LH) produced in the pituitary, develops into the *corpus luteum* or yellow body. The latter, as a result of production of the *luteotrophic hormone* (LTH) by the pituitary, produces *progesterone*, which stimulates the mucosa of the uterus to undergo further development necessary for implantation if fertilization has taken place. Physiologically, therefore, the cycle is controlled by the action and interaction of both gonadotrophic and gonadal hormones.

In the male FSH stimulates the seminiferous tubules and sper-

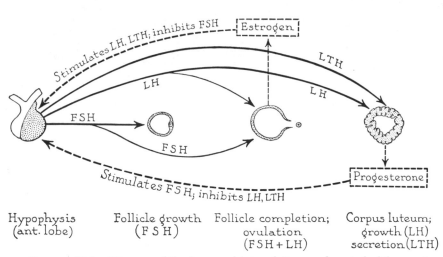

Hypophysis      Follicle growth    Follicle completion;    Corpus luteum;
(ant. lobe)        (F S H)              ovulation             growth (LH)
                                       (F S H + LH)         secretion (LTH)

**FIGURE 12.1.**   Diagram of the hormonal interrelations and control of the ovarian cycle. (Arey, L. B.: Developmental Anatomy. 7th Ed., W. B. Saunders Co., 1965.)

matogenesis, but the development of the interstitial cells in the testis is dependent upon LH. For this reason LH in the male is usually referred to as the *interstitial cell stimulating hormone* (ICSH). The interstitial cells themselves produce *testosterone*, a gonadal hormone responsible for the male secondary sexual characters.

Another hormone produced by the anterior lobe of the pituitary is the lactogenic hormone, *prolactin*. This stimulates the production of milk in mammals. In birds an increase in prolactin is associated with the termination of laying, with the development of a brood patch, and with incubation. Under laboratory conditions it has been shown that a brood patch can be induced in both the male and female of certain passerine species by injections of estradiol benzoate alone or in combination with progesterone or prolactin. In galliforms such as quail, progesterone will cause loss of feathers on the brood patch but no other histological changes, whereas prolactin causes complete incubation patch development. The production of pigeon milk in the crop of pigeons also appears to be controlled by prolactin.

## SEXUAL MATURITY

There is considerable diversity in different species of vertebrates with regard to the length of time required by the individual to attain sexual maturity. Among mammals the time involved may range from several weeks in certain kinds of rodents to quite a few years in man.

The earlier reproductive development of the female has been observed in many species of mammals. Hatfield (1935) recorded female voles (*Microtus californicus*) copulating when they were about 21 days old, although males of this species do not mate until six weeks of age. Wright (1947) found that male long-tailed weasels (*Mustela frenata*) born in April do not attain sexual maturity until the following spring, with spermatogenesis commencing in March. Females, on the other hand, come into estrus in the summer of the year of their birth and have young when they are about one year old.

Among our large North American herbivores more is known about the reproductive age of the females than the males. Pimlott (1959) found that 37 per cent of the yearling moose (*Alces alces*) that he studied in Newfoundland were pregnant. Olaus J. Murie (1951) concluded that three year old bull elk (*Cervus elaphus*) were sexually mature, but doubted that they could hold a herd of cows when in competition with older bulls. Cows of this species normally breed in the third rutting season after birth, at which time they are about two years and four months old. Adolph Murie (1944) found that ewes of the Dall sheep (*Ovis dalli*) in the Mt. McKinley region of Alaska first breed when two and one-half years of age.

The availability of food probably plays an important part in determining the reproductive age of many mammals as well as other vertebrates. Taber and Dasmann's (1957) studies on mule deer (*Odocoileus hemionus*) have shown that, in favorable areas, does often breed at the age of 17 months, but on poorer ranges some does do not breed until 29 or even 41 months of age.

Female northern fur seals (*Callorhinus ursinus*) may become pregnant in their third year, but the greatest number of breeding females are above the fifth year class. Males may produce spermatozoa at three or four years of age, but they are not regarded as sexually mature until the fifth or sixth year of life. They begin to breed at the age of seven, but do not attain their peak until they are 12 to 15 years old. They may attain an age of 30 years.

Most passerine birds in temperate regions are sexually mature and ready to breed in the spring of the year following hatching. It would appear that in certain equatorial birds that lack a specific breeding season, sexual maturity may sometimes be attained in as little as five months (Miller, 1959).

Most gulls do not breed until they are several years old. Johnston's (1956a, 1956b) detailed study of the California gull (*Larus californicus*) has shown that neither first nor second year birds breed. By the fourth year members of both sexes have reached reproductive maturity.

It is probable that the California condor (*Gymnogyps californianus*) requires more time to attain sexual maturity than any other North American bird. Koford (1953) estimated that the condor is at least five years of age before it engages in nesting activities. In this species the adult plumage is not acquired until the fifth year of life. The royal albatross (*Diomedea epomophora*) of the Antarctic, however, is reported to nest for the first time in its ninth year.

Some reptiles and amphibians attain sexual maturity in one year, but there are others that require several years. In most reptiles maturity is attained more rapidly in the tropics than in temperate climates where the growing season may amount to only four months of the year. Certain Temperate Zone lizards, such as members of the genera *Anguis* and *Ophiosaurus,* may require four or five years to become sexually mature. Size also is a factor. Small lizards and snakes mature more rapidly than large species. Even in warm temperate regions maturity may be reached the first year. Fitch (1940) concluded that fence lizards (*Sceloporus occidentalis*) become sexually mature following their second hibernation, at which time they are about two years old. Blair (1953), as a result of marking over 350 baby Mexican toads (*Bufo valliceps*), found that sexual maturity in both males and females was attained in the year following hatching. Maturity is attained more slowly in certain salamanders. Stebbins (1954) estimated that Eschscholtz salamanders (*Ensatina eschscholtzi*) do not become sexually active until the third or fourth year following hatching.

There are some caudate amphibians that fail to metamorphose completely and attain sexual maturity in the larval condition. This is known as *neoteny.* The common mud puppy (*Necturus*) that is widely used for laboratory dissection is an example of a neotenic species. It retains its gills permanently and remains in water. One of the classical examples of neoteny is presented by the tiger salamander (*Ambystoma tigrinum*) which ranges from Canada south to Mexico (Fig. 12.2). Throughout most of its range it undergoes complete metamorphosis. At the southern end, however, it reproduces in the larval form. For many years this larval form, which is capable of reproduction, was considered a separate species and known in Mexico as the *axolotl.* As a result of transfer of captive axolotls to different environments it was accidentally discovered that they were capable of metamorphosing into adult tiger salamanders.

Sexual maturity may occur at an extremely early age in some species of fishes. In others it is not attained until the life span is almost terminated. Young males of several species of viviparous perch (family Embiotocidae), including the reef perch (*Micrometrus aurora*), the dwarf perch (*Micrometrus minimus*), and the yellow shiner (*Cymatogaster aggregata*), are sexually mature immediately

FIGURE 12.2. An adult tiger salamander (*Ambystoma tigrinum*). This species is neotenic in the southern part of its range. (Photograph by Nathan W. Cohen.)

after birth (Hubbs, 1921). The males of many other kinds of fishes attain sexual maturity at an earlier age than do the females. This is well demonstrated by the Atlantic salmon (*Salmo salar*) which spawns in rivers and certain lakes from Greenland south to New England. Males of this species may become sexually mature before migrating to the sea and, when only 10 cm. in length, will spawn with the adult females many times larger that have returned to fresh water from the Atlantic Ocean.

Fishes that breed but once in a lifetime, such as salmon of the genus *Oncorhynchus* and eels of the genus *Anguilla*, are migratory. The return to the spawning ground is a part of the reproductive cycle. It has been shown that European eels (*Anguilla anguilla*) that have been prevented from migrating have been able to survive for over 50 years without attaining sexual maturity.

Most silver salmon (*Oncorhynchus kisutch*) return to spawn when they are three or four years of age. King salmon (*O. tshawytscha*) may be anywhere from two to seven years old when they spawn, although most of them do so in the fourth or fifth year of life.

Steelhead trout (*Salmo gairdnerii*) usually attain sexual maturity when they are about four years of age, having spent two years in fresh water and two years in the sea. Unlike salmon of the genus *Oncorhynchus* they do not necessarily die, but may return to the sea. They may survive several spawning seasons.

## HERMAPHRODITISM AND PARTHENOGENESIS

Hermaphroditism is not uncommon among many kinds of invertebrate organisms, but until very recently it was considered to be of rare occurrence in vertebrates except in certain cyclostomes (see Chapter 2, under "Urogenital System"). At present, however, we know of a number of kinds of marine fishes as well as a few freshwater species that exhibit intersexuality. Individuals of some species function only as one sex at a time, but there are others that are capable of fertilizing their own eggs, sometimes within the ovotestis itself.

*Protandrous* species are those that undergo a change from male to female, whereas *protogynous* species change from female to male. *Synchronous* hermaphrodites can both produce ova and fertilize them. Furthermore, there are some fishes such as certain kinds of sea bass that may be synchronous hermaphrodites yet are capable of behaving as a male or a female and engaging in courtship. This is true of *Serranus subligarius*. Sudden reversal of the sex roles may also occur in this species.

Little information is available as yet on the time required for various kinds of protandrous and protogynous hermaphrodites to undergo sex reversal. In the synbranchid genus *Monopterus* the change from female to male takes place at about three years of age. In the red grouper (*Epinephelus morio*) of the Atlantic, which is another protogynous species, the change from female to male does not occur until it is about 12 years old.

Little is known regarding the selective advantage of hermaphroditism, although the reproductive potential in species that exhibit synchronous hermaphroditism would seem to be markedly greater than in species with separate sexes.

Until fairly recently parthenogenesis was known only in invertebrates. It does occur, however, in some lizards, notably members of the New World family Teiidae. In certain populations of the genus *Cnemidophorus*, which ranges from northern United States to South America, males are rare or appear to be lacking entirely. *Cnemidophorus cozumelus cozumelus* on the Yucatan Peninsula is one of these. *C. exsanguis*, living in Chihuahua, Mexico, is another parthenogenetic species. No males have yet been found.

## PERIODICITY

The cyclic nature of reproductive activity is most apparent in vertebrates in nonequatorial regions and is more marked in females than in males. Even in equatorial regions, where many species appear to breed the year around, it seems likely that, within the individual, gonadal activity is cyclic.

Among mammals the terms *monestrous* and *polyestrous* are frequently used to describe the frequency of occurrence of the estrous cycle. Monestrous species are those having an annual cycle. Polyestrous species have two or more cycles a year. Most small rodents and lagomorphs are polyestrous. In permanent polyestrous species, one cycle succeeds the other without any prolonged period of sexual inactivity. A good example of this is to be found in the laboratory rat (*Rattus norvegicus*), which ovulates every four or five days unless interrupted by pregnancy. Seasonally polyestrous species have a period of sexual inactivity between cycles.

The adult female African elephant (*Loxodonta africana*), whose gestation period, like that of the Indian elephant (*Elephas maximus*), is approximately 22 months, produces one young about every four years. The young nurse for approximately two years. The bulls at any one time in the year exhibit marked individual differences in testosterone production. The maximum life expectancy in this species is believed to be nearly 70 years.

Most of our temperate North American bats, carnivores, large herbivores, and pinnipeds that have been studied are monestrous. The monestrous condition is often associated with species whose young require a long period of parental care or those that engage in extensive migrations or that hibernate. In a number of monestrous species, such as seals and sea lions, mating follows the birth of the young within a few days. Although the females of these marine mammals are almost continually pregnant during their reproductive life, they have delayed implantation. The fertilized ovum undergoes development to the blastocyst stage and then becomes quiescent for several months.

The majority of our North American birds exhibit definite reproductive periodicity. In general the nesting season for nonequatorial species is the spring of the year, although there are many exceptions. Actually within the United States there are certain species, such as the Anna hummingbird, that commence nesting in December, and others, such as the band-tailed pigeon, that may still be engaged in nesting in October. While most species rear one or two broods of young a year, there are some that occasionally have three, possibly more. The opposite extreme is represented by the California condor, which is reported to lay a single egg every other year.

Most North American reptiles and amphibians living away from the tropics exhibit reproductive periodicity with ovulation occurring in the spring of the year. In some species that hibernate it appears that the estrous cycle may be quite prolonged. Follicular growth usually commences in the weeks or months preceding the period of dormancy so that ovulation can take place shortly following emergence in the spring. This condition commonly prevails in most hibernating species of amphibians and probably a number of kinds of reptiles. In the red-backed salamander (*Plethodon cinereus*) of eastern North America, however, mature females require two years to accumulate sufficient yolk in their eggs for deposition. Ovulation, therefore, occurs only every other year.

Rahn's (1942) very interesting study of the prairie rattler (*Crotalus v. viridis*) suggests that, at least in certain parts of its range, this viviparous species ovulates every other year. Adult females that have not borne young during the year develop large ovarian follicles in the fall. Copulation precedes hibernation and the sperm remain viable until the next spring. When these females emerge from dormancy they are ready to ovulate. During the long period of gestation that follows, the corpora lutea in the ovary inhibit the development of new follicles prior to the beginning of the next period of hibernation. These females do not copulate in the fall. During the succeeding summer and fall follicular growth once again occurs and insemination takes place before the onset of winter. Consequently, an examination of the reproductive tracts of hibernating adult females will show one group with well developed follicles and sperm present and another with regressed ovaries in a postpartum condition and lacking sperm.

There is a great variation in the frequency of the reproductive cycle as well as the time or times of the year that it occurs in various fishlike vertebrates. As has already been noted, salmon of the genus *Oncorhynchus* and eels of the genus *Anguilla* as well as some lampreys spawn but once in a lifetime and then die. Most salmon enter the spawning streams in the late fall or early winter and deposit their eggs in winter. In larger rivers, however, there may be spring runs, but these fish usually do not spawn until sometime between August and November. Most steelhead trout (*Salmo gairdnerii*) leave the ocean and enter the home streams in late fall or winter and spawn early in the spring. Like salmon, however, there are spring runs of these fish that enter fresh water in a sexually immature condition and spawn in the fall. Some species of freshwater trout spawn in the spring, others in the fall.

The true cod of the Pacific Ocean (*Gadus macrocephalus*) spawns in shallow water close to shore in winter and early spring. Many other marine species such as the smelts and sea perches spawn during the summer months. One of the most interesting examples of reproductive periodicity is presented by the grunion (*Leuresthes tenuis*), a species inhabiting the coastal water of southern California and northern Baja California. These fish, which average between 5 and 6 inches in length, have a spawning season that extends from

**FIGURE 12.3.** Two female grunion (*Leuresthes tenuis*), surrounded by males, starting to dig in the sand. (Photograph by Joseph Brauner.)

late February to early September. Actually spawning, however, only takes place on the three or four nights following the full moon or new moon and is limited to a period of one to three hours immediately following high tide (Fig. 12.3). Adult females may spawn from four to eight times a season. A closely related species, *Leuresthes sardinia*, that occurs in the Gulf of California, has very similar breeding habits except that spawning may occur in the daytime.

The dogfish (*Squalis acanthias*) is reported to produce young every other year.

## ENVIRONMENTAL FACTORS

Although the reproductive cycle of vertebrates is known to be basically under hormonal control, it exhibits such definite seasonal periodicity in many species that numerous studies have been made on various environmental factors that might possibly function as regulators. Much of our present knowledge in this field has been derived from investigations on birds.

It has long been known that most birds in regions away from the equator nest in the spring of the year, when environmental temperatures are rising and the days are increasing in length. Temperature, however, was generally considered to be the critical factor. Professor William Rowan (1929) was the first to present evidence that gonadal *recrudescence* (renewed activity) could be induced in slate-colored juncos (*Junco hyemalis*) in the late fall and early winter months by subjecting them to increased daily amounts of light. The results were very conclusive for male birds.

Since the early work of Rowan similar results involving testicular recrudescence, following increased daily periods of illumination in late fall and winter, have been obtained in a number of other species of birds as well as in a few mammals, reptiles, and fishes. Artificial stimulation by light, however, fails to increase gonadal activity in the late summer and early fall. This has been termed the *refractory period*. In nature it represents a time of physiological rest and gonadal regression following the strenuous breeding season. Termination of this refractory period varies with different species. Miller (1948) found that in golden-crowned sparrows (*Zonotrichia atricapilla*) it ended between November 5 and 20 in central California. The same author (1954) found that the refractory period terminated in several races of white-crowned sparrows (*Zonotrichia leucophrys*) in the latter part of October. Wolfson (1954) and others, however, have shown that the refractory period may be reduced by subjecting birds to very short photoperiods daily following gonadal activity. As many as five gonadal cycles within a year have been induced by artificial light manipulation in the slate-colored junco (*Junco hyemalis*).

The significance of light as a regulator of sexual periodicity is obvious, since it is the one factor in the environment that has regular annual periodicity. The length of the period of daylight on any given day is essentially the same each year and increases steadily from the winter solstice to the summer solstice, after which it decreases steadily until the following winter solstice.

It is now generally agreed that in certain vertebrates the stimulus of light, transmitted by way of the optic nerve to the hypothalamus of the brain, in some way induces the pituitary to produce gonadotrophins which in turn affect the gonads. That the eye is the important factor has been demonstrated rather clearly. Ducks whose bodies have been masked so that only the eyes are exposed react to increased amounts of light like those whose bodies were not masked. On the other hand, varying hares and ferrets whose eyes were masked did not show testicular recrudescence when subjected to increased photoperiods although, when subjected to increased amounts of light with the eyes unmasked, both of these species of mammals showed marked gonadal activity (Bissonnette, 1936; Lyman, 1943).

Lyman (*op. cit.*) showed that both coat color and gonadal activity could be manipulated in the varying hare (*Lepus americanus*) by increasing and decreasing the daily period of illumination. He also found that there was an innate reproductive rhythm that was independent of light. Hares maintained under constant minimum amounts of light showed signs of periodic gonadal activity. Those kept under constant illumination for 18 hours a day underwent periods of gonadal regression.

It was suggested long ago that gonadal recrudescence in animals subjected to increased illumination might be the result of the increase in the period of activity rather than to light directly. This obviously is not true in a nocturnal mammal such as the varying hare. The daily period of activity in hares subjected to maximum illumination is much less than in those receiving a minimum amount of light. Similar observations have been made on the raccoon (*Procyon lotor*), another nocturnal mammal.

Several experimental studies also have been made upon birds to determine whether gonadal recrudescence correlated with increased periods of daily illumination might not be the result of increased activity in diurnal species. To date the evidence fails to support the activity theory. An interesting method of testing this was employed by Farner and Mewaldt (1953). They subjected one group of migratory white-crowned sparrows (*Zonotrichia leucophrys gambelii*) to high environmental temperatures in midwinter so that the birds became very active at night. In fact their activity was about equal to that of birds kept outdoors but subjected to increased amounts of light. The testes of the former group did not recrudesce, while those of the latter did.

It must not be inferred that light is the only environmental regulator of gonadal periodicity. Artificially induced gonadal activity outside of the normal period of reproduction as a result of increased daily photoperiods has been successful with males of a number of species of vertebrates. The results of similar experiments on females have not been nearly as conclusive. Significant perhaps in this regard is the fact that gonadal periodicity is by no means as sharply defined in most male vertebrates as in females. Spermatogenesis frequently begins long before the onset of the breeding season and may even be completed weeks or months before the time of ovulation.

In equatorial regions, where the length of daylight is essentially the same from day to day throughout the year, photoperiodicity obviously cannot control the reproductive cycle. If environment plays a role it would appear that it must do so by other means.

In his studies on the Andean sparrow (*Zonotrichia capensis*) in a constant equatorial environment in Colombia, Miller (1959) found that individual birds had two breeding cycles a year. Each extended over a period of four months followed by two months of rest. This, however, was an individual rhythm. Within a local population birds could be found breeding at any time during the year. The region in which these observations were made not only lacked significant annual light and temperature cycles but had poorly defined wet and dry seasons. This acyclic reproductive pattern in a population, however, is not characteristic of most birds in tropical equatorial regions where there are one or occasionally two rainy seasons. The onset of the rainy period is frequently correlated with reproductive activity.

Studies made on the tricolored blackbird (*Agelaius tricolor*) by Robert B. Payne (1969) have shown that this species, unlike the closely related redwinged blackbird (*A. phoeniceus*), has two breeding seasons a year. This is a highly gregarious, nomadic species belonging to the New World family Icteridae and restricted largely to the valleys of California. While the redwinged blackbird has a somewhat extended breeding season in the spring and males have territories of several hundred square feet, the tricolor has a compressed breeding season and very small territories, maintained only for a single nesting. In the redwinged blackbird the male has a series of successive matings with females, i.e., they are serially polygamous within the semi-permanent territory. The testes of the male tricolor regress during the period of feeding the young, permitting a very early refractory period and a summer molt. In many tricolors, however, there is an autumnal recrudescence as days grow shorter. Autumnal breeding activity reaches a peak in October. Rain appears to stimulate this, although it is not essential. Nesting success in autumn is low. In general autumnal breeding seems to be associated with individuals having a short and early spring period of reproduction and, to some extent, with food available with early fall rains.

There is some evidence that gonadal response to light may be innate in some birds presently living close to the equator. Passerine birds inhabiting the Galápagos Islands nest during the rainy season, which extends from December to April. This appears to be advantageous, since it is the time of year when food is most abundant. Experiments involving the transportation of several species of the native ground finches from these equatorial islands to central California resulted in these birds changing their principal period of reproductive activity in the latter region to spring and summer. This, in general, conformed to the breeding season of most of the local passerine birds in California (Orr, 1945).

Although the correlation between day length and gonadal activity is very marked in many birds inhabiting temperate regions, there are other species that begin nesting in midwinter. A good example of this is presented by the Anna hummingbird (*Calypte anna*) in California. A study of this species by Williamson (1956) showed that there is no time of year when the testes are completely inactive. In general, however, recrudescence of the testes is well under way by August, and by December the birds are in full breeding condition and ready to nest. It is suggested that recrudescence may be influenced in this species by such factors as increased rainfall, which increases the food supply, by completion of the molt, which occurs between June and January, and by the presence of females.

Although Farner and Mewaldt's (*op. cit.*) experiments on white-crowned sparrows showed that increased temperatures and the consequent increased periods of activity did not stimulate gonadal activity under minimum light conditions, temperature is a factor when the daily photoperiod is increased to 11 or 12 hours. Engels and Jenner (1956) subjected juncos to such exposures for six weeks and found testicular recrudescence much more advanced in birds kept in a warm environment than those in cold. Even under natural conditions, cold weather in the spring appears to retard reproductive activity in many kinds of birds, while warm weather has the opposite effect.

In a number of species of insectivorous bats in the Northern Hemisphere, it has been found that spermatogenesis begins in the early summer but is not completed until early autumn. Copulation in these species, however, does not occur until October and November and may take place sporadically during periods of awakening from dormancy during the winter and even in spring. By the time copulation begins, the testes have undergone involution and the spermatozoa are stored in the epididymis.

Ovulation in such bats is not synchronized with insemination. The sperm may be stored in the reproductive tract of the female all winter and remain viable until the females emerge from dormancy in the spring, at which time ovulation and fertilization occur. Maximum gonadal activity in males and females is thus at opposite times in the year. Experimental evidence indicates that in these bats temperature is an important factor. Females removed from hibernation in midwinter and placed in a warm environment may be induced to ovulate within a few days (Orr, 1954).

A similar reproductive cycle has recently been found to occur in the Australian skink, *Hemiergis peronii*. In this reptile the females ovulate in spring, at which time the males are reproductively inactive. The testes attain maximum size in late summer and decrease toward winter. Insemination occurs in the fall and sperm can be found in the oviduct all winter.

Relatively few experimental studies have been made on the effects of light and temperature on cold-blooded vertebrates. Bartholomew (1953) found that yucca night lizards (*Xantusia vigilis*) of southwestern United States and northwestern Mexico responded to these stimuli like many birds. Males subjected to 16 hours of light daily in late fall and winter showed considerably greater testicular response when maintained at temperatures of 20° or 32° C. than at 8° C. Increased periods of light and temperature had only a slight effect on the ovaries of females.

In the anole lizard (*Anolis carolinensis*) photoperiodic response of the testes is dependent upon temperature. In order that long periods of daylight stimulate testicular recrudescence, body temperature must be raised to about 32° C. during at least part of the daylight hours. Cool days with high temperature during the night have an even more retarding effect on testicular growth than continual low temperatures. In some salamanders it has been found experimentally that both increased temperature and increased daily photoperiods serve to stimulate testicular activity in the spring under natural conditions; it is likely that temperature is the more important factor in such animals that are primarily nocturnal.

Hoover and Hubbard (1937) found that by subjecting brook trout (*Salvelinus fontinalis*) to increased amounts of daily illumination in late winter and early spring, followed by a gradual decrease in the daily photoperiod to seven hours a day, spawning could be induced in July. This is about four months earlier than the natural spawning time. Medlen (1951), as a result of experiments on mosquito fish (*Gambusia affinis*) in Texas, concluded that increased water temperature stimulated reproductive activity, while light played a minor part.

Many behavioral and physiological studies have been made on sticklebacks because of their availability, small size, and ability to live in aquaria. These are small fresh, marine, or brackish water fishes of the family Gasterosteidae found in the Northern Hemisphere. The males are territorial and defend the nests that they construct for the eggs. In two genera, *Gasterosteus* and *Culaea*, increased day length in the spring appears to be an important factor stimulating gonadal activity. Temperature also is a factor. *Gasterosteus* is tolerant of high temperatures during the three-month breeding period, but *Culaea* is not. The five-spined or brook stickleback (*C. inconstans*) is one of the most northern species of freshwater fishes, ranging from northern British Columbia (east of the Rocky Mountains) to the Atlantic coast and south to northern United States. Breeding takes place in May, June, and July when the water temperature does not get above 19° C. Photostimulation in fall and winter, with 14 to 16 hours of light in water not exceeding 19° C., will artificially stimulate gonadal activity and territorial behavior in males.

Emphasis in this chapter has been placed on the significance of light and temperature on reproductive periodicity, since these can be controlled in experimental studies and considerable information has been obtained on their effects. There are many other factors, however, that influence gonadal activity. Brief mention has been made of the relationship between rainfall, food supply, and the reproductive season of many tropical birds. These are also important to many vertebrates, especially aquatic species, far from the tropics. Periods of drought or food scarcity may definitely inhibit breeding.

Tactile and visual stimuli influence ovulation in some species. The development of the follicle and ovulation follow copulation in the rabbit. Copulation is also necessary for ovulation in the ferret and the cat. In birds the tactile (by means of the feet) or visual stimulus of eggs in the nest may inhibit or prolong ovulation. Most species lay a rather definite number of eggs. When the clutch is completed ovulation ceases. However, in some species removal of some of the eggs before the clutch is completed may prolong the period of ovulation. There is one record of a yellow-shafted flicker (*Colaptes auritus*) that laid 71 eggs in 73 days when only one egg was left in the nest and the others were removed daily after they were laid.

The sight of nesting material is a prerequisite to ovulation in some birds. In others the presence or at least sight of a male appears to stimulate the production of gonadotrophins in the female. This is true of the domestic pigeon. Even auditory stimulation may induce ovulation. Female budgerigars maintained in darkness have been known to lay when they were able to hear other breeding pairs. There are a number of kinds of fishes that produce sound, including the toadfish, sea horses, and the gourami. It has been suggested that in some of these species sound acts as a sexual stimulus.

## SEX RECOGNITION, COURTSHIP, AND PAIR FORMATION

Sex recognition is an essential part of vertebrate reproduction. Courtship and pair formation represent special types of reproductive behavior that are highly evolved in certain groups of vertebrates, notably among birds, and poorly developed or essentially lacking in others.

Visual, olfactory, auditory, or even tactile stimuli may be involved in sex recognition. Visual recognition may result from sexual dimorphism or from differences in the behavior of males and females within a species. Many kinds of birds exhibit marked secondary sexual dimorphism (Fig. 12.4). In such groups as ducks, gallinaceous birds, hummingbirds, trogons, warblers, troupials, and finches the males are often much more brightly colored than the females. Plum-

**FIGURE 12.4.** The red-winged blackbird (*Agelaius phoeniceus*) shows marked sexual dimorphism in plumage. The males are black with a scarlet patch, usually bordered by buff, on the bend of the wing. The females are brownish, heavily streaked on the underparts, and have the throat tinged with pink or buff.

age frequently plays an important part in courtship display. Combs, wattles, spurs, gular pouches, even eye color may be additional secondary male characters in some species.

Pronounced secondary sexual differences are less frequent in most other kinds of vertebrates although examples may be found in each of the major groups. In some mammals there is a marked size difference between males and females. This is especially true of certain pinnipeds. Adult northern fur seals, for example, weigh from 400 to 700 pounds, while the adult females weigh 75 to 125 pounds. In most species of carnivores, toothed whales, and hoofed animals the males are larger than the females. Among North American members of the deer family (Cervidae) antlers represent a secondary sexual character of the male except in the genus *Rangifer*, in which they are present in both sexes. In horned artiodactyls the horns usually show greater development in males than in females. Obvious secondary sexual differences are lacking in most North American insectivores, bats, and rodents.

The males of some lizards are much more brightly colored than the females. Although this is a secondary sexual character it seems to serve primarily as a means of intimidating other males. There is relatively little secondary sexual dimorphism in snakes. There are some turtles and tortoises in which males differ from the females in color.

In many salamanders the males and females differ slightly in coloration, shape of tail, and shape of head. Among North American anurans, the greatest sexual dimorphism is seen in the Yosemite toad (*Bufo canorus*) of the Sierra Nevada of California (Fig. 12.5). In this species the female is not only much larger than the male but also much more brightly colored.

Although there are few or no obvious external differences between the sexes in some kinds of fishes, the most extreme examples of secondary sexual dimorphism are to be found in this group. There are certain deep sea fishes in which the male is many times smaller than the female. A classic example is found in the oceanic angular fish, *Photocorynus spiniceps*. Although the tiny males are free-swimming for a while, each ultimately becomes attached to a female by means of the mouth. Subsequently, the attachment becomes permanent and the male receives all of his nourishment from the blood stream of the female (Fig. 12.6). It has even been suggested (Regan, 1925) that the female may be able to control the reproductive activities of the male. Parasitic males are found in other members of this group of fishes which inhabit ocean depths ranging from 1500 to over 4500 feet. It would appear to be an adaptation to life in a permanently dark environment.

Size differences, though not as extreme as in the oceanic angler fishes, color differences, and differences in shape of fins, especially

female  male

FIGURE 12.5. The Yosemite toad (*Bufo canorus*) exhibits unusual sexual dimorphism for North American anurans. (From Stebbins, R. C.: Amphibians and Reptiles of Western North America. McGraw-Hill Book Co., Inc., 1954.)

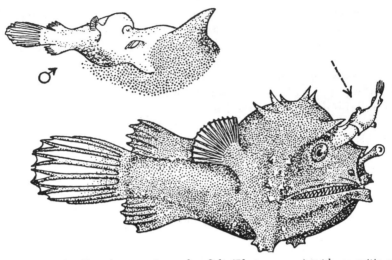

FIGURE 12.6. Female oceanic angler fish (*Photocorynus*) with parasitic male attached. (From Norman, in Young, J. Z.: The Life of Vertebrates. Oxford, Clarendon Press, 1955.)

the anal and pelvic fins, frequently serve to distinguish the sexes in some species of fishes. Less distinctive, though no less a secondary sexual character, is the protruding lower jaw of the mature male salmon.

It seems likely that visual recognition of sex in many vertebrate animals, some of which may even differ in secondary sexual characters, is often the result of differences in behavior. A male bird that has established territory at the beginning of the nesting season may attack any other member of the same species entering his territory, irrespective of sex. If it is another male that has intruded the latter may either retreat or fight back. A receptive female, however, will exhibit neither type of behavior but remain passive and thus identify herself.

Similar passive or receptive behavior on the part of the female appears to be an important means of sexual recognition in some mammals, many reptiles, amphibians, and fishes. Even though sex recognition may be the result of several types of stimuli the behavior of the female often indicates whether or not she is receptive. For example, most female mammals will repulse advances of males except when in estrus.

The behavior of the male, of course, may be equally important in sex recognition. Studies by Noble and Curtis (1939) and others on the social behavior of the jewel fish (*Hemichromis bimaculatus*), a well-known tropical cichlid from Africa, have shown that sexual selection in this species is comparable to that of many kinds of birds

(Fig. 12.7). In the breeding season members of either sex assume a reddish color. Although the females are more disposed toward brightly colored males than pale males, sex recognition appears to be a result of identification of particular movements. A female will select a pale male that is active and performing sex-stimulating movements over a brightly colored male that is sluggish in his behavior.

The sense of smell plays a very important part in sex recognition among mammals. Mention has already been made in Chapter 6 of the many kinds of scent glands that are so highly developed among members of this class of vertebrates. Although some of these glands are purely for defense or species recognition, others are associated with reproductive activity and become enlarged and active during the breeding season. The castor glands of the beaver, which secrete a substance called castoreum, are in this category. These glands increase greatly in size in the breeding season and their secretion is deposited at various places by the beavers. Presumably it aids these animals in finding mates. Among many carnivores the males may detect the scent of a female in estrus a considerable distance away.

Since birds have such a poorly developed sense of smell, it is

**FIGURE 12.7.** A male jewel fish (*Hemichromis bimaculatus*) guarding eggs that have been deposited by the female on a rock. (New York Zoological Society photograph.)

highly improbable that the olfactory organs have any part in sex recognition. Odor as well as behavior appears to be important as a means of enabling many male snakes to recognize receptive females. Although special scent glands are not well developed among reptiles, crocodilians possess musk glands on the throat and in the cloacal region which become enlarged during the breeding season. Some turtles also possess musk glands. The femoral and preanal glands of lizards have been suspected of aiding in sex recognition. The sense of smell also appears to assist in sex recognition in some caudate amphibians and certain fishes. The ovaries of the shad (*Alosa alosa*) produce a chemical which, when secreted into the water and received by the olfactory organs of the male, stimulates sexual behavior.

Sound is an important means of announcing sex in certain animals. Good examples among mammals are the bugling of male elk and the call of bull moose during the rutting season. Although song in birds serves primarily to announce territory on the part of the male, there is no doubt that it may also serve as a means of sex recognition. In nonsong birds sound may also function for sex recognition. Among members of the grouse family, the male's production of sound by various means seems to be a means of attracting females. The nightly vocalization of males of many kinds of frogs and toads at breeding ponds in the spring of the year appears definitely to be a means of attracting females to these areas.

Courtship represents one phase of the reproductive cycle that is most highly evolved among birds. There are relatively few species in other major groups of vertebrates that have developed such an elaborate type of sexual behavior. It has been suggested that the function of courtship is to effect a close bond between the members of a pair. Such a bond is decidedly advantageous in vertebrates that must build a nest to hold eggs, incubate the latter, and care for the young until they are old enough to be on their own. When both members of the pair participate in these activities, the chances of survival of the offspring, especially in altricial species, are considerably increased. In many passerine as well as some nonpasserine species of birds the male not only establishes and defends a nesting territory but may assist the female in nest construction, in incubating the eggs, and in the feeding of the young. The latter activity often requires a great expenditure of energy on the part of both members of the pair. In precocial species the care of these young, which are taught to feed themselves, is generally the responsibility of the female and the courtship ceremony performed by the males serves primarily to attract females for the brief period necessary for copulation.

Lack (1940) gives a very excellent summary of pair formation in birds in which he lists five main categories as follows: (1) sexes meeting only for copulation; (2) sexes remaining together for a few days at time of copulation; (3) pair formation established some time before copulation but ending shortly after; (4) sexes remaining together for the breeding season; (5) birds paired for life. Examples of each of these five types are to be found among North American birds.

The sage grouse (*Centrocercus urophasianus*) presents a good example of type 1. In this species the males gather on definite strutting grounds in the early morning and evening during the breeding season (Fig. 12.8). Here they space themselves, display, and fight adjacent males. Part of the display involves the production of sound by rubbing the wings against stiff breast feathers and by the actions of inflated air sacs above the breast. Females enter the strutting areas primarily for copulation.

Our North American hummingbirds come under type 2. Males and females remain together for a few days but the former take no part in the nesting activities and are generally driven away by the females.

Many species of ducks fall into type 3. Pair formation may occur in the wintering area weeks or even months in advance of nesting. On the nesting ground, however, the male will frequently leave the female after she has finished laying. This, of course, is not true of all waterfowl.

Type 4 is characteristic of the majority of our passerine birds.

**FIGURE 12.8.** Male sage grouse (*Centrocercus urophasianus*) displaying on the strutting ground. (Photograph courtesy of the Wyoming Game and Fish Commission.)

The males select territories prior to the nesting season, indulge in song, secure a mate, frequently display, and may or may not assist in nest building, incubation, and feeding the young. The members of a pair remain together for some time prior to nesting and afterward until the young are on their own. The same pair may even produce additional broods in a season.

The wrentit (*Chamaea fasciata*), whose behavior was described in Chapter 9, presents an excellent example of type 5. Once the bond between a male and female has been established, the birds remain together for life and both defend the territory the year around. In this species there is no obvious difference in the external appearance of the male and female. For most North American passerine species it has been shown that where little or no sexual dimorphism occurs there is greater participation in nesting activities by the male than in nonmonomorphic species.

In the majority of mammals the male has little or no part in the care of the young. After the mutual attraction in the breeding season necessary to effect fertilization, the male in most species of mammals has completed his family duties. The young develop within the female, who provides milk for their nourishment for some time after birth and often cares for them for a long time subsequently, depending upon the rate of growth. Elaborate courtship, if it is to establish a bond between the sexes that is to last for some time, is unnecessary. There are of course exceptions, especially among what might be termed colonial or herd mammals, such as certain pinnipeds and hoofed animals. As has been pointed out, the mature males of certain species of seals and sea lions establish harems of females which they fight for and defend. There may be some question, however, as to whether this behavior is true courtship. In those species of pinnipeds in which the male maintains a harem there is considerable sexual dimorphism in size, with the weight of the males sometimes several times that of the females. Bartholomew (1952) has suggested that this serves to further aggressive behavior in the male which in itself "determines whether or not he will be able to participate in the reproductive activities of the population." Bull elk, likewise, fight to establish harems of cows. The same is true of the pronghorn antelope. These bonds, however, are purely sexual, the males not being concerned with the birth or care of the young.

Perhaps the longest and most elaborate mammalian courtship, in an effort to secure a permanent bond, is indulged in, not always successfully, by man.

In reptiles and amphibians, courtship, if the term may properly be used, is generally brief, although the reproductive period may be fairly extended. Male fence lizards of the genus *Sceloporus* display in the presence of females by flattening the body and raising and lowering themselves rhythmically with the front legs. The blue ventral body patches are exposed at such times. Motion and color both appear to be involved. As the male approaches the female the up and down movements become more frequent.

Stebbins (1954) describes a series of body movements involved in the courtship of salamanders of the genus *Ensatina* (Fig. 12.9). The male upon approaching a female first noses her head and neck, then moves his body past and beneath her so that her throat is over his sacral region. The latter portion of the male then strokes the female's throat as a result of rotary motion. The two courting individuals then move along until the male stops and deposits a spermatophore, a gelatinous mass containing spermatozoa, on the substratum. During this act and for a time following, the female strokes the lower back of the male with her chin. They then move forward until her vent is over the spermatophore, which she picks up and takes into the cloaca. Following a writhing motion on the part of the male, during which he strokes the lower back and base of the tail of the female, they part.

Among many frogs and toads, the male, after he has contacted a female, grasps her behind the forelimbs and holds on tightly. The swollen glandular thumb pads of the male are an aid in this process and the pressure applied to the body of the female assists in the extrusion of eggs. This union, known as *amplexus*, may last for several weeks or more. In a few primitive species, the grasp is pelvic. In most anurans fertilization is external and the male deposits sperm on the eggs as they pass from the cloaca of the female. As soon as this has taken place members of a pair separate.

In some species of fishes the bond between sexes is very brief, merely long enough to insure fertilization of the eggs, while in others there may be elaborate courtship ceremonies. An example of the former is seen in the grunion, whose periodicity has already been discussed. When grunion come to shore to spawn they permit wave action to carry them up on a sandy beach. Within a period of about one-half a minute the female digs her body into the sand, extrudes from 1000 to 3000 eggs, and is ready to go back to sea with the next wave. While the female is laying, the male or males deposit milt on her body, down which it flows to the eggs where fertilization is effected. No courtship is involved and the sexual bond is only a matter of some seconds.

The blue gourami (*Trichogaster tricopterus*), on the other hand, represents a species with a fairly complicated courtship. This fish belongs to the family Anabantidae, most of which are bubble-nest builders. Studies by Hodges and Behre (1953) have shown that part of the preliminary reproductive behavior involves the construction of a floating nest of bubbles, presumably by the male. Nest building is

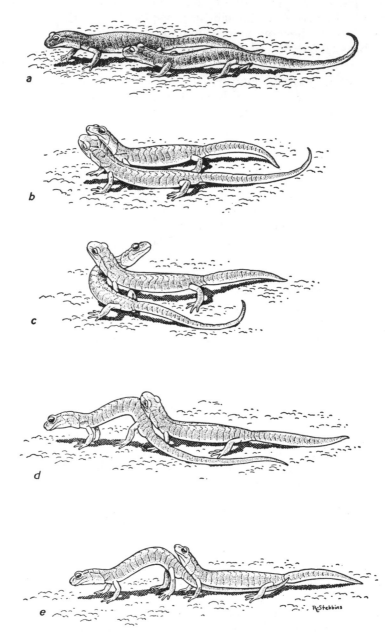

**Figure 12.9.** Part of the courtship movements of the plethodontid salamander (*Ensatina eschscholtzi*). (From Stebbins, R. C.: Natural History of the Salamanders of the Plethodontid Genus *Ensatina*. Univ. Calif. Publ. Zool., No. 54, 1954.)

an indication of breeding condition. Part of the courtship ceremony involves the spreading of the fins and the use of the mouth in nudging a prospective mate. Considerable chasing and nipping of fins also takes place until the female becomes passive. Sometimes the nest is destroyed during these activities, in which instance it is rebuilt by the male. Finally the female either swims of her own accord or is herded by the male under the nest. The male then curves his body around her so that his head and tail meet on her dorsal side. The body of the male vibrates and squeezes the female as she extrudes eggs and he deposits milt. When laying is completed both rise to the surface and the female is released and chased away. Care of the eggs and young is the responsibility of the male.

While courtship display serves primarily to stimulate the physiology of the opposite sex, it has been demonstrated in gulls and certain fishes that it may also be of protective value by reducing the likelihood of attack. In the cichlid *Etroplus maculatus*, individuals will court before mates that are larger than they more often than before those that are smaller. This reduces the probability of attack on the smaller of the two.

## References Recommended

Allen, G. M. 1925. Birds and Their Attributes. Boston, Marshall Jones Co.

Armstrong, E. A. 1965. Bird Display and Behavior. An Introduction to the Study of Bird Psychology. New York, Dover Publications, Inc.

Asdell, S A. 1964. Patterns of Mammalian Reproduction. 2nd ed. Ithaca, N.Y., Comstock Publishing Co.

Atz, J. W. 1965. Hermaphroditic Fishes. Science 150 (3697):789-797.

Barlow, G. W. 1968. Effect of Size of Mate on Courtship in a Cichlid Fish, *Etroplus maculatus*. Commun. Behav. Biol. (A) 2:149-160.

Bartholomew, G. A. 1952. Reproductive and Social Behavior of the Northern Elephant Seal. Univ. Calif. Publ. Zool. 47:369-472.

Bartholomew, G. A. 1953. The Modification by Temperature of the Photo-periodic Control of Gonadal Development in the Lizard *Xantusia vigilis*. Copeia 1953:45-50.

Bissonnette, T. H. 1936. Modification of Mammalian Sexual Cycles. V. J. Comp. Psych. 22:93-103.

Blair, W. F. 1953. Growth, Dispersal and Age at Sexual Maturity of the Mexican Toad (*Bufo valliceps* Wiegmann). Copeia 1953:208-212.

Blair, W. F., and Pettus, D. 1954. The Mating Call and Its Significance in the Colorado River Toad (*Bufo alvarius* Girard). Texas J. Sci. 6:72-77.

Briggs, J. C. 1953. The Behavior and Reproduction of Salmonid Fishes in a Small Coastal Stream. Calif. Dept. Fish and Game, Fish Bull. 94:1-62.

Brown, M. E., Ed. 1957. The Physiology of Fishes. New York, Academic Press, Inc., 2 Vols.

Collias, N. E., and Collias, E. E. 1964. Evolution of Nest-building in the Weaverbirds (Ploceidae). Univ. Calif. Publ. Zool. 73:1-162.

Darevsky, I. S. 1966. Natural Parthenogenesis in a Polymorphic Group of Caucasian Rock Lizards Related to *Lacerta saxicola* Eversmann. Jour. Ohio Herp. Soc. 5:115-152.

Enders, A. C. 1963. Delayed Implantation. Published for William Marsh, Rice University, by University of Chicago Press.

Engels, W. L., and Jenner, C. E. 1956. The Effect of Temperature on Testicular Recrudescence in Juncos at Different Photoperiods. Biol. Bull. *110*:129-137.

Farner, D. S., and Mewaldt, L. R. 1953. The Relative Roles of Diurnal Periods of Activity and Diurnal Photoperiods in Gonadal Activation in Male *Zonotrichia leucophrys gambelii* (Nuttall). Experimentia 9:219.

Fitch, H. S. 1940. A Field Study of the Growth and Behavior of the Fence Lizard. Univ. Calif. Publ. Zool. *44*:151-172.

Harrington, R. W., Jr. 1961. Oviparous Hermaphroditic Fish with Internal Self-fertilization. Science *134*:1749-1750.

Hatfield, D. M. 1935. A Natural History Study of *Microtus californicus*. J. Mamm. *16*:261-271.

Hodges, W. R., and Behre, E. H. 1953. Breeding Behavior, Early Embryology, and Melanophore Development in the Anabantid Fish, *Trichogaster trichopterus*. Copeia *1953*:100-107.

Hoover, E. E., and Hubbard, H. E. 1937. Modification of Sexual Cycle in Trout by Control of Light. Copeia *1937*:206-210.

Hubbs, C. L. 1921. The Ecology and Life-history of *Amphigonopterus aurora* and Other Viviparous Perches of California. Biol. Bull. *40*:181-209.

Jackson, A. W. 1952. The Effect of Temperature, Humidity, and Barometric Pressure on the Rate of Call in *Acris crepitans* Baird in Brazos County, Texas. Herpetologica 8:18-20.

Johnston, D. W. 1956*a*. The Annual Reproductive Cycle of the California Gull. I. Criteria of Age and the Testis Cycle. Condor *58*:134-162.

Johnston, D. W. 1956*b*. The Annual Reproductive Cycle of the California Gull. II. Histology and Female Reproductive System. Condor *58*:206-221.

Jones, R. E. 1969. Hormonal Control of Incubation Patch Development in the California Quail *Lophortyx californicus*. General and Comparative Endocrinology *13*:1-13.

Koford, C. B. 1953. The California Condor. Research Report 4, National Audubon Society, New York.

Lack, D. 1940. Pair-formation in Birds. Condor *42*:269-286.

Lack, D. 1954. The Natural Regulation of Animal Numbers. Oxford, The Clarendon Press.

Lehrman, D. S. 1959. Hormonal Responses to External Stimuli in Birds. Ibis *101*:478-496.

Licht, P. 1966. Reproduction in Lizards: Influence of Temperature on Photoperiodism in Testicular Recrudescence. Science *154*:1668-1670.

Lowe, C. H., and Wright, J. W. 1966. Evolution of Parthenogenic Species of *Cnemidophorus* (whiptail lizards) in Western North America. J. Arizona Acad. Sci. *4*:81-87.

Lyman, C. P. 1943. Control of Coat Color in the Varying Hare *Lepus americanus* Erxleben. Bull. Mus. Comp. Zool. *43*:394-461.

Marshall, A. J. 1959. Internal and Environmental Control of Breeding. Ibis *101*:456-478.

Marshall, F. H. A. 1936. Sexual Periodicity and the Causes Which Determine It. Philos. Trans. Roy. Soc. London *226*:423-456.

Maslin, T. P. 1962. All-female Species of the Lizard Genus *Cnemidophorus*, Teiidae. Science *135*:212-213.

Medlen, A. B. 1951. Preliminary Observations on the Effects of Temperature and Light upon Reproduction in *Gambusia affinis*. Copeia *1951*:148-152.

Miller, A. H. 1948. The Refractory Period in Light-induced Reproductive Development of Golden-crowned Sparrows. J. Exper. Zool. *109*:1-11.

Miller, A. H. 1954. The Occurrence and Maintenance of the Refractory Period in Crowned Sparrows. Condor *56*:13-20.

Miller, A. H. 1959. Reproductive Cycles in an Equatorial Sparrow. Proc. Nat. Acad. Sci. *45*:1095-1100.

Murie, A. 1944. The Wolves of Mount McKinley. U.S. Department of Interior, National Park Service, Fauna Ser. 5.

Murie, O. J. 1951. The Elk of North America. The Stackpole Co., Harrisburg, Pennsylvania and the Wildlife Management Institute, Washington, D. C.

Noble, G. K., and Bradley, H. T. 1933. The Mating Behavior of Lizards; Its Bearing on the Theory of Sexual Selection. Ann. N.Y. Acad. Sci. *35*:25-100.

Noble, G. K., and Curtis, B. 1939. The Social Behavior of the Jewel Fish, *Hemichromis bimaculatus* Gill. Bull. Am. Mus. Nat. Hist. 76:1-46.

Orr, R. T. 1945. A Study of Captive Galapagos Finches of the Genus *Geospiza.* Condor 47:177-201.

Orr, R. T. 1954. Natural History of the Pallid Bat, *Antrozous pallidus* (Le Conte). Proc. Calif. Acad. Sci., 4th Ser. 28:165-246.

Parkes, A. S., and Bruce, H. M. 1961. Olfactory Stimuli in Mammalian Reproduction. Science 134:1049-1054.

Payne, R. B. 1969. Breeding Seasons and Reproductive Physiology of Tricolored Blackbirds and Redwinged Blackbirds. Univ. Calif. Publ. Zool. No. 90.

Pearson, O. P., Koford, M. R., and Pearson, A. K. 1952. Reproduction of the Lump-nosed Bat (*Corynorhinus rafinesquei*) in California. J. Mamm. 33:273-320.

Pimlott, D. H. 1959. Reproduction and Productivity of Newfoundland Moose. J. Wildlife Mgmt. 23:381–401.

Rahn, H. 1942. The Reproductive Cycle of the Prairie Rattler. Copeia 1942:233-240.

Regan, C. T. 1925. Dwarfed Males Parasitic on the Females in Oceanic Angler-fishes (Pediculati Ceratioidea). Proc. Roy. Soc., Ser. B 97:386-400.

Reisman, H. M., and Cade, T. J. 1967. Physiological and Behavioral Aspects of Reproduction in the Brook Stickleback, *Culaea inconstans.* Amer. Midl. Nat. 77:257-295.

Rowan, W. 1929. Experiments in Bird Migration. I. Manipulation of the Reproductive Cycle: Seasonal Histological Changes in the Gonads. Proc. Boston Soc. Nat. Hist. 39:151-208.

Sayler, A. 1966. The Reproductive Ecology of the Red-backed Salamander, *Plethodon cinereus*, in Maryland. Copeia 1966:183-193.

Selander, R. K., and Kuich, L. L. 1963. Hormonal Control and Development of the Incubation Patch in Icterids, with Notes on Behavior of Cowbirds. Condor 65:73-90.

Shank, M. C. 1959. The Natural Termination of the Refractory Period in the Slate-colored Junco and the White-throated Sparrow. Auk 76:44-54,

Smyth, M., and Smith, M. J. 1968. Obligatory Sperm Storage in the Skink *Hemiergis peronii.* Science 161:575-576.

Stebbins, R. C. 1949. Courtship of the Plethodontid Salamander *Ensatina eschscholtzii.* Copeia 1949:274-281.

Stebbins, R. C. 1954. Natural History of the Salamanders of the Plethodontid Genus *Ensatina.* Univ. Calif. Publ. Zool. 54:47-124.

Stuart, G. R. 1954. Observations on Reproduction in the Tortoise *Gopherus agassizi* in Captivity. Copeia 1954:61-62.

Taber, R. D., and Dasmann, R. F. 1957. The Dynamics of Three Natural Populations of the Deer *Odocoileus hemionus hemionus.* Ecology 38:233-246.

Tinbergen, N. 1959. Comparative Studies of the Behaviour of Gulls (Laridae): A Progress Report. Behaviour 15:1-70.

Verner, J., and Willson, M. F. 1969. Mating Systems, Sexual Dimorphism, and the Role of Male North American Passerine Birds in the Nesting Cycle. Ornith. Monog. 9, Amer. Ornith. Union.

Walker, B. W. 1952. A Guide to the Grunion. Calif. Fish and Game 38:409–420.

Werner, J. K. 1969. Temperature-Photoperiod Effects on Spermatogenesis in the Salamander *Plethodon cinereus.* Copeia 1969:592-602.

Williamson, F. S. L. 1956. The Molt and Testis Cycle of the Anna Hummingbird. Condor 58:342-366.

Wolfson, A. 1954. Production of Repeated Gonadal, Fat, and Molt Cycles Within One Year in the Junco and White-crowned Sparrow by Manipulation of Day Length. J. Exper. Zool. 125:353-376.

Wright, P. L. 1947. The Sexual Cycle of the Long-tailed Weasel (*Mustela frenata*). J. Mamm. 28:343-352.

## Chapter Thirteen

# GROWTH AND DEVELOPMENT

The process of growth begins at the time the ovum is fertilized and may continue in some species throughout life. Furthermore, from the time of hatching or birth until maturity the developing individual undergoes changes in behavior that are associated with the process of growing up.

The parent-offspring relationship in vertebrates varies greatly. In a few birds, most reptiles and amphibians, and many fishes parental care is entirely lacking. The female completes her family obligations with the laying of the eggs and the male his with their fertilization. In other vertebrates, notably mammals and most birds, as well as some lower forms, parental care is essential to the survival of the young. The degree of dependence of the latter on the parent or parents varies as does the time involved.

To express degrees of relationship between the female and the young during the earlier stages of development the terms oviparous, ovoviviparous, and viviparous are used. As will be seen, however, it is sometimes difficult to say in which of these categories certain species belong. Animals that lay eggs that subsequently develop outside the body are said to be *oviparous*. Those that retain the eggs within the body, where they are incubated before hatching, are said to be *ovoviviparous*. Those whose developing young become attached to the reproductive tract of the female, from which they receive oxygen and nourishment, are said to be *viviparous*. Oviparity seems clear in fish such as trout that lay eggs which are even fertilized outside the body of the female and are left to develop by themselves. In birds we find that cell division leading to the development of the embryo has proceeded to a considerable extent in the few hours between fertilization and laying. Some reptiles may retain their eggs in the reproductive tract until incubation is well along, others until time of hatching. Thus we can find examples among vertebrates representing every stage between oviparity and ovoviviparity. Similarly there are species whose young develop within the body of the parent, to which the attachment may be mere contact or which may be extremely complicated. The simple contact between a vascular larval tail and the uterine epithelium, as a means of securing oxygen, might represent one extreme, and the complicated mammalian placenta and its associated fetal membranes connecting mother and young might represent the other.

To complicate matters more, there are vertebrates whose young develop in or on parts of the body other than the reproductive tract of the female, or may even develop within parts of the body of the male. For convenience, however, the terms oviparous, ovoviviparous, and viviparous are used here and examples of variations are mentioned.

## PREHATCHING OR PRENATAL PERIOD: OVIPAROUS AND OVOVIVIPAROUS SPECIES

Monotreme mammals and birds are classified as oviparous even though, as mentioned previously, some early development may occur prior to laying. Furthermore, the development of the young up to time of hatching is, with few exceptions, dependent upon heat from the body of the parent or parents during the incubation period. The majority of reptiles, amphibians, and fishes are either oviparous or ovoviviparous. The eggs of ovoviviparous species are of necessity fertilized within the body of the female. Fertilization may be either external or internal in oviparous species. External fertilization occurs in most marine and aquatic species that do not have their eggs encased in shells. In egg-laying mammals, all birds and reptiles, and some kinds of amphibians and fishes, the eggs are fertilized internally. In most, although not all, of these species, the eggs when laid are covered with leathery or calcareous shells which would prevent subsequent fertilization. An exception is seen in the tailed frog (*Ascaphus truei*) of western North America. Although the eggs of this amphibian are not encased in shells they are frequently deposited in swift-flowing mountain streams, where external fertilization would be very difficult.

### MAMMALS

The only egg-laying mammals living in the world today are the platypus (*Ornithorhynchus anatinus*) of Australia and the spiny anteaters, belonging to the genera *Tachyglossus* and *Zaglossus*, which occur in parts of New Guinea, Australia, and Tasmania. The female platypus constructs a nest of vegetation in a chamber within the breeding tunnel. The whitish, soft-shelled eggs, varying from one to three in number, are believed to require about 14 days of incubation before hatching (Burrell, 1927).

## BIRDS

All birds are oviparous and the majority of them construct nests of one sort or another in which the eggs are laid. There are some species, however, that do not lay their eggs in nests, and others that do not incubate their own eggs.

The fairy tern (*Gygis alba*) of the Pacific and southern Atlantic oceans lays a single egg on a rock or a horizontal branch without any nesting material whatsoever. Many birds belonging to the order Caprimulgiformes, such as the whip-poor-will, chuck-will's-widow,

FIGURE 13.1.   The western gull (*Larus occidentalis*) of North America builds a simple nest of available vegetation and feathers.

FIGURE 13.2.   The nest of the killdeer (*Charadrius vociferus*) frequently consists only of pebbles.

the poor-will, and nighthawks, lay their eggs on bare ground. Although the eggs of many members of this group are white, the protective coloration of the incubating adult makes them difficult to find. The killdeer (*Charadrius vociferus*) is a well known North American bird that frequently lays its egg on bare ground. Occasionally a small amount of nesting material such as bits of vegetation or pebbles are used (Fig. 13.2). The eggs of this species are protectively colored with dark blotches on a background of buff. They blend so well with the substratum that it is often difficult to distinguish them a few feet away. This is true of most shorebirds.

There are a few parasitic birds that lay their eggs in the nests of other birds and leave the problem of incubation as well as the rearing of the young to the host. The brown-headed cowbird (*Molothrus ater*) is a well known North American species that has this habit. Most remarkable, as regards their nesting habits, are the megapodes or brush turkeys inhabiting parts of Australia and the East Indies. These birds bury their eggs in mounds of decaying vegetation or in warm volcanic ash, and the heat of the nest itself does the incubating.

The majority of birds, however, construct some sort of nest in which the eggs are laid and incubated by one or both of the parents. Nesting sites are many and varied. Most petrels and shearwaters nest in holes in the ground or under rocks. Some kingfishers and swallows nest in holes in banks. Some hawks and cormorants nest on rocky ledges. Certain swifts and swallows attach their nests to vertical cliffs or even to man-made structures (Fig. 13.3). Many birds, including most woodpeckers, nest in holes in trees (Fig. 13.4). Other kinds nest on the ground, in low vegetation, or in trees. Grebes build floating nests on the water.

Not only do the nesting sites selected by birds vary greatly with the species but so also do the materials used in nest construction, as well as the shape and form of the nest. Many small passerine species construct an open, cup-shaped nest out of grass stems, plant fiber, and rootlets with other materials such as plant down, feathers, and animal hair frequently incorporated into the lining. Larger birds building cup-shaped nests use larger materials. Jays and crows make nest of twigs (Fig. 13.5). Golden eagles use small limbs to construct their bulky nests, which are sometimes used repeatedly year after year. Most members of the tit family (Paridae) nest in holes but some that do not, such as the bushtit (*Psaltriparus mimimus*), construct a long, covered, pendant nest with a hole on the side. The American

FIG. 13.3                  FIG. 13.4                  FIG. 13.5

FIGURE 13.3.   The nest of the barn swallow (*Hirundo rustica*) is made of mud mixed with straw, feathers, and sometimes horsehair.

FIGURE 13.4.   Nest of a red-breasted nuthatch (*Sitta canadensis*) in a cavity in an aspen (*Populus tremuloides*). A section of the trunk has been removed.

FIGURE 13.5.   The open, cup-shaped nest of a Steller jay (*Cyanocitta stelleri*) in a Douglas fir (*Pseudotsuga Menziesii*).

robin (*Turdus migratorius*) strengthens its cup-shaped nest with mud. Some swallows build cup- or retort-shaped nests almost entirely of mud. Most swifts glue their nesting material together with saliva, but some species belonging to the genus *Collocalia* of Asia and the eastern Pacific area make their nests entirely out of salivary secretions. These are the nests from which bird's nest soup, a delicacy in the Orient, is made.

In general, among birds there tends to be a direct correlation between body size and egg size. Hummingbirds produce the smallest eggs and the ostrich the largest. However, the eggs of species whose young are hatched in an advanced state of development are usually larger than those of comparable sized species whose young are entirely dependent on their parents when hatched. Birds that nest in holes generally have eggs that are white or whitish. Most, though not all, birds that nest in the open tend to have colored eggs. The eggs may be of a solid color or they may be strikingly decorated with spots, blotches, or lines.

The number of eggs laid may vary from one every other year by the California condor to 20 or more in a single clutch in some gallinaceous birds. The period of incubation ranges from as little as 11 days in some passerine birds to 11 or 12 weeks for the royal albatross (*Diomedea epomophora*). In certain birds, such as humming-birds, only the female incubates. In others, such as phalaropes, only the male incubates. In a great many species, however, both sexes participate in incubation.

## REPTILES

Turtles and tortoises are oviparous and generally bury their eggs in sand, mud, or soil in an open situation (Fig. 13.6). The eggs tend to be spherical or oval, and the number laid varies with the species and with the age and size of the female. They are usually hard-shelled although in some species the shell may be quite thin.

The female of the common snapping turtle (*Chelydra serpentina*) may travel a mile or more from the home pond or stream until she finds a suitable spot in which to lay her eggs. These are deposited in a cavity which she digs with her hind legs. After covering the eggs she returns to the water. The eggs hatch in from 12 to 16 weeks.

The western pond turtle (*Clemmys marmorata*) lays five to 11 eggs. These are buried in a hole which the female digs with her hind legs. Open, sunny ground is selected for this purpose.

The loggerhead turtle (*Caretta caretta*) generally comes to shore at night to lay. The eggs, numbering 50 to 1000, are deposited in a hole dug with the front flippers in the sand above high tide. Imme-

FIGURE 13.6.   A green turtle (*Chelonia mydas*) returning to the sea after laying her eggs. (By Fritz Goro, Life Magazine.)

diately after the eggs are covered, the female returns to the sea. The eggs hatch in about two months or slightly more.

Observations made by Stuart (1954) on the desert tortoise (*Gopherus agassizi*) in captivity showed that the female digs a hole with her rear feet, using them one at a time, from one to four days before laying. After the two to nine eggs are laid, the hind legs are again employed in covering the hole. Digging and laying have been observed in the late afternoon, sometimes continuing until after dark. Artificially incubated eggs hatched 80 to 118 days after being laid.

The females of some species of turtles are capable of retaining viable spermatozoa within their reproductive tract for a very long time. The diamondback terrapin (*Malaclemys terrapin*) has been reported as laying fertile eggs for four years following a single mating.

All crocodilians lay eggs which are deposited in either a hole or a nest. The nest of an American alligator (*Alligator mississippiensis*), reported by Ditmars (1922), consisted of a water-soaked mound of twigs, moss, and debris about 5 feet in diameter and 2 feet high. It contained 38 white, oval, hard-shelled eggs measuring 3¼ inches by 1¾ inches. They were arranged in two layers at the bottom of the nest. The incubation period is about nine or 10 weeks. Sometimes the female returns to the nest during incubation.

Although most lizards and snakes are oviparous, there are some that are ovoviviparous and even a few that are definitely viviparous (see p. 463). The eggs of oviparous lizards are white or cream-colored and usually have tough but flexible shells. In some of the geckos the eggs are hard-shelled.

Size in reptiles seems to be a determining factor as far as the number of young in a brood or litter is concerned. Large snakes and lizards tend to have more young than small species. Furthermore, the young are relatively larger in small broods or litters than in larger ones. Snakes are more prolific than lizards, usually having 2 to 16 young at a time, whereas most lizards, according to Fitch (1970), have only 2 young per litter or brood. There are always exceptions, however: certain snakes, especially boas and pythons, have been known to lay 80 to 100 eggs; and some of the large iguanid lizards may produce clutches of 30 to 60 eggs. Age also is a factor, older females producing more eggs or young than *primiparous* females. In *Python reticulatus*, females from 10 to 11½ feet long have clutches ranging from 14 to 16 eggs, whereas those of maximum length, in excess of 25 feet, may have as many as 100 in a clutch.

Many lizards bury their eggs in earth or sand. Some deposit them in rotten logs, under rocks, or in crevices. The females of

FIGURE 13.7.   The common gecko (*Gecko gecko*) of southeastern Asia may attach its eggs to rocks. Three such eggs, with one hatching, are seen in the above picture. (Photograph courtesy of Steven C. Anderson.)

several North American species of skinks (genus *Eumeces*) guard
their eggs in subterranean chambers until the young have hatched.
This requires five to six weeks. The Gila monster (*Heloderma suspectum*) lays four or five very large eggs, ranging from about 2½ to 3
inches in length. These are deposited in a hole, 3 to 5 inches deep,
which the female digs in damp sand in a sunny location. The incubation period for this species is said to be about a month. The eggs of
the western whiptail (*Cnemidophorus tigris*) laid in captivity have
hatched in 80 to 82 days. The incubation period for oviparous
species of horned toads (genus *Phrynosoma*) is two months or slightly more.

The eggs of snakes, like those of lizards, are white or cream-
colored. They usually have flexible shells and are quite elongate
(Fig. 13.8). The number of eggs laid varies from one to more than
100, depending on the species. Sometimes the eggs adhere to one
another. Many snakes bury their eggs in soft, often moist soil. Sometimes they are deposited in rotten logs or under stones. The eggs of
the common whipsnake (*Masticophis flagellum*) of southern United
States and Mexico have been found buried in soil to a depth of 11
inches.

Few snakes care for their eggs once they have finished laying.

FIGURE 13.8.   Blue racer (*Coluber constrictor*) laying eggs. (By Fritz Goro, Life
Magazine.)

The female python, however, after laying is completed, arranges the eggs in the form of a pyramid about which she coils her body. The head is placed over the top of the eggs so as to partly cover them from above. The body temperature of the female python is reported to increase from 5.5 to 7.5° F. during incubation. During the six week period that she incubates she occasionally leaves to drink. The mud snake (*Farancia abacura*) of North America has also been found to coil around its eggs, although it is not known to be able to increase its body temperature.

The incubation period for the eggs of many snakes ranges from two to three months under natural conditions. Under conditions of captivity it may be shorter. The eggs of captive western ring-necked snakes (*Diadophis amabilis*) have hatched in six weeks. There are records of the eggs of the smooth green snake (*Opheodrys vernalis*) hatching in from four to 23 days, although this suggests that incubation has progressed quite far before the eggs were laid (Stebbins, 1954).

All our North American rattlesnakes are ovoviviparous. The eggs have a parchment-like covering around them and are retained in the female reproductive tract until they are ready to hatch.

## AMPHIBIANS

Although the majority of amphibians are oviparous, there are some ovoviviparous species and a few that might be classified as viviparous.

Amphibian eggs are usually covered with one or more gelatinous coats. Sometimes the outer layer is quite sticky so that the eggs may remain attached to one another or to some object. The number of eggs laid by a female varies greatly with the species. The diminutive Cuban frog (*Sminthillus limbatus*) lays a single egg. The red-backed salamander (*Plethodon cinereus*) of eastern North America may lay as few as three eggs. The opposite extreme is found in the Woodhouse toad (*Bufo woodhousei*) of the Great Plains which is reported to lay as many as 25,000 eggs. The anurans in general tend to produce more eggs than the caudate amphibians.

Amphibian eggs may be laid singly, in clusters of various sizes, in flat masses, in strands, or strung together like beads. They may float freely in water, they may be attached to stones, aquatic plants, or other submerged objects, they may be buried in soil, placed in holes in the ground or in cavities in trees. Some species build various sorts of nests in or out of the water. The developing eggs may even be attached to or contained within parts of the bodies of either sex. Eggs that are deposited in well lighted situations are usually pigmented to some extent, whereas those that are laid in dark places or retained within the body generally lack pigment.

Some remarkable adaptations have developed in this class of vertebrates for the protection of the eggs and the young. In certain South American tree frogs of the genus *Hyla*, the male builds a circular enclosure of mud along the shallow edges of ponds. The female lays in these inclosures or cells and the eggs are thereby protected from attack by aquatic predators. There are Asiatic tree frogs of the genus *Rhacophorus* that build a nest of foam. The foam is produced as a result of both the male and female beating the egg mass with their hind feet. When the outer layer of foam dries it forms a protective crust, but the air contained within the bubbles inside provides oxygen for the developing young. Another interesting type of nest is constructed by the tree frog (*Phyllomedusa sauvagei*) of Paraguay. The female of this species at first lays empty egg capsules which she uses to glue the tips of several living leaves together, thus forming a cuplike chamber suspended from a branch. The eggs are laid in this nest and then covered with more empty capsules giving them protection from above. The South American leptodactylid frog, *Engystomops pustulosus*, deposits its eggs in a white, frothy mass that floats on the water but usually is attached to some object above the surface. Sexton and Ortleb (1966) discovered, as a result of a series of experiments, that in this species the oviposition site seems to be dependent upon the presence of a solid vertical surface intersecting the water.

There are other South American anurans that carry the eggs or the young on or in the body (Figs. 13.9 to 13.12). A well known example is the Surinam toad (*Pipa pipa*). In this species the female has an elongated ovipositor which permits her to lay eggs on her own back. The male then fertilizes them and pushes them down into the dorsal skin of the female. A crust soon forms over each egg so that it is inclosed in a small cell of its own. In the hylid frog, *Ceratohyla bubalus*, the eggs and young are also carried on the back of the female (Fig. 13.9). The female Riobama pouched frog (*Gastrotheca riobamae*) of South America has a pouch on her back which has a small posteriorly directed opening. The eggs are incubated in this pouch (Fig. 13.11). The Chilean frog (*Rhinoderma darwini*) is unique in that the male incubates the eggs in his vocal pouch until the young hatch as small, fully formed frogs after completing metamorphosis.

The great majority of our North American amphibians lay their eggs in water, where the young develop unattended by either parent. There are some species of salamanders, however, that do guard the eggs. The arboreal salamander (*Aneides lugubris*) of California, for example, lays its eggs in cavities in trees or rotten logs or in holes in

FIGURE 13.9.   A female South American hylid (*Ceratohyla bubalus*) with young
on her back. (Photograph of specimen in the American Museum of Natural History
by Alan Leviton.)


the ground and one or both parents brood them. The eggs of Eschscholtz's salamander (*Ensatina eschscholtzi*) are guarded by the female under litter or rotten bark or in an animal burrow. In the red
salamander (*Plethodon cinereus*), another plethodontid species in
which larval development occurs prior to hatching, the female
guards her eggs from the time of laying in June until August, when
the young emerge. She may even stay with the hatchlings until they
leave the nest two or three weeks later. A number of other species
have similar habits.

The incubation period for caudate amphibians is generally longer than for anurans. The eggs of *Amphiuma* require about four
months to hatch. Those of *Necturus* hatch in two to two and one-half
months. The crested triton (*Triton cristatus*) of Europe has a relatively short incubation period of 13 days. The eggs of the red-legged
frog (*Rana aurora*) of North America hatch about six weeks after they
are laid, whereas those of spadefoot toads (*Scaphiopus*) may hatch in
36 hours. The eggs of the black toad (*Atelopus*) of Uruguay hatch
within 24 hours.

FIGURE 13.10.    In some of the South American brachycephalid frogs the males
guard the eggs and transport the tadpoles on their backs. The above picture is that
of a male nurse frog (*Phyllobates bicolor*) with larvae attached. (Photograph taken
near Tingo Maria, Peru, by Edward S. Ross.)

FIGURE 13.11.    A female Riobama pouched frog (*Gastrotheca riobamae*). The
bulges on the back are caused by eggs in the pouch. (Photographed in Ecuador by
Edward S. Ross.)

FIGURE 13.12. A dorsal view of a female Ecuadorian hylid frog (*Amphignatho-don guntheri*) showing the slitlike entrance to the pouch on the back in which the eggs are incubated. This is the only frog with true teeth in both jaws. (Photograph by Alan Leviton.)

## FISHES

Fishlike vertebrates exhibit even more diversity than amphibians with regard to eggs and their care during the developmental period.

The majority of fishes are oviparous and their eggs are laid in water or at least where they will be kept wet by wave action. Some species lay eggs that have a relatively low specific gravity and float freely. These are referred to as *pelagic*. Others produce eggs of higher specific gravity that sink. These are called *demersal*. Demersal eggs are usually laid at the bottom and frequently have a sticky outer coat that causes them to adhere to objects or to each other.

The eggs of oviparous fishes usually have some protective outer covering. Those of salmon, for example, are encased in a tough but permeable outer membrane called the chorion. Those of some sharks are surrounded not only by layers of albumen but also by a shell. In general, fishes that lay their eggs in some type of nest or that guard them produce far fewer ova than those that do not. The stickleback

(*Gasterosteus*) lays less than 100 eggs. Sea horses may produce 100 to 600 eggs. By way of contrast, sturgeon and cod have been estimated to lay several million eggs in a season. The number of eggs laid also varies with age, older females producing more than younger females.

There is great variation in the size of the eggs of various fishes. Those of some oviparous sharks may be several inches in length, while it is estimated that it would take about half a million eggs of the white bass (*Roccus chrysops*) of central United States to fill a quart jar.

Although there are many kinds of oviparous fishes that do not care for the eggs once they have been laid and fertilized, there are others that have developed various methods of providing the eggs with protection. They may be guarded by one or both of the parents, they may be buried, they may be placed in nests or carried on or in parts of the body.

The eggs of the lingcod (*Ophiodon elongatus*) are deposited in adhesive masses in protected rocky situations below the low tide level and guarded zealously by the male. The bluntnose minnow (*Pimephales notatus*) lays its eggs under stones or other objects in the water, where they are guarded by the male. The male bowfin (*Amia calva*) clears a circular area in quiet water and guards the eggs

FIGURE 13.13.   The male smallmouth bass (*Micropterus dolomieui*) guards the eggs until they hatch.

that the female deposits here. The males of the smallmouth bass (*Micropterus dolomieui*) and the largemouth bass (*M. salmoides*) both dig a circular depression with the tail for the eggs. The latter are placed in the bottom of the nest and guarded by the male. The eggs of the sand goby (*Gobius minutus*) are both guarded by the male and aerated as a result of his movements.

There are other fishes that bury their eggs. Salmon and trout come under this category. In this group a depression or nest called a *redd* is dug in gravel or sand by the female. After the eggs have been fertilized the female moves a short way upstream and churns up the sand and gravel so that it is carried down with the current and covers the eggs. Lampreys also bury their eggs in gravel. Some of the killifishes of South America and Africa live in small disconnected ponds that may dry up during periods of drought. Before this happens, however, the adults spawn and the tough-coated eggs are capable of surviving in the mud until the arrival of the rainy season.

The stickleback (*Gasterosteus*) makes an elaborate nest, somewhat like that of a bird, in which the female lays less than 100 eggs. The nest and eggs are guarded by the male. The male blue gourami (*Trichogaster trichopterus*) builds a nest of bubbles. This is done by coming to the surface, drawing air into the mouth, where it is enveloped in mucus, and releasing the bubbles by blowing them out so

FIGURE 13.14. A male jawfish (*Opisthognathus*) with its head protruding from its burrow. The mouth is full of eggs. This is one of the very few marine mouth-breeders. (Photographed at Steinhart Aquarium.)

they will become attached to some floating aquatic vegetation. After the female lays, the male picks up the eggs and blows them with numerous bubbles into the nest and then guards the nest. There are several families of fishes that are known as mouth-breeders. In these fishes one or sometimes both parents incubate the eggs in their mouths (Fig. 13.14).

As was noted in Chapter 9, p. 328, the bitterling (*Rhodeus amarus*) has developed the unique habit of laying its eggs in the siphon of a mussel where they develop (Fig. 13.15). In order to do this the female, under hormonal influence, develops an elongate tubular ovipositor.

There are fishes that have developed methods of carrying the eggs on the body much like certain amphibians. They may adhere to the underside of the body or underside of the tail or be enclosed in some sort of pouch. The pouch may be abdominal or subcaudal. It

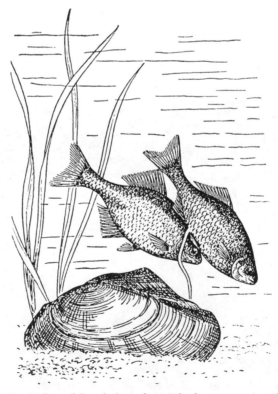

FIGURE 13.15. Male and female bitterling (*Rhodeus amarus*) with the ovipositor of the latter inserted into a pond mussel. (After Norman, in Young, J. Z., The Life of Vertebrates. Oxford, Clarendon Press, 1950.)

FIGURE 13.16. A male sea horse (*Hippocampus hudsonius*) showing the slitlike opening of the abdominal pouch in which the eggs are incubated.

may consist of lateral folds on either side which do or do not join. If they join they may or may not be fused. In the male sea horse (Fig. 13.16) the lateral abdominal folds fuse throughout much of their length, leaving a narrow opening in front for the insertion of the eggs.

The incubation period for the eggs of fishes varies with the species. The eggs of the blue gourami hatch from 22 to 24 hours after fertilization. Those of the white bass require about two days. The eggs of the muskellunge and pickerel (genus *Esox*) hatch in one to two weeks. Those of rainbow trout require about 35 days, while Atlantic salmon eggs require about 68 days' incubation. The sea horses incubate their eggs from four to six weeks.

## PREHATCHING OR PRENATAL PERIOD: VIVIPAROUS SPECIES

### MAMMALS

All mammals except the monotremes are viviparous. The period of intrauterine life, however, varies greatly in different groups. It is shortest in marsupials. The gestation period for the opossum (*Didel-phis marsupialis*) is about 13 days. For many mice it is approximately three weeks and for the domestic rabbit it is 32 days. In general there is a direct correlation between body size and the length of the gestation period although, as will be seen, there are many exceptions to this rule. The gestation period for the white-tailed deer (*Odocoileus virginianus*) is about 200 days, for the moose (*Alces alces*) approximately eight months, and for the elephant 20 months. Some of our small insectivorous bats whose bodies weigh less than that of a house mouse have a gestation period of several months and, as has been pointed out in the previous chapter, fertilization in certain of these species may occur some months after insemination. The proportionately long period of development may be associated with the heterothermic condition characteristic of these animals. It seems likely that embryonic development is retarded when the temperature of the female drops to approximately that of the environment during the daily periods of rest as happens with some insectivorous bats.

There are some mammals in which there is delayed implantation which greatly prolongs the gestation period. Studies by Wright (1942) and Wright and Rausch (1955) have shown that certain mustelids, such as weasels, marten, and wolverine, breed during the summer months and the embryos develop up to the blastocyst stage. Development then is arrested until midwinter or slightly later, at which time the blastocyst becomes implanted in the uterine wall and resumes growth, with the young being born in the spring. To all appearances, the gestation period would seem to be about three-fourths of a year, but actual embryonic growth occupies less than half of this time.

Most large herbivores as well as cetaceans, pinnipeds, and many kinds of bats have one or, less commonly, two young at a time. The number of young in a litter is quite variable in insectivores, carnivores, and rodents. Most shrews have three to seven young, wolverines two to four, weasels as many as nine, wolves generally five to eight, bears two to four, voles five to seven, occasionally more, and some ground squirrels may have 10 young in a litter. The opossum may successfully rear 12 young.

The number of young in a litter as well as the number of litters per year is greater in species in which the mortality rate is high. Meadow voles and white-footed mice are prolific breeders in many parts of North America. Females may produce a number of good sized litters a year. These small rodents, however, are continually being preyed upon by carnivores, hawks, owls, and snakes. This necessitates a high reproductive potential.

Another factor that may affect not only the number of litters produced per year but also the number of young in a litter is the length of the breeding season. Species living in regions where mild climatic conditions permit an extended breeding season often have smaller litters, but more litters per year, than closely related species inhabiting areas where the reproductive season is short because of unfavorable climate throughout much of the year. By way of example, the Audubon cottontail (*Sylvilagus audubonii*), which lives in the lower, arid or semiarid parts of western North America where the winters are relatively mild, has been found to breed throughout the year, although there are peaks, locally (Ingles, 1941). In California, where evidence indicates that many females bear more than two litters of young a year, the average number per litter is slightly more than three. The Nuttall cottontail (*Sylvilagus nuttallii*), another western North American species inhabiting parts of the Great Basin and Rocky Mountain regions, where the summers are relatively short and the winters long and rigorous, has one or at most two litters a year, but the number of young per litter is about six (Orr, 1940).

The arctic shrew (*Sorex arcticus*), whose range is principally confined to Alaska and Canada, is reputed to have four to 10 young per litter (Asdell, 1964). On the other hand, the Trowbridge shrew (*Sorex trowbridgii*), a Pacific coastal species, was found by Jameson (1955) to have from three to six young per litter in California and was suspected of having two or three litters a year.

As in many other kinds of vertebrates it appears that age has a bearing on fertility. Hatfield (1935) found that young meadow voles (*Microtus californicus*) bear two to four young at a time, while older females may have five to seven in a litter.

## REPTILES AND AMPHIBIANS

There are no viviparous birds, but true viviparity has been found in some lizards and snakes and there are amphibians in which the developing embryos depend upon maternal tissue for respiration. The presence of chorioallantoic and yolk-sac placentae in certain reptiles is significant in that it possibly suggests the manner in which mammalian placentation may have developed.

Weekes (1935), in her comprehensive study of this phase of reptilian reproduction, divides placentation in this class of vertebrates into three types. In the first and simplest type the eggs show no reduction in yolk content. During development in the oviduct, however, there is a certain amount of degeneration of both the embryonic and maternal epithelium to permit close apposition of the blood systems of parent and embryo. In the second type the maternal blood vessels are raised into folds with glandular tissue between the folds.

The chorion of the embryo which contacts the uterine wall is also frequently thick and glandular. In the most advanced type of reptilian placenta a specialized part of the oviduct develops vascular and glandular folds which interdigitate with comparable embryonic folds. The placental contact may be of a yolk-sac or allantoic type, not unlike that of many mammals having *indeciduate placentae.* The latter term implies that there is no loss of maternal tissue at time of birth. Most viviparous reptiles in this third category have a marked reduction in the yolk content of the eggs, suggesting that much of the embryonic nourishment is derived from the female.

Viviparity is best known in some of the Australian lizards of the family Scincidae and snakes of the genus *Denisonia.* However, it is by no means confined to Australian reptiles. The adder (*Vipera berus*) of Europe as well as certain African and South American reptiles is viviparous. In North America the developing young of the common garter snake (*Thamnophis s. sirtalis*) are attached to the reproductive tract of the female. Clark, Florio, and Heerowitz (1955) concluded that in this species the placenta serves not only for respiration but also for the transfer of amino acids from the parent to the embryo. Common garter snakes generally mate between the middle of March and late May and produce young between July and early September. In Michigan the period of embryonic development has been found to range from 87 to 116 days. There is one record of a female giving birth to 73 young. The northwestern garter snake (*Thamnophis ordinoides*) produces three to 15 young but has an intrauterine period of development of only about nine weeks. Most horned lizards of the genus *Phrynosoma* are oviparous, a primitive character retained by many reptiles. However, two species (*P. douglassii* and *P. orbiculare*) are viviparous. *Phrynosoma douglassii* is known to have produced 31 young in a litter, although the average is about 15. *P. douglassii* is of relatively northern distribution as well as montane, and *P. orbiculare* is montane. Other iguanid lizards that are viviparous are also generally either montane or northern in occurrence. This is of adaptive significance in species living in regions where the growing season is short. The value of viviparity to reptiles also includes the elimination of necessity to find a suitable place for egg deposition and to protect the vulnerable eggs from numerous predators.

Although there is no true placental formation in amphibians, the East African toad (*Nectophrynoides vivipara*) might be considered as viviparous. The young, sometimes numbering over 100, develop within the reproductive tract of the female. The tails of the embryonic larvae are very vascular and are kept in contact with the uterine wall by means of which they obtain their oxygen.

## FISHES

As in reptiles and amphibians there are some fishes that border between the ovoviviparous and viviparous. Others, however, are definitely viviparous, with an intimate contact existing between the developing embryo and the female. In certain cartilaginous as well as bony fishes there is a placental attachment between the yolk sac of the embryo and the wall of the oviduct enabling the former to obtain nourishment and oxygen from the mother. The gestation period of the smooth dogfish (*Mustelus canis*) of the Atlantic is reported to be 10 months. In some rays a placental attachment is lacking, but the uterine wall becomes glandular and secretes an albuminous milk for the nourishment of the embryo.

In certain bony fishes the embryo obtains nourishment by means of various attachments to the wall of the ovary. Ovarian folds may grow into the mouth or become attached to the urogenital region or to the dorsal fin of the embryo. The latter condition occurs in the viviparous sea-perches of the family Embiotocidae along the shores of the North Pacific.

There are other species of fish in which the ova are fertilized while still within the ovarian follicle. The egg may be released from the follicle after it has completed the early cleavage stages or it may remain and undergo complete development with rupture occurring when the young is ready to be born.

## HATCHING AND BIRTH

The process of emerging from the egg capsule by the young is referred to as *hatching*. In some vertebrates this may occur at a very early stage of development, while in others the young hatch in an advanced state. It may result from mechanical or chemical action. The young platypus is thought to break through the thin shell by applying pressure with the forepaws and the caruncle or protuberance on the top of the bill.

Young birds by the time they are ready to hatch have a horny protuberance called an *egg tooth* present on the tip of the upper mandible. By means of this egg tooth they are able to fracture or "pip" the shell anywhere from several hours to several days prior to hatching. The weakening of the shell as a result of this action combined with the pressure applied by the young ultimately results in its emergence. Sometimes the parent lends some assistance. In species whose young remain in the nest any length of time, the parent removes the shells immediately after hatching. This probably serves to remove odors that might attract predators. The egg tooth is lost usually a few days after hatching.

The young of oviparous reptiles develop an egg tooth or else a horny projection on the front of the head to assist in breaking or rupturing the shell. The egg tooth of lizards and snakes is a true tooth that is attached to the premaxillary although it is generally lost a few days after hatching. Young turtles and alligators have a spine on the front of the mouth that is shed when they are one to several weeks old (Fig. 13.17).

In most ovoviviparous reptiles the egg tooth is poorly developed or lacking, since the egg membrane is very thin and easily broken by pressure on the part of the young. In some species the female assists in the process. Cowles (1944) observed the female of a yucca night lizard (*Xantusia vigilis*) to tear the egg membranes as they protruded from her cloaca. After the young hatched, the female proceeded to swallow the empty membranes.

A few amphibians whose eggs are not laid in water develop an egg tooth, somewhat comparable to that of birds and certain reptiles, which assists the young in rupturing the egg capsule. Most amphibians, however, prior to hatching develop glands on the snout which secrete a chemical that digests the egg capsule, thus freeing the young. In certain frogs these glands also occur on the back. The mechanical action of pressure from within is also probably an impor-

FIGURE 13.17. A young snapping turtle (*Chelydra serpentina*) emerging from the shell. (Photograph by Hal H. Harrison, National Audubon Society.)

tant factor assisting in the liberation of the young of many amphibians.

In many kinds of oviparous fishes, the egg capsule becomes soft and easily ruptured around the time of hatching as a result of the action of enzymes secreted by glands which are located principally on the head of the embryo. The enzymes mix with the perivitelline fluid beneath the capsule, and when the latter is sufficiently weakened muscular action on the part of the young results in its liberation.

Although, as has already been indicated, there are some cold-blooded vertebrates that are viviparous this is primarily a mammalian character except in the few living monotremes. At the termination of this gestation period the process of birth or *parturition* results in the emergence of the young from the female genital tract. This is accomplished by uterine contractions which result in the rupture of certain fetal membranes. During the process the female usually bites the umbilical cord. Following the birth of the young, the afterbirth, consisting of the placenta and associated membranes, is expelled.

The birth of only relatively few wild mammals has been observed. In most instances the young emerge head first but there are exceptions. Breech presentation with the rump and tail appearing first appears to be normal among bats. Bottlenosed dolphins as well as some rodents and carnivores have a breech presentation. The females of most land mammals that have been observed during parturition carefully lick the young as soon as they are born and also eat the afterbirth. This reduces the chances of predators finding the young and preserves the cleanliness of the nest or den in species having such special places for their offspring. It is probable also that the placenta and fetal membranes provide the female with hormones that stimulate lactation.

## POSTHATCHING OR POSTNATAL PERIOD

### MAMMALS

Young monotremes are very immature at time of hatching. Although capable of movement they are blind, naked, and entirely dependent on the mother. The platypus is said to be six weeks old before the eyes are opened, at which time the fur is about ¼ inch in length.

Among viviparous mammals there is great diversity in the degree of development of the young at birth. The terms *altricial* and *precocial* are sometimes used to express this. Altricial animals are those that are helpless at birth or hatching and require parental care for some time. Precocial animals are those that are able to move about and care for themselves to a considerable extent immediately following birth or hatching.

Marsupial mammals are definitely altricial. Young opossums (*Didelphis marsupialis*) are born in a very premature condition. Only their front limbs show advanced development. By means of these appendages combined with a *negative geotropism,* or desire to move opposite to the pull of gravity, they are able to climb of their own accord up into the mother's pouch or *marsupium.* By the time a young opossum is 50 days old it is sparsely haired and about the size of a house mouse. It is about 70 days old before it ventures out of the pouch on small excursions (McCrady, 1938).

Most insectivores, bats, carnivores, and many rodents are altricial (Fig. 13.18). The young of some species are naked at birth. The eyes and ears may not be functional for some days. Hatfield's (1935) observations on California voles (*Microtus californicus*) showed that the eyes of the young did not open until they were 10 days old. The ears were closed at birth and captive-born young did not react to sound until they were from five to seven days old. Dermal pigment first appeared on top of the head eight hours following birth. Later it spread to other parts of the body and hair made its appearance on the head at 36 hours. The juvenal pelage was complete by the fifth day and was retained until the end of the third week, when it was

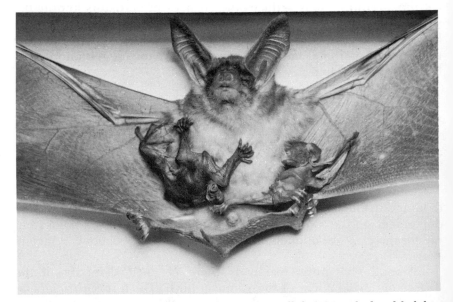

**Figure 13.18.** Young pallid bats (*Antrozous pallidus*) are naked and helpless when born.

replaced by the postjuvenal pelage. The latter was succeeded by the adult pelage at the eighth or ninth week following birth.

Not all rodents are altricial. In fact the porcupine (*Erethizon dorsatum*) is one of the most precocial of North American mammals. The gestation period in this species is long, averaging about 30 weeks. Furthermore, the single young (rarely two) is proportionately large at birth. An adult female weighing 12 pounds may produce a young that weighs 1 pound. The latter is actually, as well as proportionately, larger than a newborn black bear. The period of suckling, if any, is only a matter of several days and the young with well developed quills, sense organs, and teeth is prepared to defend itself and secure its own food.

Young pinnipeds are active immediately after birth and are even capable of swimming, although they cannot survive long in the water at this early age. In many of the Phocidae the pups are weaned when they are a few weeks old. In the Otariidae, however, the young may be dependent upon milk from the mother for nearly a year.

In colonial mammals, such as certain seals and sea lions that gather in large numbers at rookeries during the reproductive season, parent and offspring recognition is very important since females will only suckle their own young. The location of the young by a female returning from the sea appears to be largely by vocal utterances. Among Steller sea lions, for example, the female after being away some time returns to the general rookery area and calls. She is soon answered by her own pup, and after coming out of the water works her way over to the young animal. The final confirmation is effected by smelling the young. Sound and smell, therefore, play a major part in parent-offspring recognition. As the pups grow older and gather into groups this method of recognition becomes more important.

Rapid growth is sometimes essential for survival. Mayer (1953) found that young arctic ground squirrels (*Citellus undulatus*) attain the weight of adults in less than two months and are prepared to enter hibernation no later than October, when they are less than four months old (Fig. 13.19).

Among North American lagomorphs, hares of the genus *Lepus* are precocial (Fig. 13.20), whereas rabbits of the genus *Sylvilagus* are altricial. Young hares are born in what is termed a "form," which consists essentially of a depression in the ground that may or may not be partly protected by overhanging shrubbery. They are fully furred, have their eyes and ears open, and are capable of hopping about shortly after birth. Young rabbits are born in nests which are frequently lined with fur from the mother's body. They are essentially naked at birth and their eyes and ears do not function for some days.

The young of large herbivores are well advanced at time of birth and some may be able to follow the mother within a short period of

**FIGURE 13.19.** Young arctic ground squirrels (*Citellus undulatus*) grow very rapidly. (Photograph by John Koranda.)

**FIGURE 13.20.** Young varying hares (*Lepus americanus*) are fully furred at time of birth. (Photograph by Gayle Pickwell.)

time. In some members of the deer family the young are spotted. This probably serves to make them less conspicuous during the early, vulnerable period of life. Growth, however, is quite rapid. Young moose (*Alces alces*) are said to gain 1 or 2 pounds a day the first month and during the second month the growth rate increases to 3 to 5 pounds per day. White-tailed deer (*Odocoileus virginianus*) fawns double their weight by the time they are about 15 days old and may quadruple it by the time they attain an age of 30 or 40 days.

The young of some pinnipeds grow very rapidly and can increase their weight at a very remarkable rate. This is correlated with a very high fat content, which may exceed 50 per cent in some species. Young northern elephant seals (*Mirounga angustirostris*) may gain 200 pounds during the first four to six weeks following birth. The southern elephant seal (*Mirounga leonina*) is reported to gain as much as 20 pounds a day during the latter part of the nursing period, which is reported to be less than one month.

Unlike elephant seals, sea lions grow rather slowly. In most species the young nurse for nearly one year, and quite often one sees female California sea lions (*Zalophus californianus*) as well as female Steller sea lions (*Eumetopias jubata*) nursing both newborn young and yearling pups in harem areas.

FIGURE 13.21. A white-tailed deer (*Odocoileus virginianus*) fawn showing how the spotted juvenal pelage presents the illusion of sunlight and shadow. (Photograph courtesy of the U. S. Forest Service.)

**Figure 13.22.** Elephant seals at time of birth weigh 50 to 60 pounds. By the time they are 4 to 6 weeks old, as seen here, their weight may exceed 250 pounds.

## BIRDS

Birds, like mammals, may be altricial or precocial. These, how-ever, are relative terms and the degree of dependence of the young on the adults may range from total to none. The majority of birds are helpless at time of hatching and remain in the nest for days, weeks, or even months, depending on the species concerned. The nestling period, as this is called, is less than two weeks for many small passerine birds. For some hummingbirds it is slightly more than three weeks. Young golden eagles (*Aquila chrysaëtos*) remain in the nest from nine to 10 weeks, while California condors (*Gymnogyps californianus*) have a nestling period of five months. Young altricial birds may be entirely naked at time of hatching, as is true of most woodpeckers, pigeons, pelicans, and their relatives; they may have a sparse coat of down, as most sparrows and finches; they may be heavily clothed with down, as are birds of prey, albatrosses, and petrels.

During the early nestling stage most young birds lack the ability to control body temperature and therefore must be brooded by the adults. Thermoregulation comes with the development of the feathers as well as other body systems. Nice (1937) in her studies on song sparrows (*Melospiza melodia*) found that the young are brooded more than half of the time during the first two days of life, although they are ready to leave the nest when 10 days old, having increased their weight tenfold since hatching.

Once the young have left the nest they are referred to as *fledglings*. Their dependence on the parents, however, continues for some time. In the song sparrow the parents care for the young another 18 to 20 days. Erickson (1938), in her studies of the wrentit (*Chamaea fasciata*), found that the young left the nest when about 16 days old but did not become completely independent of the parents until nine or 10 weeks of age, when they finally began to leave the home territory. According to Richdale (1942), the royal albatross (*Diomedea epomophora*) may be unable to fly, and therefore unable to secure its own food, until it is 243 days old.

In communal species such as the smooth-billed ani (*Crotophaga ani*) most members of a flock or colony may assist in the care of the young and the young of a first brood may assist in feeding a second brood.

Loons, grebes, ducks, geese, swans, rails, and their relatives, as well as shorebirds and gallinaceous birds, are among the major groups of precocial birds (Fig. 13.23). The young are well covered with down which is often protectively colored. They are usually able to leave the nest very shortly after hatching and are therefore said to be *nidifugous* as opposed to *nidicolous* species that remain in the nest for some time. Most birds of this type stay with the parent or parents in family groups for some weeks or even for many months. Young California quail (*Lophortyx californica*), for example, are able to run about within a few minutes after hatching. During their first few days they depend upon their protective coloring and ability to squat down and remain motionless to avoid detection when danger threatens. They are able to fly when about 10 days old. By midsummer several family groups may join together to form a flock or covey which may remain intact until the next breeding season.

The existence of a very interesting adaptation to provide water for the young of a precocial type of bird has recently been verified for sandgrouse of the genus *Pterocles*, birds of Old World desert regions questionably related to doves and pigeons. Studies by Cade and Maclean (1967) on these birds nesting in the Kalahari Desert region of South Africa have shown that the male is capable of transporting water to the young by means of feathers on his belly. These feathers have unusual water-retaining properties. After soaking them at a water hole, the male may carry the liquid as much as 20 miles to

**FIGURE 13.23.** Young avocets (*Recurvirostra americana*), like the young of many kinds of shorebirds, are able to move about and forage very shortly after hatching.

the young, who strip the feathers through their bills. This was reported at the end of the last century but hardly believed for many years.

Many young birds of this type develop fixations within the first few hours following hatching. This is referred to as *imprinting*. Incubator hatched young of gallinaceous birds, waterfowl, and certain other kinds frequently develop the habit of following their keeper or feeder. They may even become attached to inanimate objects. No doubt in the wild, imprinting serves a number of purposes, including parent recognition and early recognition of food seen eaten by or presented by the parents.

In many species the appearance of the parent bird at the nest triggers the young to extend their necks and open their mouths. In altricial species movement of the nest or the presence of any object above it will evoke the same reaction during the first few days following hatching. With gulls it has been found that bill recognition and contact is most important in the feeding process. In some species the adults have red or black spots on their bills. When the parent returns to the nest with food in its digestive tract, the young recognizes the bill and pecks at the spot. This in turn stimulates the adult to regurgitate food upon which the young feeds. In many passerine birds the young have a brightly colored mouth which, when opened as a response to the presence of the parent, stimulates the adult to place the food that it is carrying within the gape. Those with the widest gape usually receive the food.

Many oceanic birds nest in great colonies, with the young sometimes numbering hundreds of thousands. One might wonder how a parent returning from the sea ever finds its own young. Both visual and auditory recognition appear to be employed. This is well illustrated in nesting colonies or *crèches* of penguins. A returning adult comes to the general vicinity of the young, depending on sight to determine the location. It then calls and the chick is seemingly able to distinguish its own parents' voices from those of others.

Gulls and their relatives, though well clothed with protectively colored down and out of the nest a few days after hatching, are dependent upon the adults for food until they are able to fly (Fig. 13.24). The most precocial birds known are the megapodes of Australia, whose peculiar nesting habits have already been referred to. The young are entirely on their own and receive no parental care whatsoever.

FIGURE 13.24. Young gulls leave the nest a few days after hatching but are dependent on their parents for food until they are able to fly. (Photograph by O. J. Heinemann. Courtesy of the California Academy of Sciences.)

## Reptiles and Amphibians

Few reptiles or amphibians provide any parental care for the young. Among North American reptiles, the alligator (*Alligator mississippiensis*) is the only known species in which the female cares for her offspring. According to Oliver (1955) newly hatched alligators may be assisted in emerging from the nest mound by the mother, who is attracted by their sounds. She may stay with her young for from one to three years. Most reptiles, however, are on their own once hatched or born. They usually seek concealment immediately. Young horned toads will bury themselves in the ground.

Some young reptiles rely on the unabsorbed yolk sac for a considerable time. Studies by Bellairs, Griffiths, and Bellairs (1955) have shown that in the adder (*Vipera berus*), which is a viviparous species, the yolk sac is still large at time of birth and represents 8 per cent of the body weight. Although it is retracted into the body cavity, it is believed to provide the young with nourishment during their approximately four-week period of activity in the autumn prior to hibernation, as well as during the subsequent winter months when they are dormant.

While some amphibians of the New and Old World tropics care for their young, none of our North American species do so. Unlike mammals, reptiles, and birds, most newly hatched amphibians differ markedly from the adults and are referred to as larvae until they complete their metamorphosis. The degree of difference between the larval form and the adult is not as great among salamanders as it is among frogs and toads. Larval salamanders resemble the adults in many respects except that they possess external gills. As has previously been mentioned there are certain salamanders that remain permanently in the larval state. Larval frogs and toads differ so much from the adults in form that they are commonly referred to as tadpoles.

The time required from hatching to metamorphosis, when the adult form is assumed, varies greatly with the species and even within a single species may vary throughout its range, possibly because of temperature. Increased temperature appears to speed up development. The bullfrog (*Rana catesbiana*) has a larval period of one year in Louisiana. In New York it is two years and in Nova Scotia it may extend over three summers. The shortest larval period of any North American amphibians is to be found in the spadefoot toads of the genus *Scaphiopus*. *S. holbrooki* is said to have a larval period of about 12 days, and *S. couchi* young have developed legs in the laboratory 10 days after hatching. This rapid larval development in these amphibians is seemingly an adaptation to an arid or semiarid environment where breeding ponds are of irregular occurrence and of short duration for the most part. The young of some plethodontid salamanders, such as the red salamander (*Plethodon cinereus*), are first abroad about two or three weeks after hatching but do not attain maturity until the beginning of their third year.

## FISHES

The majority of fishes provide no care for their young. This is especially true of species laying large numbers of free-floating eggs and of those that bury their eggs. Those that guard the eggs, however, frequently care for the young for some time after hatching. Frequently this duty falls to the male.

The male blue gourami (*Trichogaster trichopterus*) tries to keep the young in the bubble nest after they hatch. Much of his time is spent in blowing the constantly straying young back into this nest. The males of both smallmouth and largemouth bass (*Micropterus dolomieui* and *M. salmoides*) guard the young. The male bowfin (*Amia calva*) tends the young until the end of their first summer.

Many young fish at time of hatching still possess a large yolk sac which provides them with the food necessary for growth during their first few days or weeks. Gradually the yolk sac is absorbed. In the muskellunge (*Esox masquinongy*) this requires about two weeks.

Certain fishlike vertebrates pass through a larval stage somewhat comparable to that of amphibians. Young lamprey eels differ so markedly from the adults that they were first thought to be separate species. Even today they are referred to as ammocoete larvae. Since the lamprey eggs are buried in sand and gravel, the larval young must work their way up to the surface. Once this is accomplished they drift with the current until they come to a soft substratum, into which they burrow and remain in the larval state for one or more years, depending on the species.

Young eels of the genus *Anguilla* hatch as larval forms called *Leptocephali*. American eels require one year to develop into elvers, whereas their European counterparts do not become elvers until their third autumn or winter.

## AGE CRITERIA

Growth in vertebrates may be either determinate or indeterminate. *Determinate* growth implies a steady increase in size up to maturity, after which the process slows down and essentially ceases for the remainder of life. *Indeterminate* growth implies a continuous increase in size throughout life.

Birds and mammals in general have determinate growth, while many reptiles, amphibians, and fishes have indeterminate growth. Even in indeterminate species, however, the rate of growth is not necessarily constant. There may be times when favorable food and climatic conditions speed up growth and other times when it slows down or stops entirely as a result of food shortage or drought, or during hibernation. Furthermore, the process of growth generally slows down as an animal becomes older.

There are many different ways to determine growth rate and age. Some of these are applicable to all vertebrates; others are specific to certain groups. The most accurate method is to capture, mark, and take standard measurements of newly hatched or born individuals. Similar measurements made subsequently as a result of repeated recaptures during succeeding days, weeks, months, or years provide accurate growth records under natural conditions. This procedure is, of course, not only difficult but very time-consuming. It is less difficult but also less accurate to obtain data from the observation and measurement of captive-reared animals.

The length-frequency method is often employed, especially in the study of animals having indeterminate growth. This is based upon the theory that if a representative and adequate sample of an entire population of one species is taken, certain individuals of the same age will tend to approach a mean length. Consequently, when the data are plotted there will be aggregations of individuals of successive ages in successive length groups. In this way age groups may be segregated.

Teeth often provide much information on the age of a mammal. Deciduous teeth may erupt through the gums in definite sequence and at definite times following birth. This also is true of the permanent teeth. Furthermore, the degree of wear of the latter in some species provides fairly accurate data on the age of adults.

Studies by Scheffer (1950) on marked fur seals (*Callorhinus ursinus*) have shown that annual growth ridges on the roots of the canine teeth are accurate indicators of age for the first few years of life (Fig. 13.25). These ridges are accounted for by the fact that dentine deposition is much more rapid during the winter months, when these animals are at sea where there is an abundance of food, than during the summers at the breeding rookeries. Similar annual growth rings have been noted on the teeth of certain other pinnipeds, on the claws of the Greenland seal, and on the horns of mountain sheep.

Chittleborough (1959) has summarized the most recent methods that have been developed to determine the age of whales. These include counts of the number of corpora albicantia in the ovary, study of the cortical layer of baleen plates which show seasonal

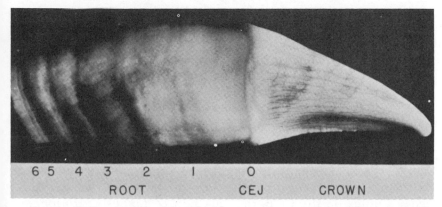

**FIGURE 13.25.** Annuli, or growth ridges, on the root of the right upper canine tooth of a 6-year-old male northern fur seal, June 17, 1947. Numbers identify the mid-summer fasting periods. *CEJ*, Cemento-enamel junction. About 3.4 times natural size. (From Scheffer, V. B.: Seals, Sea Lions, and Walruses. Stanford University Press, © 1958 by the Board of Trustees of Leland Stanford Junior University.)

variation in growth rate, and determination of the number of laminations, which increase by two each year, in the ear plug.

It has been shown that the weight of the lens of the eye, which increases up to a certain age, is another means of determining age in some mammals.

In many fishes that live in water where marked seasonal variations in temperature occur, growth may be greatly retarded in winter. This is often reflected in the growth rings or annuli on the scales and serves as a ready means of determining age (Fig. 13.26). There are also some fish bones, notably the vertebral centra, the otoliths or ear bones, and the operculars that show zones of bone depositions which

**FIGURE 13.26.** Scale of salmon: *1,* marks the edge of the smolt scale; *2,* the first winter zone formed in the sea; *3,* the spawning mark; *4,* the second winter zone formed after return to the sea. (After Regan, in Young, J. Z.: The Life of Vertebrates. Oxford, Clarendon Press, 1950.)

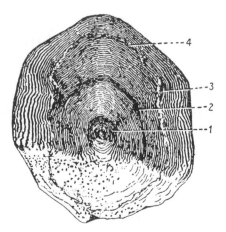

may be correlated with annual growth (Fig. 13.27). Other osteological features are also of value in determining age. By cross-sectioning the dorsal spine and preparing thin sections which may be cleared in 1% KOH, one can readily study the annual growth rings in such fishes as carp (*Cyprinus carpio*).

In mammals and to some extent in reptiles, the separation of the epiphyseal plates from many bones is an indication of immaturity. Among North American bats the joints of the bones in the wing remain swollen until ossification is completed during the first fall. Some mammals, especially the males of certain kinds of carnivores and pinnipeds, develop prominent ridges and crests on the skull with age. This is generally more marked in males than females. Recent studies on the California sea lion (*Zalophus californianus*) have demonstrated that the sagittal crest that is so prominent on the skulls of adult males does not manifest itself until the fifth year of life and does not attain full development until the tenth year.

The pelage of mammals as well as the plumage of birds also provides information regarding age up to maturity. Most young mammals, at time of birth or shortly thereafter, possess a juvenal pelage which is readily distinguishable from succeeding pelages by its color and by the proportionately fewer number of guard hairs. This is followed by a postjuvenal pelage which is more like that of the adult but still usually recognizable. These pelages are worn for rather definite periods of time by the young of any one species. The same applies to the various plumages described in Chapter 5 for birds. Although some young birds may acquire a plumage identical with

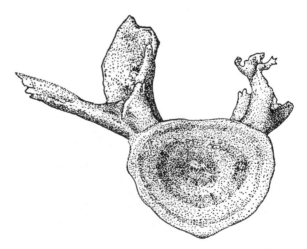

FIGURE 13.27. Centrum of trunk vertebra of a specimen of *Schilbeodes mollis*, 92 mm. in standard length, showing two dark bands (×24). (From Hooper, F. E.: Copeia, No. 1, April 15, 1949.)

that of the adult by the first winter, there are other species that require two, three, or even five years to attain adult plumage. Each of the various *subadult* plumages can be recognized and serves to indicate a particular age.

## References Recommended

Angel, F. 1947. Vie et Moeurs des Amphibiens. Payot, Paris.

Angel, F. 1950. Vie et Moeurs des Serpents. Payot, Paris.

Asdell, S. A. 1964. Patterns of Mammalian Reproduction. Ithaca, N.Y., Comstock Publishing Co., Inc.

Bellairs, R., Griffiths, I., and Bellairs, A. d'A. 1955. Placentation in the Adder, *Vipera berus.* Nature 176:657-658.

Blair, W. F. 1953. Growth, Dispersal and Age at Sexual Maturity of the Mexican Toad (*Bufo valliceps* Wiegmann). Copeia 1953:208–212.

Briggs, J. C. 1953. The Behavior and Reproduction of Salmonid Fishes in a Small Coastal Stream. Calif. Dept. Fish and Game, Fish Bull. 94:1-62.

Brinley, F. J., and Eulberg, F. J. 1953. Embryological Head Glands of the Cichlid Fish *Aequidens portalegrensis.* Copeia 1953:24-26.

Burrell, H. 1927. The Platypus. Sydney, Australia, Angus and Robertson, Limited.

Cade, T. J., and Maclean, G. L. 1967. Transport of Water by Adult Sandgrouse to Their Young. Condor 69:323-343.

Carlander, K. D. 1953. Handbook of Freshwater Fishery Biology with the First Supplement. Dubuque, Iowa, W. C. Brown Co.

Carlton, W. G., and Jackson, W. B. 1964. The Use of Spines for Age Determination of Fish. Turtox News 42:282-283.

Chittleborough, R. G. 1959. Determination of Age in the Humpback Whale, *Megaptera nodosa* (Bonnaterre). Australian J. Marine and Freshwater Res. 10:125-143.

Clark, H., Florio, B., and Heerowltz, R. 1955. Embryonic Growth of *Thamnophis s. sirtalis* in Relation to Fertilization Date and Placental Function. Copeia 1955:9-13.

Cole, J. E., and Ward, J. A. 1970. An Analysis of Parental Recognition by the Young of the Cichlid Fish, *Etroplus maculatus* (Block). Zeitschrift für Tierpsychologie 27:156-176.

Cooper, E. L. 1951. Validation of the Use of Scales of Brook Trout, *Salvelinus fontinalis,* for Age Determination. Copeia 1951:141-148.

Cowles, R. B. 1944. Parturition in the Yucca Night Lizard. Copeia 1944:98-100.

Ditmars, R. L. 1922. Reptiles of the World. New York, The Macmillan Co.

Erickson, M. M. 1938. Territory, Annual Cycle, and Numbers in a Population of Wrentits (*Chamaea fasciata*). Univ. Calif. Publ. Zool. 42:247-334.

Fitch, H. S. 1970. Reproductive Cycles in Lizards and Snakes. Univ. Kansas Mus. Nat. Hist. Misc. Publ. 52.

Hamilton, W. J., Jr. 1933. The Weasels of New York. Am. Mid. Nat. 14:289-344.

Hatfield, D. M. 1935. A Natural History Study of *Microtus californicus.* J. Mamm. 16:261-271.

Hooper, F. E. 1949. Age Analysis of a Population of the Ameiurid Fish *Schilbeodes mollis* (Hermann). Copeia 1949:34-38.

Huxley, J. 1932. Problems of Relative Growth. New York, Dial Press, Inc.

Ingles, L. G. 1941. Natural History Observations on the Audubon Cottontail. J. Mamm. 22:227-250.

Jameson, E. W., Jr. 1955. Observations on the Biology of *Sorex trowbridgei* in the Sierra Nevada, California. J. Mamm. 36:339-345.

Lagler, K. F. 1956. Freshwater Fishery Biology. Dubuque, Iowa, W. C. Brown Co.

Lindsay, S. T., and Thompson, H. 1932. Biology of the Salmon (*Salmo salar*) Taken in Newfoundland Waters in 1931. Repts. Newfoundland Fishery Research. Comm. 1, No. 2, pp. 1-80.

Longhurst, W. M. 1964. Evaluation of the Eye Lens Technique for Ageing Columbian Black-tailed Deer. J. Wildlife Mgmt. 28:773-784.

Lord, R. D., Jr. 1959. The Lens as an Indicator of Age in Cottontail Rabbits. J. Wildlife Mgmt. 23:358-360.

Lord, R. D., Jr. 1961. The Lens as an Indicator of Age in the Gray Fox. J. Mamm. 42:109-111.

Lorenz, K. 1970. Studies in Animal and Human Behavior. Vol. 1. Cambridge, Mass., Harvard Univ. Press.

Low, W. A., and Cowan, I. McT. 1963. Age Determination of Deer by Annular Structure of Dental Cementum. J. Wildlife Mgmt. 27:466-471.

Mayer, W. V. 1953. Some Aspects of the Ecology of the Barrow Ground Squirrel, *Citellus parryi barrowensis*. Stanford University Publ., Univ. Ser. 11:48-55.

McCrady, E., Jr. 1938. The Embryology of the Opossum. Am. Anat. Memoirs, 16. Philadelphia, Wistar Institute of Anat. and Biol.

Miller, L. 1955. Further Observations on the Desert Tortoise, *Gopherus agassizi*, of California. Copeia 1955:113-118.

Nice, M. M. 1937. Studies in the Life History of the Song Sparrow. I. Trans. Linnean Soc. N.Y.

Oliver, J. A. 1955. The Natural History of North American Amphibians and Reptiles. Princeton, N.J., D. Van Nostrand Co., Inc.

Oppenheimer, J. R., and Barlow, G. W. 1968. Dynamics of Parental Behavior in the Black-chinned Mouthbreeder, *Tilapia melanotheron* (Pisces: Cichlidae). Zeitschrift für Tierpsychologie 25:889-914.

Orr, R. T. 1940. The Rabbits of California. Calif. Acad. Sci., Occasional Paper 19.

Orr, R. T., Schonewald, J., and Kenyon, K. W. 1970. The California Sea Lion: Skull Growth and a Comparison of Two Populations. Proc. Calif. Acad. Sci. (4) 37:381-394.

Payne, R. B. 1961. Growth Rate of the Lens of the Eye of House Sparrows. Condor 63:338-340.

Richdale, L. E. 1942. Post-egg Period in Albatrosses. Dunedin. Biol. Monog. No. 4. (Otago Daily Times and Witness Newspapers Co., Ltd.)

Rongstad, O. J. 1966. A Cottontail Rabbit Lens-growth Curve from Southern Wisconsin. J. Wildlife Mgmt. 30:114-121.

Saunders, A. A. 1956. Descriptions of Newly Hatched Passerine Birds. Bird-Banding 27:121-128.

Scheffer, V. B. 1950. Growth Layers on the Teeth of Pinnipedia as an Indication of Age. Science 112:309-311.

Sexton, O. J., and Ortleb, E. P. 1966. Some Cues Used by the Leptodactylid Frog, *Engystomops pustulosus*, in Selection of the Oviposition Site. Copeia 1966:225-230.

Shadle, A. R. 1951. Laboratory Copulations and Gestations of Porcupine, *Erethizon dorsatum*. J. Mamm. 32:219-221.

Stebbins, R. C. 1954. Amphibians and Reptiles of Western North America. New York, McGraw-Hill Book Co., Inc.

Stuart, G. R. 1954. Observations on Reproduction in the Tortoise *Gopherus agassizi* in Captivity. Copeia 1954:61-62.

Van Tyne, J., and Berger, A. J. 1959. Fundamentals of Ornithology. New York, John Wiley & Sons, Inc.

Weekes, H. C. 1935. A Review of Placentation Among Reptiles with Particular Regard to the Function and Evolution of the Placenta. Proc. Zool. Soc. London 1935:625-645.

Wright, P. L. 1942. Delayed Implantation in the Long-tailed Weasel (*Mustela frenata*), the Short-tailed Weasel (*Mustela cicognani*) and the Marten (*Martes americana*). Anat. Rec. 83:341-353.

Wright, P. L., and Rausch, R. 1955. Reproduction in the Wolverine, *Gulo gulo*. J. Mamm. 36:346-355.

Zimmermann, A. A., and Pope, C. H. 1948. Development and Growth of the Rattle of Rattlesnakes. Fieldiana, Zool. 32:355-413.

# Chapter Fourteen

# POPULATION DYNAMICS

Theoretically if the reproduction and mortality rates in a nonmobile population were equal and constant, the number of individuals constituting the population would also remain constant. Natural populations of animals, however, are not static but constantly subject to change and motion because of many variable factors both in the environment and within the organisms themselves. These aggregate changes, therefore, that take place during the life of a population are referred to as *population dynamics*.

In studying an animal population it is important to know not only its density but also its composition. Most vertebrate populations show marked fluctuations in numbers of individuals as well as in age composition from season to season. These are the result of such factors as reproduction, invasion, emigration, migration, and mortality. Furthermore, there may be cyclic fluctuations of considerably greater length and magnitude involving several years or more.

To obtain population data many of the techniques of trapping, marking, or otherwise identifying individuals, previously described in Chapter 9, are employed. Special methods are sometimes used for particular species to obtain information on numbers. Aerial photographs frequently prove useful in obtaining population estimates on some colonial mammals such as seals and sea lions. Photography is also an aid in determining numbers of wintering waterfowl and of marine birds that nest in large aggregations. Fecal pellet counts may provide an index as to numbers of deer, elk, or lagomorphs in a given area. Fish may be seined or stunned by rotenone (derris root) in ponds and streams to obtain population data.

## REPRODUCTIVE CAPACITY

A population increase may be effected as a result of immigration or migration, but basically it is dependent upon reproduction. In order to avoid extinction a species must produce new individuals in numbers sufficient to replace those that die. This also applies to populations within a species.

Biologists frequently use the term *reproductive potential* which, in the absolute sense, refers to the maximum number of individuals that a population could produce. In any group of animals, however, the number of new individuals that could be produced exceeds the number that actually is produced. For some species this realized reproductive capacity is not greatly exceeded by the potential reproductive capacity, but in others it represents only a very small percentage. For example, the survival rate of the single young produced once a year, sometimes less often, by certain large mammals is high, while the percentage of individuals that develop to maturity from the several million eggs produced annually by a female sturgeon or cod is extremely small. The latter have a high reproductive potential, but the survival rate of the eggs and young is low because of environmental resistance.

Just how a species in the course of evolution attained a reproductive rate that is adjusted to balance mortality has been a subject of considerable speculation. Among birds the reproductive rate is determined by the size of the clutch and the number of clutches produced annually. Within any one species the number of eggs that comprise a clutch tends to be fairly uniform. This does not mean that every female lays exactly the same number. To cite a specific example, four is the usual number of eggs laid by a killdeer (*Charadrius vociferus*), but occassionally clutches of three or five are produced. Furthermore, there may be geographic variation in the average number of eggs laid in some species.

Lack (1954*b*) has summarized the four principal theories that have been presented to account for clutch size. The first assumes that birds produce as many eggs as they are physiologically capable of producing. This, however, is contradicted by the fact that removal of eggs from the nest of a laying bird frequently stimulates the production of more eggs to complete the clutch. The second theory presumes that the clutch size is determined by the number of eggs that can satisfactorily be incubated by a setting bird. There seems to be no reason, however, why a pigeon could not easily incubate three or four eggs instead of two. The third theory is that clutch size has, as a result of natural selection, become adjusted to balance mortality. One of the arguments presented by Lack in opposition to this is that within a given population it is difficult to see how natural selection would favor individuals that produced fewer young over those that produced more young. This objection is valid, however, only if one assumes that selection is concerned with individuals and not with whole populations. It seems reasonable to believe that species that increase beyond the carrying capacity of their environment possibly have less chance of survival over long periods of time than those that do not. This might also apply to populations within a species where they are isolated from one another. One would expect that selection would favor those whose reproductive rate is adjusted to balance mortality over those that had either too high or too low a reproductive rate.

The fourth hypothesis listed by Lack and one which he favors is that, as a result of natural selection, birds produce as many young as the parents, on the average, can satisfactorily provide food for. There are a number of weaknesses to this theory. Mortality is considered primarily insofar as it affects the young of individuals producing more progeny than they can properly feed. It fails to explain the population equilibrium that exists in so many species. Likewise, it does not account for clutch size in precocial species that do not feed their young.

As will be noted later in this chapter, there are various environmental factors that may affect the reproductive rate and cause it to fluctuate from year to year. Furthermore, in some species of vertebrates the number of eggs or young produced varies with age.

## REPRODUCTION AND MORTALITY

A species with a high mortality rate must likewise have a high reproductive rate if it is to survive. Animals that are long-lived usually reproduce slowly, whereas those whose life span is brief generally are prolific. Most large herbivorous mammals as well as pinnipeds and cetaceans produce one or occasionally two young a year. Few of these animals attain sexual maturity the first year of life, and some require a number of years to reach the peak of sexual activity. Their life expectancy is relatively long.

Many northern insectivorous bats produce only one or two young a year, but banding has shown that some individuals under natural conditions may attain an age of more than 20 years. This, of course, is far beyond the average life expectancy of a bat, but it indicates potential longevity in this group.

California condors (*Gymnogyps californianus*) may not breed until they are about five years old and then generally produce only one young every other year. One captive condor in the National Zoological Park in Washington, D.C., is known to have reached the

FIGURE 14.1. The California condor (*Gymnogyps californianus*) is an avian species with a very low reproductive capacity. (Photograph courtesy of Ed N. Harrison.)

age of 45 years, although Koford (1953) has suggested that the average life expectancy of wild condors might be somewhere near 12 years.

Among many small mammals we find a high reproductive rate associated with species whose life expectancy is about one year or slightly more. Voles may begin breeding at three to six weeks of age. The period of gestation as well as that of parental care is brief. There is one record, cited by Hamilton (1939), of a captive vole producing 17 litters of young within one year. These small rodents, however, are heavily preyed upon by many other vertebrates.

A study of the small mammals (*Microtus, Reithrodontomys,* and *Mus*) inhabiting a grassland area along the central California coast by Pearson (1964) revealed a population peak among rodents in June 1961. At this time of year, which would normally be the height of the breeding season, reproduction essentially ceased and was not resumed until the following spring. The total population was determined by live-trapping. An analysis was made of the droppings of all the small carnivores, which included feral house cats as well as native skunks, raccoons, and foxes that also occurred there. Predation by these species alone accounted for 88 per cent of the 4400 voles, 33 per cent of the 1200 harvest mice, and 7 per cent of the 7000 house mice during this nonreproductive period.

Blair (1940*b*) found that there was an almost complete annual turnover in the population of deer mice (*Peromyscus maniculatus*) on the Edwin S. George Reserve in Michigan. Most of the individuals breeding in the spring had been born the previous fall. The spring population was small, but a maximum was reached by the end of July. This was followed by a decline in numbers until autumn breeding commenced. Following this the winter population declined to the spring low.

Between 1939 and 1944, the author maintained an aviary occupying 400 square feet of floor space on the tar and gravel roof of one of the buildings of The California Academy of Sciences in San Francisco. The presence of food soon attracted house mice (*Mus musculus*), which could only hide in nest boxes, under an adjacent narrow board walk, or in a few spaces beneath the flashing at the edge of the roof. Persistent trapping, which rarely resulted in the capture of less than one mouse a night, over this period failed to reduce the mouse population. An attempt was made to eliminate the nests in which the females had their young. It was soon found that the elimination of one nest only resulted in the rapid construction of another in the same place or in some other situation where it might not be discovered for some time. Following this the nests were left and the young removed when found. During the summer and autumn of 1944 records were kept of the numbers and approximate ages of the young that were removed from some of the nests. From one of these alone, 10 young belonging to two age groups were taken on July 3. Four were naked and six had fur appearing on the body. There were two adults present when the nest was first examined. On July 5 four newly born young were removed. Twelve more newly born young were removed on July 7. When next examined on July 24, 24 young belonging to three age groups were found and two adults ran away. This was but one of several nests that obviously were used at the same time by more than one female to house young. The amazing reproductive capacity of this species is one of the reasons why it has successfully established itself over so many parts of the world.

## DENSITY AND AGE COMPOSITION

In most natural populations of animals the density and age composition fluctuate during the year. For species that produce their young in the spring and summer, population density will be at a low in the months immediately preceding and will rise as the reproductive season progresses. Likewise, the ratio of young to adults will also increase at the same time.

Both density and age composition may also affect the reproductive capacity of a population. Rapidly increasing populations usually have a higher percentage of young breeding individuals than stationary or declining populations. However, as will be discussed later in this chapter, if the population density becomes too high it may inhibit reproduction.

Davis (1951) found fewer pregnancies in stationary populations of Norway rats (*Rattus norvegicus*) than in increasing or decreasing populations, although the prevalence of lactation was approximately the same at the three levels. He assumed that in increasing populations there was a greater parturitional mortality but better survival on the part of the young that were weaned. No explanation was apparent for the increase in pregnancies in a decreasing population.

Data recently compiled on the black-tailed jackrabbit (*Lepus californicus*) of western North America indicate that in this species when the population density increases there may be a decrease in fertility, resulting in a shortening of the breeding season and a reduction in the number of young per litter. Conversely, with a declining population the opposite trend in length of breeding season and litter size occurs.

Blair (1957) observed that severe drought greatly reduced the adult population of rusty lizards (*Sceloporus olivaceus*) on a 10 acre plot in Travis County, Texas. There was, however, a higher survival of the offspring, although they were fewer in number. Since yearling females breed but produce fewer eggs than those in higher age classes, the number of breeding individuals in the population underwent little change although the breeding potential was reduced and the age composition changed. Cricket frogs (*Acris crepitans*) in the same area produced fewer young, which, however, had a higher survival rate and matured more rapidly than in normal years. Two other species, the anole (*Anolis carolinensis*) and the bullfrog (*Rana catesbiana*) were eliminated.

## FOOD

As pointed out by Lack (1954*b*), the reproductive rate, as well as many other factors that influence animal numbers, is the result of natural selection. This leads to a distinction between what are termed proximate and ultimate causes. Thus in some species increased day length may stimulate reproductive activity in the spring, when food is sufficiently abundant for the raising of the young. Light may be considered as the proximate factor here, but it is the food supply which is necessary for survival that is the ultimate factor.

Food supply seems sometimes to have a very direct effect upon reproductive behavior. Jameson (1955) found that whereas reproduction in deer mice (*Peromyscus maniculatus*) in the Sierra Nevada of northern California was essentially the same in the spring over a three year period, fall breeding was dependent on food supply. In years when there was an abundance of conifer seeds, manzanita berries, and acorns, breeding continued until December with a consequent high population of mice the following year. In the absence of such food, breeding ceased in June and a marked decline in the population followed. The lull in reproduction—which was correlated with food scarcity—rather than increased mortality was believed responsible for the population decline.

Taber's (1956) observations on deer (*Odocoileus hemionus*) in California showed that does over two years old averaged 1.65 fawns apiece in open shrubland when food was abundant, as compared with an average of 0.77 fawn for those inhabiting areas of dense chaparral, where less food was available (Fig. 14.2).

In some birds food supply seems to influence clutch size. Rendall (1925) commented on the abnormally large clutches of eggs of the short-eared owl (*Asio flammeus*) during a year in which voles were at peak numbers in Canada. In Sweden, nutcrackers (*Nucifraga caryocatactes*) regularly have an average of three eggs to a clutch, but in years when nuts are abundant four eggs are usually laid (Lack, 1954*b*).

It has been suggested that food is the mechanism that controls the populations of sea birds in tropical waters. Since many species of oceanic birds wander very far outside of the breeding season, it is only during this season that food is an effective control. Among

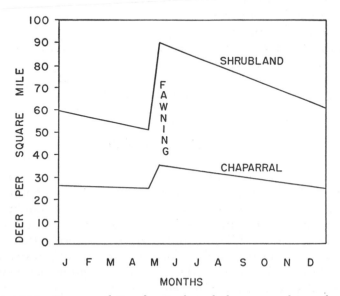

**FIGURE 14.2.** Deer population density through the year on chaparral and shrubland range. (From Taber, R. D.: Transactions of the Twenty-First North American Wildlife Conference, 1956.)

species that aggregate in large numbers during the nesting season, the competition for food in the surrounding waters may be very great. This means that when the population reaches a certain maximum many adults may fail to succeed in successfully rearing their young.

## CLIMATE

Climate may also affect reproduction in many kinds of vertebrates. This has been discussed to some extent in Chapter 12. Cold weather in spring delays nesting in certain species of birds. Drought inhibits the breeding activities of many kinds of amphibians and some reptiles. Martin (1956), in his study of the prairie vole (*Microtus ochrogaster*) in Kansas, found that subnormal rainfall appeared to inhibit reproduction, whereas heavy rainfall killed large numbers of juveniles. It has been observed that in certain regions the production of mallards (*Anas platyrhynchos*) is high in years when there is a high spring rainfall and low in years when there is a spring drought. According to Mayhew (1955), high relative humidity and periodic wetting of the eggs is essential for a good hatch.

Rogers (1959) found that when, as a result of drought, the water level was drastically reduced in the potholes of southwestern Manitoba, there was an abrupt decline in the production of lesser scaup (*Aythya affinis*). This resulted partly from the drying up of many of the nesting ponds. However, predation was higher than in wet years, possibly because of the greater accessibility of the nests to predators. There was also greater loss from agricultural activities including cattle grazing and mowing. Furthermore, and possibly of greater importance, was the fact that there was a marked failure to renest on the part of birds whose nests were destroyed.

## CARRYING CAPACITY

The carrying capacity or maximum number of individuals that a given area can satisfactorily accommodate is another important factor that influences populations. This is primarily determined by the availability of food and the presence of shelter and breeding sites. For example, Taber (1956) found that in California the carrying capacity of dense chaparral for deer was considerably less than that of open shrubland where herbaceous plants grow between well spaced shrubs (Fig. 14.2). In December the chaparral supported about 25 deer per square mile, whereas the shrubland, which provided more food, supported about 60 per square mile. Blair (1940a)

FIGURE 14.3. Vast numbers of small lakes in northern United States and the prairie provinces of Canada provide breeding grounds for many kinds of waterfowl. (Photograph taken in southern Manitoba.)

found 11.9 meadow voles (*Microtus pennsylvanicus*) per acre in moist grassland where there was good cover, while only 2.6 voles per acre were found in dry grassland in Michigan.

In Africa it has been shown that the tropical savanna ecosystem can support a very high population of animals even though the land is not rich and the rainfall is low and irregular. This is the result of a combination of burning by man and grazing by native mammals. The former prevents overencroachment by woody species and the latter permits vigorous growth of grasses and forbs of many kinds. Provided this balance is maintained the animal population can remain high. If burning does not occur, woodland ultimately replaces the grassland and the carrying capacity is greatly reduced. Likewise, since different kinds of native grazing and browsing animals select different species of plants for food, there is relatively little interspecific competition along with efficient utilization of available food. The carrying capacity of such areas for domestic stock is much lower than for native herbivores, since the former are capable of utilizing only a small amount of the nutrient material present.

In territorial species the spatial separation of paired individuals or even colonies is a means of limiting populations within an environment and tends to prevent overpopulation. However, in any group of animals, if the carrying capacity of the area occupied is exceeded overpopulation results.

Since food and cover are often subject to seasonal fluctuations, the carrying capacity of the environment at the most unfavorable time of the year is of critical importance. A stream may be capable of accommodating a large trout population in spring and early summer, but its condition in late summer and fall when the runoff reaches a minimum may determine the number of fish that can satisfactorily be accommodated on a year-round basis.

In many areas winter food supply determines the carrying capacity of an area for large herbivores even though many more individuals can be accommodated during the summer months.

One of the best known examples of a population exceeding the carrying capacity of its range was presented some years ago by the deer on the Kaibab Plateau in northern Arizona. In 1906 this area, comprising about three-fourths of a million acres, was set aside as a game refuge by President Theodore Roosevelt. The deer population at that time was estimated at about 4000. Hunting by both Indians

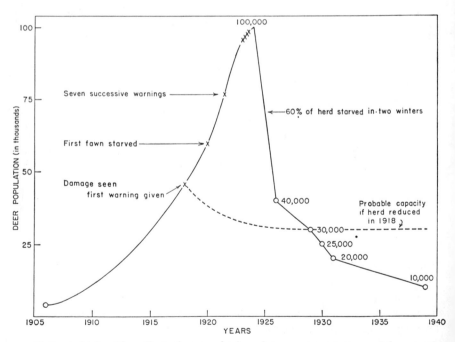

**FIGURE 14.4.** The effect of removal of predators on populations of deer on the Kaibab Plateau in Arizona (727,000 acres). Six hundred pumas were removed in 1907–1917, 74 in 1918–1923, and 142 in 1924–1939. Eleven wolves were removed in 1907–1923 and all exterminated by 1926. Three thousand coyotes were removed in 1907–1923 and 4388 in 1923–1939. (Redrawn from Leopold, after Allee et al.: Principles of Ecology. W. B. Saunders Co., 1949.)

and white men was prohibited and predator control was carried on. Six hundred mountain lions were eliminated by 1917. The small wolf population was exterminated by 1926, and over 7000 coyotes had been killed by 1939. As a result of the reduction in mortality rate the deer population rapidly increased until it far exceeded the carrying capacity of the range, which was determined by the winter food supply. By 1920, 100,000 deer were estimated to be present, but starvation during the following two winters killed 60 per cent of the population (Fig. 14.4).

The decline continued on a depleted range until the population reached 10,000 in 1939, which was estimated to be about one-third of the normal carrying capacity of the area.

Similar situations, although not of this magnitude, in which deer have increased beyond the carrying capacity of their winter range have occurred in other parts of North America.

## COMPETITION

Sometimes competition is a major factor in limiting populations of vertebrates. In areas where red squirrels and gray squirrels come into contact with each other, the former are dominant and tend to drive the latter away. Red squirrels are equally intolerant of fox squirrels (Hatt, 1929).

Martin (1956) found prairie voles (*Microtus ochrogaster*) reaching a population peak in the autumn and a population low in January. Competition with cotton rats (*Sigmodon hispidus*) rather than predators was considered the significant factor. Rapid reverses in populations, however, could occur. In the fall of 1951, in northeastern Kansas, where this study was made, a decrease in the vole population was correlated with an increase in the numbers of cotton rats. The following February a sudden cold spell decimated the cotton rats, and the vole population immediately began to rise.

Bardach (1951) presents some interesting information on the effect of both competition and disease on the population of yellow perch (*Perca flavescens*) with respect to size and numbers of individuals in Lake Mendota, Wisconsin. During the first five decades of this century there was a very marked reduction in the perch population in this lake. However, the average size and weight of these fish increased. In 1916 the average catch per 100 feet of gill net during the spawning season was 15.5 fish per hour, whereas in 1947 it was down to 3.6. The average length of individuals caught rose during this time from 162 to 243 mm., and the average weight rose from 50 to 180 grams. Fishing pressure was not considered to be a major factor, but disease and competition from the cisco (*Leucichthys artedi*) were. In 1939 an epizootic caused by the one-celled myxosporidian parasite, *Myxobolus,* was very severe. The following year it affected the cisco, which is a competitor of the perch, causing large numbers to die off. Subsequently there was a marked acceleration in the growth rate of perch.

## CYCLES

Most vertebrates living away from the equator reach a population high during the reproductive season and a low during the winter months. The production of young combined with normal mortality is responsible for this annual cycle. There are some species, however, that exhibit definite cyclic fluctuations over longer periods of time with high years and low years. The numbers during high years may occasionally reach even epidemic proportions.

Man's knowledge of the spectacular rise and fall in the populations of some of these animals goes back several thousand years. Plagues or outbreaks of small rodents are recorded in the Old Testament, described in detail by Aristotle, and form a part of the history of Europe. A summary of these historical accounts has been compiled by Charles Elton (1942).

In North America somewhat similar fluctuations in the populations of certain mammals and birds have been known for the past 200 years. It was really not until the present century, however, that their cyclic nature was realized. One of the first comprehensive accounts of population cycles was presented by Elton (1924). Since then many detailed studies on the subject have been made, although we still know little about the causative factors.

The most pronounced cycles are to be found among certain animals inhabiting the tundra and the northern boreal forests. The principal groups exhibiting such cyclic periodicity are: (1) the lemmings (both *Lemmus* and *Dicrostonyx*), voles (*Microtus*), and the varying hare (*Lepus americanus*); (2) birds and mammals that prey upon these animals; (3) certain gallinaceous birds, notably ptarmigan and grouse. In species inhabiting the tundra or the tundra-coniferous forest ecotone, the cycle tends to be a relatively short-term one of three to four years, whereas species of the northern coniferous forests have a long-term cycle of nearly 10 years. Many theories have been presented to account for these rhythmic fluctuations in populations. They have been attributed to sunspot cycles, climatic cycles, food supply, disease, and parasites, to mention a few. Food obviously plays a major role in the regulation of predators that feed upon other birds or mammals whose members are subject to cyclic fluctuations. As the prey becomes more abundant the predators increase. A decline in the former results in a decline in the latter. The predator cycle, however, usually lags behind that of the prey by about one year.

The factors controlling nonpredatory species are much less clearly understood. In arctic North America, the brown lemming (*Lemmus trimucronatus*) has a three or four year cycle (Fig. 14.5). Rausch (1950) comments on the rapid decline in these lemmings from peak numbers in the region of Pt. Barrow, Alaska, in the spring of 1949. These animals were heavily preyed upon by snowy owls, jaegers, and arctic foxes. Vast numbers, nevertheless, died from other causes and reproductive activity ceased. By early June of that year the lemmings had essentially disappeared, although snowy owls were still abundant in October. No obvious parasitic infections were noted in the dead lemmings and, while the tundra vegetation was reduced, there was no reason to conclude that actual starvation was responsible for the decline.

The most classic example of a North American mammal that is subject to long-term periodic fluctuations in numbers is provided by the varying hare. The fur of this animal has been of commercial value for many generations, and data kept by the Hudson's Bay Company on the annual catch made by trappers since the end of the eighteenth century show that this species has a cycle somewhat less than 10 years in length (Fig. 14.6). MacLulich (1937) summarized

FIGURE 14.5. Captive brown lemmings (*Lemmus trimucronatus*) photographed at Pt. Barrow, Alaska. (By G. Dallas and Margaret M. Hanna.)

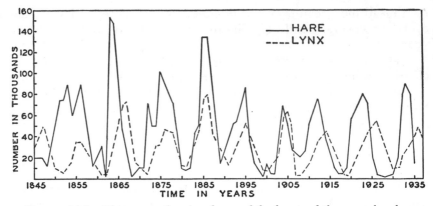

**FIGURE 14.6.** Changes in the abundance of the lynx and the snowshoe hare, as indicated by the number of pelts received by the Hudson's Bay Company. This is a classic case of cyclic oscillation in population density. (Redrawn from MacLulich, 1937. From Villee, Biology. 5th Ed., W. B. Saunders Co., 1967.)

this information and supplemented it with data obtained from the literature, questionnaires, and personal field work. The results showed that the peak years varied in different parts of Canada and that there was no correlation between the hare cycle and the sunspot cycle. During peak years, hares occupied nearly every available habitat but when the populations were down they were restricted to their favorite habitat. The reproduction rate declined as the population declined, but this was not considered to be responsible for the cycle. Nonspecific epidemics appeared to be the cause of the rapid die-off or crash that occurred after a population peak. The fact that the disease may vary with different declines, however, suggests that it may be a secondary factor. The cycle of the Canada lynx (*Lynx canadensis*) was definitely correlated with that of the varying hare (Fig. 14.6). The latter provides the principal food for this carnivore.

In western North America in the northern part of the Great Basin, the montane vole (*Microtus montanus*) is known to have a three to four year population cycle. Occasionally very high peaks are attained locally (Kartman, Prince, and Quan, 1959). An outbreak of these voles occurred in Humboldt County, Nevada, in 1908–1909. Following population highs in 1949 and 1954 an irruption occurred in the latter part of 1957 and the spring of 1958 in parts of Washington, Oregon, Idaho, Nevada, and California. This was unquestionably the greatest vole outbreak recorded in North America. Population estimates (Spencer, 1959) indicated maximum numbers in favorable localities may have been as high as 2000 to 3000 individuals per acre, although the average was considerably below this. When the decline began, both laboratory and field studies showed a high incidence of tularemia, which is caused by *Pasteurella tularensis.* Approximately 30 per cent of the voles captured from various localities within this area died from tularemia in the laboratory within 25 days. Another 40 per cent of those captured, however, died but not from any identified rodent disease (Kartman, Prince, and Quan, 1959).

There is considerable evidence to indicate that even if disease is present in rapidly declining populations it is not the direct cause of the decline. Chitty (1954), in a study of voles (*Microtus agrestis*) in England found that, although the incidence of tuberculosis was higher than usual during periods of abnormal mortality, advanced tuberculosis was comparatively rare. He suggested that the chances of survival of individuals during a crash year are reduced because most of them are abnormal from birth. This may be influenced by the physiological condition of the parents as a consequence of strife during the previous breeding season. Data obtained by Godfrey (1955) on this same species tends to support this view. This idea was presented earlier by Christian (1950), who suggested that cyclic declines in rodents may be the result of exhaustion of the adrenopituitary system, which in turn is caused by the stress of a high population level.

Lack (1954a) suggested that the basic cause of the cycles in the three major groups concerned "is the dominant rodent interacting with its vegetable food to produce a predator-prey oscillation. When the rodents decline in numbers, their bird and mammal predators become short of food, prey upon and cause the decrease of the gallinaceous birds of the same region and themselves die of starvation and/or emigrate."

The significance of the effect of a decline in the populations of rodents or hares on their predators is apparent, as already pointed out. It is quite possible also to see that this may cause the predators to turn to ptarmigan or grouse as the next source of food and in turn reduce the populations of these gallinaceous birds. There is some evidence to indicate that the population cycles of these birds correspond to those of the mammals inhabiting the same region. The willow ptarmigan (*Lagopus lagopus*) in Norway has a four year cycle like that of the lemmings with which it is associated. This same species has a 10 year cycle on the north shore of the Gulf of St. Lawrence in eastern North America, where it occurs within the range of the varying hare.

Another interpretation of the possible relationship between food and animal populations is the nutrient-recovery hypothesis which was first presented on the basis of studies made in northern Europe and, later, on microtines in the American Arctic. It is suggested that the nutritional value of the food of northern animals, rather than its

FIGURE 14.7. The spruce grouse (*Canachites canadensis*) of northern North America is rare in years when the varying hare population is low.

actual quantity, is the important factor. Many northern plants have been observed to have seed cycles. A certain number of years are required for a plant to build up sufficient nutrient material to produce seeds. Following seed production the plants are depleted and must once again gradually accumulate nutrient material for the next seed cycle. Most seed cycles are about three and one-half years, which is very similar to the population cycle of many small mammals.

As is pointed out, one would expect animals that eat foliage or herbage to have a population peak about a year earlier than spermophiles, since the production of seeds removes nutrient from other parts of the plant long before the seeds mature. Observations made in Finland have shown that populations of grouse, which depend to a large extent upon buds for food, reach their peak a year before those of seed-eating rodents such as squirrels.

Few biologists believe that sunspots or other extraterrestrial influences control population cycles in animals. Most favor the idea that these oscillations are correlated with environmental variables. The fact that these environmental correlates have not yet been discovered for many species that undergo regular periodic fluctuations in numbers still leaves the possibility that they do not exist. This has led to the proposal of a new idea, presented by Palmgren in 1949 and later developed by Cole (1951, 1954), that these cycles "are essentially random fluctuations with serial correlations between the populations of successive years." This theory does not disregard the environmental variables nor discourage further investigation in this field as one might at first be led to believe. Instead it suggests that because of the many variable factors that affect animal populations random oscillations necessarily result and, as Cole (1954) says, "produce a short basic cycle superimposed on a longer cycle three times the length of the basic cycle."

In reviewing all the various theories that have been presented to account for population cycles in warm-blooded vertebrates in the north, especially rodents and hares, we find that they may be grouped into three major categories: (1) theories based on an external environmental cause, (2) those based on an interaction between the population and its biological environment, and (3) those based on an inherent self-regulation within the population itself.

The first category of theory, if it could be proved, would show a correlation between population numbers and some regularly fluctuating physical or biological factor in the environment such as temperature, rainfall, sunspots, parasites, or disease-causing organisms. The second supposes that a population upon reaching a certain peak would adversely affect its environment, for example, its food supply, and thereby effect a decline in its numbers. The third theory presupposes an innate self-regulating mechanism which maintains regular cycles within populations.

There are advocates of all three theories, although the second and third are most strongly supported at present. It would seem likely that, in most instances at least, population cycles are the result of a number of factors rather than being attributable to a single one.

## References Recommended

Allee, W. C., Emerson, A. E., Park, O., Park, T., and Schmidt, K. P. 1949. Principles of Animal Ecology. Philadelphia and London, W. B. Saunders Co.

Ashmole, N. P. 1963. The Regulation of Numbers of Tropical Oceanic Birds. Ibis 1036:458–473.

Bardach, J. E. 1951. Changes in the Yellow Perch Population of Lake Mendota, Wisconsin, Between 1916 and 1948. Ecology 32:719–728.

Barnaby, J. T. 1944. Fluctuations in Abundance of Red Salmon, *Oncorhynchus nerka* (Walbaum), of the Karluk River, Alaska. U.S. Fish and Wildlife Ser., Fishery Bull. 50:237–295.

Beverton, R. J. H., and Holt, S. J. 1957. On the Dynamics of Exploited Fish Populations. London, Ministry of Agriculture, Fisheries and Food. Fisheries Invest., Ser. 2, Vol. 19.

Blair, W. F. 1940a. Home Ranges and Populations of the Meadow Vole in Southern Michigan. J. Wildlife Mgmt. 4:149–161.

Blair, W. F. 1940b. A Study of Prairie Deer-mouse Populations in Southern Michigan. Am. Mid. Nat. 24:273–305.

Blair, W. F. 1941. Techniques for the Study of Mammal Populations. J. Mamm. 22:148–157.

Blair, W. F. 1957. Changes in Vertebrate Populations Under Conditions of Drought. Cold Spring Harbor Symposia on Quantitative Biology 22:273–275.

Chitty, D. 1954. Tuberculosis Among Wild Voles: With a Discussion of Other Pathological Conditions Among Certain Mammals and Birds. Ecology 35:227–237.

Chitty, D. 1960. Population Processes in the Voles and Their Relevance to General Theory. Canadian J. Zool. 38:99–113.

Christian, J. J. 1950. The Adreno-Pituitary System and Population Cycles in Mammals. J. Mamm. 31:247–259.

Christian, J. J., and Davis, D. E. 1964. Endocrines, Behavior, and Population. Science 146:1550–1560.

Clarke, C. H. D. 1949. Fluctuations in Populations. J. Mamm. 30:21-25.

Clough, G. C. 1965. Lemmings and Population Problems. Am. Scientist 53:199-212.

Cole, L. C. 1951. Population Cycles and Random Oscillations. J. Wildlife Mgmt. 15:233-252.

Cole, L. C. 1954. Some Features of Random Population Cycles. J. Wildlife Mgmt. 18:2-24.

Davis, D. E. 1951. The Relation Between Level of Population and Pregnancy of Norway Rats. Ecology 32:459-461.

Davis, J., and Williams, L. 1964. The 1961 Irruption of Clark's Nutcrackers in California. Wilson Bull. 76:10-18.

Dice, L. R. 1952. Natural Communities. Ann Arbor, University of Michigan Press.

Elton, C. S. 1924. Fluctuations in the Numbers of Animals: Their Cause and Effect. J. Exp. Biol. 2:119-163.

Elton, C. S. 1942. Voles, Mice and Lemmings: Problems in Population Dynamics. Oxford, at the Clarendon Press.

Elton, C., and Nicholson, M. 1942. The Ten-year Cycle in Numbers of the Lynx in Canada. J. Animal Ecol. 11:215-244.

Emlen, J. T., Jr., Hine, R. L., Fuller, W. R., and Alfonso, P. 1957. Dropping Boards for Population Studies of Small Mammals. J. Wildlife Mgmt. 21:300-314.

Errington, P. L. 1951. Concerning Fluctuations in Populations of the Prolific and Widely Distributed Muskrat. Am. Nat. 85:273-292.

Errington, P. L. 1954. On the Hazards of Overemphasizing Numerical Fluctuations in Studies of "Cyclic" Phenomena in Muskrat Populations. J. Wildlife Mgmt. 18:66-90.

Errington, P. L. 1956. Factors Limiting Higher Vertebrate Populations. Science 124:304-307.

Errington, P. L. 1957. Of Population Cycles and Unknowns. Cold Spring Harbor Symposia on Quantitative Biology 22:287-300.

Evans, F. C. 1960. Population Dispersion. In McGraw-Hill Encyclopedia of Science and Technology. New York, McGraw-Hill Book Co., Inc., pp. 500-503.

Frank, F. 1957. The Causality of Microtine Cycles in Germany. J. Wildlife Mgmt. 21:113-121.

French, N. R., McBride, R., and Detmer, J. 1965. Fertility and Population Density of the Black-tailed Jackrabbit. J. Wildlife Mgmt. 29:14-26.

Godfrey, G. K. 1955. Observations on the Nature of the Decline in Numbers of Two *Microtus* Populations. J. Mamm. 36:209-214.

Green, R. G., and Evans, C. A. 1940. Studies on a Population Cycle of Snowshoe Hares on the Lake Alexander Area. III. Effects of Reproduction and Mortality of Young Hares on the Cycle. J. Wildlife Mgmt. 4:347-358.

Hairstrom, N. G., Smith, F. E., and Slobodkin, L. B. 1962. Community Structure, Population Control, and Competition. Am. Nat., Vol. 94, No. 879.

Hall, E. R. 1927. An Outbreak of House Mice in Kern County, California. Univ. Calif. Publ. Zool. 30:189-203.

Hamilton, W. J., Jr. 1937. The Biology of Microtine Cycles. J. Agric. Research 54:779-790.

Hamilton, W. J., Jr. 1939. American Mammals. New York and London, McGraw-Hill Book Co., Inc.

Hatt, R. T. 1929. The Red Squirrel: Its Life History and Habits, with Special Reference to the Adirondacks of New York and the Harvard Forest. Roosevelt Wild Life Annals, Vol. 2, No. 1.

Hoffman, R. S. 1958. The Role of Reproduction and Mortality in Population Fluctuations of Voles (*Microtus*). Ecol. Monog. *28*:79-109.

Jameson, E. W., Jr. 1955. Some Factors Affecting Fluctuations of *Microtus* and *Peromyscus*. J. Mamm. *36*:206-209.

Kartman, L., Prince, F. M., and Quan, S. F. 1959. Epizootiologic Aspects. *In* The Oregon Meadow Mouse Irruption of 1957-1958. Oregon State College, Fed. Coop. Ext. Ser., pp. 43-54.

Kendeigh, S. C. 1944. Measurement of Bird Populations. Ecol. Monog. *14*:67-106.

Klein, D. R. 1968. The Introduction, Increase, and Crash of Reindeer on St. Matthew Island. J. Wildlife Mgmt. *32*:350-367.

Koford, C. B. 1953. The California Condor. Res. Rept. No. 4, Nat. Aud. Soc. N.Y.

Krebs, C. J. 1964. The Lemming Cycle at Baker Lake, Northwest Territories, During 1959-62. Arctic Inst. N. Am., Tech. Pap. 15, pp. 1-104.

Lack, D. 1954a. Cyclic Mortality. J. Wildlife Mgmt. *18*:25-37.

Lack, D. 1954b. The Natural Regulation of Animal Numbers. Oxford, Clarendon Press.

Lauckhart, J. B. 1957. Animal Cycles and Food. J. Wildlife Mgmt. *21*:230-234.

MacLulich, D. A. 1937. Fluctuations in the Numbers of the Varying Hare (*Lepus americanus*). Univ. Toronto Studies in Biol. Ser. *43*:1-136.

MacLulich, D. A. 1957. The Place of Chance in Population Processes. J. Wildlife Mgmt. *21*:293-299.

Martin, E. P. 1956. A Population Study of the Prairie Vole (*Microtus ochrogaster*) in Northeastern Kansas. Univ. Kansas Publ. Mus. Nat. Hist. 8:361-416.

Mayhew, W. W. 1955. Spring Rainfall in Relation to Mallard Production in the Sacramento Valley, California. J. Wildlife Mgmt. *19*:36-47.

Odum, E. P. 1959. Fundamentals of Ecology. 2nd Ed. Philadelphia and London, W. B. Saunders Co.

Orr, R. T. 1944. Communal Nests of the House Mouse (*Mus musculus* Linnaeus). Wasmann Collector 9:35-37.

Palmgren, P. 1949. Some Remarks on the Short-term Fluctuations in the Numbers of Northern Birds and Mammals. Oikos *1*:114-121.

Pearson, O. P. 1960. A Mechanical Model for the Study of Population Dynamics. Ecology *41*:494-508.

Pearson, O. P. 1963. History of Two Local Outbreaks of Feral House Mice. Ecology *44*:540-549.

Pearson, O. P. 1964. Carnivore-Mouse Predation: An Example of Its Intensity and Bioenergetics. J. Mamm. *45*:177-188.

Pitelka, F. A. 1957. Some Characteristics of Microtine Cycles in the Arctic, pp. 153-184. *In* Hansen, H. P. [ed.], Arctic Biology. 18th Ann. Biol. Colloq., Corvallis, Oregon State Univ. Press.

Pitelka, F. A. 1964. The Nutrient-recovery Hypothesis for Arctic Microtine Cycles. *In* Grazing in Terrestrial and Marine Environments. Oxford, Blackwell Scientific Publications, pp. 55-56.

Rausch, R. 1950. Observations on a Cyclic Decline of Lemmings (*Lemmus*) on the Arctic Coast of Alaska During the Spring of 1949. Arctic 3:166-177.

Rendall, T. E. 1925. Abnormally Large Clutches of Eggs of Short-eared Owl (*Asio flammeus*). Canadian Field Nat. *39*:194.

Rogers, J. P. 1959. Low Water and Lesser Scaup Reproduction near Erickson, Manitoba. Trans. N. Am. Wildlife Conference *24*:216-224.

Rogers, J. P. 1964. Effect of Drought on Reproduction of the Lesser Scaup. J. Wildlife Mgmt. *28*:213-222.

Rowan, M. K. 1965. Regulation of Sea Bird Numbers. Ibis *107*:54-59.

Schultz, A. M. 1964. The Nutrient-recovery Hypothesis for Arctic Microtine Cycles. *In* Grazing in Terrestrial and Marine Environments. Oxford, Blackwell Scientific Publications, pp. 57-68.

Shelford, V. E. 1951. Fluctuations of Non-forest Animal Populations in the Upper Mississippi Basin. Ecol. Monog. *21*:149-181.

Siivonin, L. 1948. Structure of Short-cycle Fluctuations in Numbers of Mammals and

Birds in the Northern Parts of the Northern Hemisphere. Helsinki, Finnish Foundation for Game Preservation, Papers on Game Research, No. 1.

Slobodkin, L. B. 1954. Cycles in Animal Populations. Am. Scientist *42*:658-660.

Solomon, M. E. 1949. The Natural Control of Animal Populations. J. Animal Ecol. *18*:1-35.

Spencer, D. A. 1959. Biological and Control Aspects. *In* The Oregon Meadow Mouse Irruption of 1957-1958. Oregon State College, Fed. Coop. Ext. Ser., pp. 15-25.

Taber, R. D. 1956. Deer Nutrition and Population Dynamics in the North Coast Range of California. Trans. N. Am. Wildlife Conference *21*:159-172.

Taber, R. D., and Dasmann, R. F. 1957. The Dynamics of Three Natural Populations of the Deer *Odocoileus hemionus hemionus*. Ecology *38*:233-246.

Talbot, L. M. 1964. The Biological Productivity of the Tropical Savanna Ecosystem. *In* Ecosystems and Biological Productivity. IUCN Publ., New Ser. No. 4, Pt. 11, pp. 88-97.

Vorhies, C. T., and Taylor, W. P. 1933. The Life Histories and Ecology of Jack Rabbits, *Lepus alleni* and *Lepus californicus* ssp., in Relation to Grazing in Arizona. Univ. Arizona College Agric. Exp. Sta., Tech. Bull. *49*:471-587.

# INDEX

In the following index **boldface** type is used to indicate the page or pages on which a term or subject is specifically defined.